KATIE
HAFNER
AND
JOHN
MARKOFF

A TOUCHSTONE BOOK
Published by Simon & Schuster

CYBERPUNK

Outlaws

and

Hackers

on the

Computer

Frontier

▲ NEW YORK ▼ LONDON ▲ TORONTO
▼ SYDNEY ▲ TOKYO ▼ SINGAPORE ▲

TOUCHSTONE
Rockefeller Center
1230 Avenue of the Americas
New York, New York 10020

First Touchstone Edition 1995

TOUCHSTONE and colophon are registered trademarks
of Simon & Schuster Inc.

Designed by Caroline Cunningham

Manufactured in the United States of America

10   9   8   7   6

Library of Congress Cataloging-in-Publication Data is available.

ISBN: 0-684-81862-0

Portions of the Epilogue appeared in slightly different form in
Esquire.

TO OUR PARENTS

CONTENTS

CYBERPUNK

Introduction

We set out to investigate a computer underground that is the real-life version of cyberpunk, science fiction that blends high technology with outlaw culture. In cyberpunk novels high-tech rebels live in a dystopian future, a world dominated by technology and beset by urban decay and overpopulation. It's a world defined by infinitely powerful computers and vast computer networks that create alternative universes filled with electronic demons. Interlopers travel through these computer-generated landscapes. Some of them make their living buying, selling and stealing information, the currency of a computerized future. The television character Max Headroom, who lived in a network of mass media, popping up in computers and television sets everywhere, was considered pure cyberpunk. So was the 1982 movie *Blade Runner*, which portrays a slick, dark and dangerous world in which technology has triumphed and life is grim.

The inspiration for this book came when we began to see a change in the way computers were being used. We found harbingers of cyberpunk, young people for whom computers and computer networks are an obsession, and who have carried their obsession beyond what computer professionals consider ethical and lawmakers consider acceptable. They were called hackers. As the public viewed them, these computer hackers posed a sinister, if also somewhat vague, threat. This is our attempt to explain who they are and what drives them.

This book tells three stories. Kevin Mitnick fit the public's perception of an archetypal "dark-side" computer hacker. He was thought to be able to manipulate credit ratings, tap telephones and take complete control of distant computers. He saw himself as a brilliant computer renegade, and he proved to be a formidable adversary for one of the world's leading computer makers. In the end he was trapped by his own arrogance and compulsion.

The computer culture of the 1980s was as global as the youth culture of the 1960s. A young West Berliner who called himself Pengo discovered computers in his early teens. Because his parents had no understanding of the electronic world he had entered, no one knew that he was doing anything wrong when he spent hours at a time in front of a computer screen. To play out his outlaw fantasies, Pengo joined a group that sold the fruits of its wanderings through international computer networks to the Soviets. Eventually Pengo and his gang were torn apart in a series of betrayals.

Kevin and Pengo represent something close to the cyberpunk idea of the computer "cowboy" who lives outside the law. Robert Tappan Morris was different. The young Cornell University graduate student became notorious when he wrote a program that brought down a nationwide computer network. The son of a leading computer security researcher, he grew up as the ultimate insider, a member of an insular and elite community of computer scientists. Shy and thoughtful, Robert was hardly a rebel. By releasing a program that crippled several thousand computers in a matter of hours, he permanently altered the course of his life and confirmed everyone's worst fears about what hackers could do. The event marked a turning point: the private world of computer networks was suddenly of concern to the general public.

More than stories about computers or technology, this book is about the social consequences of computer networks and the communities that have grown up around them. As the world's computer networks became more closely linked in the 1980s, it was suddenly possible for anyone to travel the electronic corridors that were once the preserve of a small group of researchers. All three young men were seduced by the thrill of exploring these international computer networks, but they all went too far. Each drew national attention and contributed to a growing sense of public unease about the risks that arise from society's increasing dependence on computer networks.

In the 1960s and 1970s, to be a computer hacker was to wear a badge of honor. It singled one out as an intellectually restless soul compelled to stay awake for forty hours at a stretch in order to refine a program until it could be refined no more. It signified a dedication to computers that was construed as fanatical by outsiders but was a matter of course to the hackers themselves. The hackers from the Massachusetts Institute of Technology in particular adhered to what has been called the Hacker Ethic, which Steven Levy described in his 1984 book *Hackers* as a code of conduct that championed the free sharing of information and demanded that hackers never harm the data they found. Hacking also meant anything either particularly clever or particularly wacky, with or without a computer, as long as the manipulation of a complex system was involved. Some hacks are legendary. There was the computer program that determined the minimum amount of time it took to cover the entire New York City subway system. After the program was written, a bunch of MIT hackers actually went to New York and tried it out. And there was the legendary hoax during the 1961 Rose Bowl game between Washington and Minnesota, when Cal Tech students made substitutions for the letters on the cards to be held up by the Washington Huskies fans at halftime. Instead of *Washington,* the cards spelled out *Cal Tech.* A palindromic music composition was considered a good hack (thus making Haydn, with his Palindrome Symphony, an honorary hacker). So was anything done to establish not merely a new record but a new category altogether in the *Guinness Book of World Records.*

In the 1980s, a new generation appropriated the word "hacker" and, with help from the press, used it to define itself as password pirates and electronic burglars. With that, the public's perception of hackers changed. Hackers were no longer seen as benign explorers but malicious intruders.

These hackers are significant because of what our fear of them says about our unease with new technologies. Arthur C. Clarke once said, "Any sufficiently advanced technology is indistinguishable from magic." For many in this country, hackers have become the new magicians: they have mastered the machines that control modern life. This is a time of transition, a time when young people are comfortable with a new technology that intimidates their elders. It's not surprising that parents, federal investigators, prosecutors and judges often panic when confronted with something they believe is too complicated to understand.

The fallout from this fear is already apparent. As we were finishing the book, something of a hacker hysteria was sweeping the nation. After a two-year combined federal and state investigation, in the spring and summer of 1990 more than thirty raids on young computer users took place across the country, followed by a second wave of searches and arrests a few months later. In an attempt to root out the high-tech tools being used by this new breed of "criminal," law-enforcement agents confiscated computers, modems, answering machines, telephones, fax machines and even stereo equipment. Federal agents have gone after computer hackers in 1990 as if they are the next scourge after Communism.

Do young people who illegally enter computers really represent such a menace? We hope that from reading the following stories readers will learn that the answer isn't a simple one. All three of the young men we write about were caught up in what society views as criminal activities, yet none saw himself as a criminal. Each felt he was an explorer in a remarkable electronic world where the rules aren't clear. And each paid a price for his actions.

It is possible that once computer networks become as commonplace as our national highway system, we will learn to treat them in much the same way. Rules of the road will emerge and people will learn to respect them for their own safety and for the common good. We hope that the stories of Kevin Mitnick, Pengo and Robert Morris illustrate not just the risks of computer networks but also their allure.

—Katie Hafner
kmh@well.sf.ca.us

—John Markoff
markoff@nyt.com

PART ONE

Kevin:

The Dark-Side Hacker

The Roscoe Gang

It was partnership, if not exactly friendship, that kept the group together. Each member possessed a special strength considered essential for what needed to be done. Roscoe was the best computer programmer and a natural leader. Susan Thunder prided herself on her knowledge of military computers and a remarkable ability to manipulate people, especially men. Steven Rhoades was especially good with telephone equipment. And aside from his sheer persistence, Kevin Mitnick had an extraordinary talent for talking his way into anything. For a while, during its early days in 1980, the group was untouchable.

Susan was infatuated with Roscoe, but she never cared much for his constant companion, Kevin Mitnick. For his part, Kevin barely gave Susan the time of day. They learned to tolerate one another because of Roscoe. But for all their mutual hostility, Susan and Kevin shared a fascination with telephones and the telephone network; it was a fascination that came to dominate their lives. Susan, Kevin, Roscoe and Steven were "phone phreaks." By their own definition, phreaks were telephone hobbyists more expert at understanding the workings of the Bell System than most Bell employees.

The illegality of exploring the nooks and crannies of the phone system added a sense of adventure to phreaking. But the mechanical components of telephone networks were rapidly being replaced by computers that switched calls electronically, opening a new and far more captivating world for the telephone underground. By 1980, the members of this high-tech Los Angeles gang weren't just phone phreaks who talked to each other on party lines and made free telephone calls. Kevin and Roscoe, in particular, were taking phone phreaking into the growing realm of computers. By the time they had learned how to manipulate the very computers that controlled the phone system, they were calling themselves computer hackers.

Kevin was the only one of the original group to go even deeper, to take an adolescent diversion to the point of obsession. Susan, Roscoe and Steve liked the control and the thrill, and they enjoyed seeing their pranks replayed for them in the newspapers. But almost a decade later it would be Kevin, the one who hid from publicity, who would come to personify the public's nightmare vision of the malevolent computer hacker.

▲ ▼ ▲

Born in Altona, Illinois, in 1959, Susan was still an infant when her parents, struggling with an unhappy marriage, moved to Tujunga, California, northeast of the San Fernando Valley. Even after the move to paradise, with the implicit promise of a chance to start afresh, Susan's family continued to unravel. Susan was a gawky, buck-toothed little girl. Rejected and abused, at age eight she found solace in the telephone, a place where perfect strangers seemed happy to offer a kind word or two. She made friends with operators, and began calling random numbers in the telephone book, striking up a conversation with whomever she happened to catch. Sometimes she called radio disc jockeys.

After her parents divorced, Susan dropped out of the eighth grade, ran away to the streets of Hollywood and adopted the name Susy Thunder. Susan didn't make many friends, but she did know how to feed herself. Before long, she was walking Sunset Boulevard, looking for men in cars who would pay her for sex. She cut a conspicuous figure next to some of the more diminutive women on the street. Barely out of puberty, Susan was already approaching six feet.

When she wasn't walking the streets, she was living in a hazy, drug-filtered world as a hanger-on in the L.A. music scene, a rock-star groupie. Susan was a bruised child developing into a bruised adult. Quaalude was her medium of choice for spiriting her away from reality, and when Quaalude was scarce, she switched to alcohol and heroin. Her mother finally put her into a nine-month rehabilitation program; she was abruptly thrown out midcourse. Conflicting stories of Susan's ouster were in keeping with the blurry line between fact and myth that described her life. As Susan was to tell it, the adulation of power she developed as a groupie compelled her to single out the most powerful male staff member at the treatment center and seduce him. Another story, circulated by Susan's detractors, is that the male staff member for whom she left the program "sold" her services to a brothel.

Susan found an apartment in Van Nuys and retreated once again to the telephone, taking comfort in knowing that with the telephone she could gain access to a world of her own conjuring and shut it out whenever she chose. She began calling the telephone conference lines that were springing up all over Los Angeles in the late 1970s. By dialing a conference-line number, Susan could connect herself to what sounded like cross talk, except that she was heard by the others and could join in the conversation. Some conference-line callers were teenagers who dialed up after school; others were housewives who stayed on all day, tuning in and out between household chores but never actually hanging up the phone. By nightfall, many of the conference lines turned into telephonic sex parlors, the talk switching from undirected chitchat to explicit propositions.

One day in early 1980 Susan discovered HOBO-UFO, one of the first "legitimate" conference lines in Los Angeles in that its owners used their own conferencing equipment instead of piggybacking on the phone company's facilities. Drawing hundreds of people every day, HOBO-UFO was run from the Hollywood apartment of a young college student who called himself Roscoe. A friend of Roscoe's named Barney financed the setup, putting up the money for the multiple phone lines and other equipment while Roscoe provided the technical wherewithal. Susan decided she couldn't rest until she had met Roscoe, the power behind it all. But to achieve that goal, Susan knew she would have to abandon her disembodied telephone persona. She liked describing herself to men over the telephone. She knew from experience that all she had to do was mention that she was a six-foot-two blond and she wouldn't have to wait long for a knock at the door. She was right. No sooner did she deliver the description than Roscoe came calling.

The woman who greeted Roscoe was exactly as she had described herself. Susan had dressed up and made her face up carefully for the big date. But she could not conceal certain physical oddities. Her long face displayed a set of teeth so protrusive as to produce a slight speech impediment. And there was something incongruous about her large frame: her upper torso was narrow and delicate, but it descended to a disproportionate outcropping of hips and heavy thighs. Roscoe, for his part, was thin and pale. His brown-framed glasses met Susan's chin. But if either Susan or Roscoe was disappointed in the other's looks, neither showed it. They went to dinner, and when Roscoe asked Susan about her line of work she told him she was a therapist and then quickly changed the subject.

A business student at the University of Southern California, Roscoe was one of the best-known phone phreaks around Los Angeles. When a reporter from a local newspaper began researching a story about conference lines, he told a few HOBO-UFO regulars that he wanted to meet Roscoe. The next day a caller greeted him by reeling off the billing name on his unlisted phone number, his home address, the year and make of his car, and his driver's license number. Then the caller announced himself: "This is Roscoe."

When Susan and Roscoe met in 1980, phone phreaking was by no means a new phenomenon. Phone phreaks had been cheating the American Telephone and Telegraph Company for years. They started out with "blue boxes" as their primary tool. Named for the color of the original device, blue boxes were rectangular gadgets that came in a variety of sizes. Sometimes they were built by electronic hobbyists, at other times by underground entrepreneurs. Occasionally they were even used by the Mafia. One of Silicon Valley's legendary companies even has its roots in blue box manufacturing. Stephen Wozniak and Steven Jobs, who co-founded Apple Computer in 1976, got their start in the consumer electronics business several years earlier, peddling blue boxes in college dormitories.

A blue box was universally useful because it could exploit a quirk in the design of the nation's long-distance telephone system. The device emitted a high-pitched squeal, the 2600-hertz tone that, in the heyday of the blue boxers, controlled the AT&T long-distance switching system. When phone company equipment detected the tone, it readied itself for a new call. A series of special tones from the box allowed the blue box user to dial anywhere in the world. Using these clever devices, phone phreaks navigated through the Bell System from the palms of their hands. Tales abounded of blue boxers who routed calls to nearby pay phones through the long-distance lines of as many as fifteen countries, just for the satisfaction of hearing the long series of clicks and kerchunks made by numerous phone companies releasing their circuits.

Blue boxes were soon joined by succeeding generations of boxes in all colors, each serving a separate function, but all designed to skirt the computerized record-keeping and switching equipment that the phone company uses for billing calls.

The phone phreaking movement reached its zenith in the early 1970s. One folk hero among phreaks was John Draper, whose alias, "Captain Crunch," derived from a happy coincidence: he discovered that the toy whistle buried in the Cap'n Crunch cereal box matched the phone company's 2600-hertz tone perfectly.

Tending to be as socially maladroit as they were technically proficient, phone phreaks were a bizarre group, driven by a compulsive need to learn all they could about the object of their obsession. One famous blind phreak named Joe Engressia discovered the telephone as a small child; at age eight he could whistle in perfect pitch, easily imitating the 2600-hertz AT&T signal. Joe's lips were his blue box. After graduating from college, in tireless pursuit of knowledge about the phone company, Joe crisscrossed the country by bus, visiting local phone company offices for guided tours. As he was escorted around, he would touch the equipment and learn new aspects of the phone system. Joe's ambition was not to steal revenue from the telephone company but to get a job there. But he had made a name for himself as a phreak, and despite his vast store of knowledge, the phone company could not be moved to hire him. Eventually, Mountain Bell in Denver did give him a job as a trouble-shooter in its network service center and his whistling stopped. All that he had wanted was to be part of the system.

The Bell System needed people like Joe on its side. By the mid-1970s, AT&T estimated it was losing $30 million a year to telephone fraud. A good percentage of the illegal calls, it turned out, were being placed by professional white-collar criminals, and even by small businesses trying to cut their long-distance phone bills. But unable to redesign its entire signaling scheme overnight, AT&T decided to ferret out the bandits. Using monitoring equipment in various fraud "hot spots" throughout the telephone network, AT&T spent years scanning tens of millions of toll calls. By the early 1980s automated scanning had become routine and Bell Laboratories, AT&T's research arm, had devised computer programs that could detect and locate blue box calls. Relying on increasingly sophisticated scanning equipment, detection programs embedded in its electronic switches and a growing network of informants, AT&T caught hundreds of blue boxers.

In 1971, phone phreaking ventured briefly into the sphere of politics.

The activist Abbie Hoffman, joined by a phone phreak who called himself Al Bell, started a newsletter called *Youth International Party Line*—or *YIPL* for short. With its office at the Yippie headquarters on Bleecker Street in New York City's Greenwich Village, *YIPL* was meant to be the technical offshoot of the Yippies. Hoffman's theory was that communications were the nerve center of any revolution; liberating communications would be the most important phase of a mass revolt. But Al Bell's outlook was at odds with Hoffman's; Al saw no place for politics in what was essentially a technical journal. In 1973, Al abandoned *YIPL* and Hoffman and moved uptown to set up shop as *TAP*, the *Technological Assistance Program*.

Much of the information contained in *TAP* was culled from AT&T's various in-house technical journals. It was information that AT&T would rather have kept to itself. And that was the point. Whereas the original phreaks like Captain Crunch got their kicks making free phone calls, *TAP*'s leaders, while steering clear of a hard political line, believed that the newsletter's mission was to disseminate as much information about Ma Bell as it could. By 1975, more than thirteen hundred people around the world subscribed to the four-page leaflet. For the most part, they were loners by their own admission, steeped in private technical worlds. *TAP* was their ultimate handbook. Written in relentlessly technical language, *TAP* contained tips on such topics as lock picking, the manipulation of vending machines, do-it-yourself pay phone slugs and free electricity. *TAP* routinely published obscure telephone numbers; those of the White House and Buckingham Palace were especially popular. And in 1979, during the hostage crisis in Iran, *TAP* published the phone number of the American embassy in Tehran. Every Friday evening, a dozen or so *TAP* people held a meeting at a Manhattan restaurant, many still cloaked in the ties and jackets that betrayed daytime lives spent toiling away at white-collar jobs. After work, and inside the pages of *TAP*, they adopted such names as The Professor, The Wizard and Dr. Atomic.

In the late 1970s, a phone phreak who called himself Tom Edison took over *TAP*, bringing in another telephone network enthusiast, Cheshire Catalyst, a self-styled "techie-loner-weirdo science fiction fanatic," as one of *TAP*'s primary contributors. Tall and dark with the concave, hollow-cheeked look of someone rarely exposed to sunlight, Cheshire had been phreaking since the sixties. He discovered the telephone at age twelve, and learned to clip the speaker leads of the family stereo onto a telephone plug so that he could put the handset to his ear and listen to the radio while doing his homework. If his mother entered the room, he just had to hang up the receiver. By the time he was nineteen, Cheshire had become a telex maven, having programmed his home computer to simulate a telex machine. Before long he was sending telex pen-pal messages around the world. In his twenties, already a veteran TAP reader, Cheshire moved to Manhattan, got a job at a bank in computer support and joined the TAP inner circle.

TAP wasn't exactly a movement. It was an attitude, perhaps best described as playful contempt for the Bell System. One elderly woman from the Midwest sent her subscription check along with a letter to Tom Edison saying that although she would never do any of the things described in TAP, she wanted to support those people who were getting back at the phone company.

As one responsible for keeping TAP unassailable, Cheshire didn't sully his hands with blue boxes, and he paid his telephone bills scrupulously. Out of corporate garb, he and his friends stayed busy irking Ma Bell through their constant wanderings inside the phone system. As Cheshire and his friends would explain to outsiders, they loved the telephone network—it was the bureaucracy behind it they hated.

People like Cheshire lived not so much to defraud corporate behemoths as to home in on their most vulnerable flaws and take full playful advantage of them. Beating the system was a way of life. Flying from New York to St. Louis, for instance, was not a simple matter of seeking an inexpensive fare; it meant hours of research to find the cheapest route, even if it meant taking advantage of a special promotional flight from New York to Los Angeles and disembarking when the flight made a stop in St. Louis. And getting the best of AT&T, the most blindly bureaucratic monopoly of all, embodied a strike against everything worth detesting in a large corporation.

As private computer networks proliferated in the late 1970s, there came a generation of increasingly computer literate young phreaks like Roscoe and Kevin Mitnick. If the global telephone network could hold a phone phreak entranced, imagine the fascination presented by the proliferating networks that began to link the computers of the largest corporations. Using a modem, a device that converts a computer's digital data into audible tones that can be transmitted over phone lines, any clever interloper could hook into a computer network. The first requirement was a valid user identification—the name of an authorized user of the network. The next step was to produce a corresponding password. And in the early days, anyone could root out one valid password or another.

The rising computer consciousness of the phone phreaks was inevitable as technology advanced. Electromechanical telephone switches were rapidly giving way to computerized equivalents all over the world, suddenly transforming the ground rules for riding the telephone networks. The arrival of computerized telephone switches dramatically increased the risks and the sense of danger, as well as the potential payoff. The phone company's automatic surveillance powers grew by orders of magnitude, served by silent digital sentinels that sensed the telltale signals of the electronic phone phreaks' microelectronic armory. As the peril grew, so did the sense of adventure. Whenever an outsider gained control of a central office switch or its associated billing and maintenance computers, by evading often inadequate security barriers, the control was absolute. Thus the new generation of phone phreaks could go far beyond placing free telephone calls. Anything was possible: eavesdropping, altering telephone bills, turning off an unsuspecting victim's phone service, or even changing the class of service. In one legendary hack a phone phreak had the computer reclassify someone's home phone as a pay phone. When the victim picked up the telephone, he was startled to hear a computerized voice asking him to deposit ten cents. For phone phreaks, the temptation to step up from the simpler technology of the telephone to the more complicated and powerful technology of computers was irresistible.

In 1983, just as TAP was coming to symbolize the unification of computers and telephones, the journal came to an untimely end. Tom Edison's two-story condominium in suburban New Jersey went up in flames, the object of simultaneous burglary and arson. The burglary was professional: Tom's computer and disks—all the tools for publishing TAP—were taken. But the arson job was downright amateurish. Gasoline was poured haphazardly and the arsonist failed to open the windows to feed the fire. For years, Tom and Cheshire speculated that the phone company had engineered the fire, but proving it was another matter. Tom had a real-world name, a respectable job and a reputation to maintain. Cheshire rented a truck and hauled what remained of the operation —including hundreds of back issues of TAP, all of which escaped unsinged—over to his place in Manhattan. But in the end, TAP didn't survive the blow. A few months after the fire, Cheshire printed its final issue.

Meanwhile, the Southern California phreaks had been holding their equivalent of TAP meetings. Once a month or so, a group of phreaks, including Roscoe and occasionally Kevin, would get together informally at a Shakey's Pizza Parlor in Hollywood to talk and exchange information. But the L.A. phreaks weren't a particularly sociable bunch to begin with, and their meetings were far less organized than those on the East Coast, with fewer political overtones.

Though an avid TAP reader, Roscoe shunned blue boxes and most of the other electronic crutches of phone phreaking. They were just too easy to trace, he thought. Roscoe preferred to exploit flukes, or holes, that he and his friends found in the newly computerized telephone system. Modesty was not one of Roscoe's virtues. He claimed that he had acquired as much knowledge about the telephone system and the computers that controlled it as anyone else in the country. He kept notebooks filled with the numbers of private lines to corporations like Exxon and Ralston Purina, along with access codes to scores of computers operated by everything from the California Department of Motor Vehicles to major airlines. He boasted that he could order prepaid airline tickets, search car registrations and even get access to the Department of Motor of Vehicles' computer system to enter or delete police warrants. Whereas most pure phreaks viewed their art as a clever means of bypassing the phone company, Roscoe saw it as a potential weapon. With access to phone company computers, he could change numbers, disconnect phones or send someone a bill for thousands of dollars. Most of the numbers in his extensive log came from hours of patient exploration on a computer terminal at school. And many of the special tricks he learned from Kevin Mitnick.

Roscoe met Kevin in 1978, over the amateur radio network. When Roscoe was tuned in one day, he was startled to hear a nasty fight in progress between two hams. The control operator of the machine was accusing a fellow ham named Kevin Mitnick of making illegal long-distance telephone calls over the radio using stolen MCI codes. At the time, Roscoe knew nothing of telephones or computers. But given the vituperative tone of the angry ham, either Kevin Mitnick had done something truly terrible or he was being unjustly accused. Suspecting the latter, Roscoe switched on his tape recorder and recorded the invectives as they flew through the air. Then he got on the radio to tell Kevin that he had a tape recording of the accusations if Kevin wanted it. Sensing a potential ally, Kevin gave Roscoe a telephone number to call so that they could speak privately. As Roscoe was to find out later, Kevin had given him the number of a telephone company loop line, a number hidden in the electronic crevices of the telephone network and reserved for maintenance workers in the field who are testing circuits. Roscoe was immediately taken with this teenager, a good three years his junior, who evidently knew so much about telephones. He drove out to the San Fernando Valley to meet Kevin, gave him the tape and cemented a new friendship. Sometimes Kevin called Roscoe directly at his home in Hollywood, a toll call from the San Fernando Valley, and they talked for hours. When Roscoe asked him how a high schooler could afford it, Kevin just laughed.

▲ ▼ ▲

When Susan met Roscoe in 1980, he had been phreaking for about a year. She fell in love with him almost at once. He was the first man she had met who displayed some intelligence and whose life didn't revolve around drugs and the drug scene. She found Roscoe's interest in computers charming, even fascinating. Roscoe was taking phone phreaking to a new level, combining his knowledge of the phone system with his growing knowledge of computers. Susan saw this as a brilliant next step for someone with a phone obsession. What was more, they were both talented at employing their voices to desired ends. They shared a faith in how much could be accomplished with a simple phone call. As a teenager, Susan had employed the technique she called psychological subversion, otherwise known as social engineering, to talk her way into backstage passes at dozens of concerts. Posing as a secretary in the office of the head of the concert production company, she could get her name added to any guest list. Susan prided herself on those skills. If she and Roscoe had anything in common, both lacked the mechanism that compels most people to tell the truth.

Roscoe and Susan started to date each other. Roscoe was attending the University of Southern California and his schedule there, he told her, let him see her only on certain nights. But that was fine with Susan, as she was holding down two jobs. One was as a switchboard operator at a telephone answering service. The more lucrative job was something she knew how to make a lot of money at: she worked for a small bordello in Van Nuys. Her counselor story didn't last long. Roscoe made it his business to learn all he could about people, and Susan was no exception. When the truth emerged about Susan's profession, Roscoe found it more amusing than scandalous.

In her head, Susan was living out a romance of her own quirky invention. In fact, the relationship between Susan and Roscoe was oddly businesslike, hardly distracted by passion. Often their dates consisted of an excursion to the USC computer center, where Roscoe would set Susan up with a computer terminal and keep her occupied with computer games while he "worked." Susan soon realized that he was using accounts at the university's computer center to log on to different computers around the country. Susan lost interest in the games and turned her attention to what Roscoe was doing. Before long, she became his protegée.

Susan developed her own talent for finessing her way into forbidden computer systems. She began to specialize in military computer systems. The information that resides in the nation's military computers isn't just any data. It represents the nation's premier power base—the Pentagon. And in digging for military data, as Susan saw it, she, a high school dropout and teenage runaway, was just a silo away from the sort of control that truly mattered. Still, she was a beginner, far short of mastering the Defense Department's complex of computers and communications networks. What she couldn't supply in technical knowledge she compensated for with other skills. One of her methods was to go out to a military base and hang around in the officers' club, or, if she was asked to leave, in bars near the base. She would get friendly with a high-ranking officer, then act flirtatiously with him. Deploying her womanly charms, and implying that sexual favors might be granted, she persuaded these men to give her access to their systems. She would report each new success to Roscoe, who praised her profusely while making careful note of the specifics.

From her job at the bordello, Susan was taking home about $1,200 a week, and all of it came in handy. She invested every spare cent in computer and phone equipment. She installed a phone line for data transmission, and an "opinion line" she named "instant relay." Whoever dialed the "instant relay" number got Susan's commentary on topics of her own choosing. At the same time, she taught herself to use RSTS, Resource Sharing Time Sharing, the standard operating system for Digital Equipment Corporation's PDP-11 minicomputers. (Operating systems are programs that control a computer's tasks the way an orchestra conductor controls musicians. Operating systems start and stop programs and find and store files.) For computer intruders the fascination with a computer's operating system is obvious: it is not only the master controller but also the computer's gatekeeper, regulating access and limiting the capabilities of users.

Roscoe often employed Susan's Van Nuys apartment as a base of operations. He was usually accompanied by his younger cohort, the plump and bespectacled Kevin Mitnick. Kevin was the kind of kid who would be picked last for a school team. His oversize plaid shirts were seldom tucked in, and his pear-shaped body was so irregular that any blue jeans would be an imperfect fit. His seventeen years hadn't been easy. When Kevin was three, his parents separated. His mother, Shelly, got a job as a waitress at a local delicatessen and embarked on a series of new relationships. Every time Kevin started to get close to a new father, the man disappeared. Kevin's real father was seldom in touch; he remarried and had another son, athletic and good-looking. During Kevin's junior high school years, just as he was getting settled into a new school, the family moved. It wasn't surprising that Kevin looked to the telephone for solace.

Susan and Kevin didn't get along from the start. Kevin had no use for Susan, and Susan saw him as a hulking menace with none of Roscoe's charm. What was more, he seemed to have a malicious streak that she didn't see in Roscoe. This curiously oafish friend of Roscoe's always seemed to be busy carrying out revenge of one sort or another, cutting off someone's phone service or harassing people over the amateur radio. At the same time, Kevin was a master of the soothing voice who aimed at inspiring trust, then cooperation. Kevin used his silken entreaties to win over even the most skeptical keepers of passwords. And he seemed to know even more about the phone system than Roscoe. Kevin's most striking talent was his photographic memory. Presented with a long list of computer passwords for a minute or two, an hour later Kevin could recite the list verbatim.

Roscoe and Kevin prided themselves on their social engineering skills; they assumed respect would come if they sounded knowledgeable and authoritative, even in subject areas they knew nothing about. Roscoe or Kevin would call the telecommunications department of a company and pose as an angry superior, demanding brusquely to know why a number for dialing out wasn't working properly. Sufficiently cowed, the recipient of the call would be more than glad to explain how to use the number in question.

While Kevin's approach was more improvisational, Roscoe made something of a science out of his talent for talking to people. He kept a separate notebook in which he listed the names and workplaces of various telephone operators and their supervisors. He noted whether they were new or experienced, well informed or ignorant, friendly and cooperative or slow and unhelpful. He kept an exhaustive list of personal information obtained from hours of chatting: their hobbies, their children's names, their ages and favorite sports and where they had just vacationed.

Roscoe and Kevin didn't phreak or break into computers for money. Secret information, anything at all that was hidden, was what they prized most highly. They seldom if ever tried to sell the information they obtained. Yet some of what they had was eminently marketable. Roscoe's notebooks, filled with computer logins and passwords, would have fetched a tidy sum from any industrial spy. But phreaking to them was a form of high art that money would only cheapen. Roscoe especially thrived on the sense of power he derived from his phreaking. Presenting a stranger with a litany of personal facts and watching him or her come unhinged gave Roscoe his greatest pleasure.

Another frequent visitor to Susan's apartment was Steve Rhoades, a puckish fifteen-year-old from Pasadena with straight brown hair that cascaded down to the middle of his back. His timid intelligence and easy manner had a way of catching people off guard. He was an expert phreak who had earned a grudging respect from the Pacific Bell security force— the very people he loved to taunt. So adept was he at manipulating his telephone service from the terminal box on the telephone pole outside his house that the phone company removed the footholds. Like Kevin and Roscoe, Steve was an amateur radio buff. Two-way radios, in fact, often figured in their phreaking. When foraging through phone company trash for manuals, discarded papers containing passwords and whatever else might help them in their phreaking endeavors— a method they called trashing—they communicated by two-way radio.

Whatever the personality differences in the L.A. gang, their various talents brought results. Working together, they were able to get the numbers of lost or stolen telephone credit cards using highly imaginative methods: for instance, they would divert cardholders' toll-free calls to report lost cards to a pay phone of their choosing and answer with, "Pacific Bell, may I help you?" Their vast store of knowledge came from months of diligent research: they obtained manuals however they could get them and joined as many company tours as they could, mostly to familiarize themselves with building layouts. The group routinely got credit information from the computer at TRW's credit bureau. Sometimes it was a matter of talking an unsuspecting employee out of a password; sometimes a job required "physical research," such as a late-night excavation of TRW's Dumpsters.

By brainstorming together, the group would be inspired to ever more audacious stunts. On one occasion, Steve Rhoades figured out a way to override directory assistance for Providence, Rhode Island, so that when people dialed for information, they got one of the gang instead. "Is that person white or black, sir?" was a favorite line. "You see, we have separate directories." Or: "Yes, that number is eight-seven-five-zero and a half. Do you know how to dial the half, ma'am?"

A few months into her infatuation with Roscoe, Susan noticed unhappily that he was spending less time with her. Then someone let her in on the reason: she had been deceived. Roscoe, it turned out, was all but engaged to someone else, a law student as prim and straight in her ways as Susan was errant in hers. The other girlfriend apparently provided Roscoe's tie to respectable society. Susan despaired. He had been displaying affection for Susan when what he really wanted to do was exploit her eagerness to hack. When she confronted him, he just laughed. So she tried a veiled threat: she told him it was quite likely that FBI agents would come knocking on her door, and she would have to talk to them. Roscoe feigned puzzlement. Susan refused to believe that he wasn't at least torn between the two women. She decided to ask him, once and for all, for the truth. "Don't you feel anything for me at all?" He just smiled and said she had been misled. What Roscoe had failed to take into account was that the person he had just crushed was a woman who had been bruised once too often. By rejecting her so heartlessly, Roscoe was inviting trouble. He had yet to experience Susan's dark side.

▲ ▼ ▲

Eddie Rivera, a free-lance writer in Los Angeles, wasn't sure what he had stumbled into. As he was getting out of his car one day on Sunset Boulevard in late April of 1980, he saw a flyer that simply read: "UFO CONFERENCE CALL NOW." His curiosity stirred, the young reporter dialed HOBO-UFO and heard three people chatting idly, engaged in what seemed to be an interminable telephone conversation. A television murmured in the background. After listening to five more minutes of prattle, Eddie sensed the possibility of a story far different from what he usually produced as a rock-and-roll critic. Like Susan Thunder before him, Eddie focused on meeting the person running the conference. "Excuse me," he interrupted. "If you guys on this line know who runs it, have him call me." The line went silent; he felt as if he had switched on the kitchen light late at night and seen dozens of cockroaches running for cover. But it worked. Within a day of putting the word out that he wanted to meet Roscoe, Eddie received his first call. Eddie got a story assignment from the L. A. *Weekly*, an alternative newspaper, and went to work.

Their first meeting took place at an electronics store on Santa Monica Boulevard owned by Barney, Roscoe's pudgy, disheveled benefactor. When not in school or overseeing the HOBO-UFO line from his home, Roscoe was often at Barney's place. With not a customer in sight, the store was littered with old televisions in various stages of disrepair. For his part, Barney derived no small amount of pleasure from the venture he was financing; he used the conference line to meet adolescent girls.

Eddie hadn't known quite what to expect. Roscoe's appearance was surprisingly neat, but there was something amiss: his pale blue polyester pants with a slight flare at the bottom, and his dark polyester print shirt with an oversize collar, were already at least five years out of date. The twenty-year-old Roscoe seemed more to resemble an electrical engineering student than a telephone outlaw.

For the first interview, Barney put a "closed" sign in the shop window and, in what would become a routine preamble to the interviews, the three went out for doughnuts. Roscoe lived on junk food, as did, it seemed, all his fellow phreaks. A patina of doughnut glaze frequently rested on Roscoe's lips. In the afternoons, Roscoe moved on to Doritos and cheeseburgers. And Barney's store was strewn with Winchell's Do-nuts coffee cups whose contents suggested that Barney might be using them as petri dishes.

Roscoe made it clear from the start that he would be in complete control of the nature and quantity of information he imparted to Eddie. He enjoyed telling Eddie just how much he knew about telephones—far beyond the body of information known to the average telephone company employee.

There was something oddly mechanized about Roscoe's language. Eddie was struck by the young man's formal, almost bureaucratic way of speaking. In response to a question, Roscoe usually answered as indirectly as possible. He had a curious affection for the passive construction. Roscoe didn't simply make phone calls. Instead, telephone conversations were initiated. Perhaps, Eddie thought, Roscoe's tangled locution resulted from reading too many of those phone company manuals he kept talking about. In any case, his manner of speech distanced him from whatever he was talking about. Perhaps it made him feel more important.

Roscoe told Eddie that he had a friend at the phone company who could get him into the switching room that housed the powerful computer controlling all the telephones in Hollywood. During one recent late-night visit there, Roscoe told Eddie, he had walked over to a wall that was blanketed with telephone company switching equipment and watched as his friend flicked a switch. At the sound of a female voice, the friend announced proudly, "That's Farrah Fawcett." Bored telephone company employees, Roscoe's friend claimed, monitored people's calls all the time. Roscoe told the story in such precise detail that Eddie had no doubt about its truth.

In the early reporting stages of Eddie's article, Roscoe was highly secretive about his whereabouts, guarding his daily comings and goings like a fugitive. If Eddie wanted to speak with him, he would have to wait for him to call. After a few weeks, Roscoe gave him a number where he could leave a message. It was a month before Roscoe invited the reporter to his home, one of ten units in a plain two-story white stucco building located in a shabby neighborhood on the southern edge of Hollywood that was cluttered with similar apartment buildings. Roscoe lived with his mother in a ground-floor two-bedroom apartment. Adult bookstores dominated the neighborhood.

There seemed to be little in Roscoe's life besides school and telephones. Eddie heard that Roscoe had a girlfriend, but he never saw any sign of her. Roscoe had once introduced Eddie to someone named Susan, a bizarre and cranky young woman, eccentrically tall and with unusually wide hips, but she and Roscoe appeared to be just friends. His world was the telephone, and from his small bedroom in the back he operated his HOBO-UFO conference. His phone rang constantly as people called the line. Roscoe continuously monitored the conference through a speakerphone, which created a constant low level of conversation in the room, and he could pick up his telephone any time and interrupt. Another line attached to an answering machine rang frequently as well. Many of those calls came from giggling teenage girls to whom Roscoe had given his private number.

There was something about the way Roscoe reacted to telephone tones that made Eddie suspect he had a musical ear. He could recite a phone number just by hearing it dialed. Spying an upright piano and a large stereo in the modest quarters, Eddie decided he was right. Scholastic awards from Belmont High School lined Roscoe's bedroom wall; now he was attending USC on a scholarship. And from what Eddie could see of Roscoe's interaction with his mother, an immigrant from Argentina who appeared to speak no English, Roscoe was a model son. Eddie was taken aback to hear Roscoe's perfect Spanish, spoken with the unself-conscious ease of a native. Roscoe's complexion was so pale and his English so flatly American that Eddie would have put his roots far north of Argentina, possibly in Iowa.

If Roscoe had a side that was less restrained and formal, he displayed it to Eddie just once. On a drive through Hollywood, down a section of Western Avenue elevated above the freeway, Eddie and Roscoe were stopped at a red light a few doors away from a storefront church, where a small congregation of Hispanics lingered outside. "Slow down!" Roscoe blurted out suddenly from the passenger's seat. "I'm going to freak these people out!" He rolled down his window and leaned his torso out the window. "Ay Dios mío!" he screamed in perfect Spanish, in the evangelical wail of a recent convert. As they accelerated away from the horrified worshipers, Roscoe was beside himself with laughter. "That gets 'em every time!" he cried.

For the most part, the people who called Roscoe's conference lived for their telephonic encounters. Many were blind, Roscoe told Eddie, or otherwise handicapped. Others were housewives or single mothers. The majority were overweight. Their names—Rick the Trip, Regina Watts Towers, Dan Dual-Phase, Mike Montage—suggested to Eddie a group of shy folk making sad attempts to add mystery and intrigue to their lives. Eddie's suspicions were confirmed when, several weeks into his reporting, he attended a phreak party at Dan Dual-Phase's house. For many members of the group, it couldn't be a simple drop-in affair. So few of the conference line callers could drive that the transportation logistics alone had lent an unusual air to the party. At the event itself, the conference line callers sat in scattered clutches of embarrassed silence, too shy to speak to one another in person.

As part of the education process, Roscoe presented Eddie with literature. He gave the reporter several back issues of TAP, referring to it as if its circulation matched that of Newsweek. After a mystifying perusal of TAP's pages, with its proclamations that no code can be completely secure, Eddie could only conclude that it carried the voice of true outlaws. Roscoe also showed him a legendary Esquire magazine story about phone phreaks. One of the featured phone phreaks in the story was Captain Crunch. Roscoe angrily denounced Crunch as an idiot whose blue box was a crutch. Phone phreaks should be like Houdinis, able to cruise the telephone network as if by magic, without the visible aid of tools. Roscoe did have one hardware crutch—a Touch-Tone dialer, a square, gray plastic box small enough to fit in one's palm. On the front was a dialing pad with ten numbers; batteries were taped to the back. The dialer was elegantly simple yet indispensable. Its function was to send Touch-Tones into the mouthpiece of a rotary-dial phone. These were precisely the tones that he needed to get access to corporate phone networks for his free telephone calls. Once he reached a company's private telephone system, a friendly digitized voice asked for a code. After receiving a correct code punched into the dialer, the phone system stood open for dialing anywhere at all.

One day during Eddie's reporting assignment, Roscoe had with him a slightly younger friend named Kevin. Eddie had already heard about Kevin from Roscoe. If Eddie found Roscoe's knowledge of telephones impressive, Roscoe had told him in a reverential tone, he should meet Kevin. Kevin lived about forty-five minutes away from Roscoe in the San Fernando Valley, and until Kevin got his driver's license, Roscoe regularly retrieved him and then took him home.

Kevin was overweight and exceedingly shy. Roscoe, in fact, was downright garrulous compared with his laconic friend, who assiduously avoided eye contact. Where Eddie had seen glimpses of normalcy in Roscoe, in Kevin he saw nothing but a life steeped in telephones and computers. When he joined Roscoe and Kevin for an afternoon of phreaking, Eddie noticed that Roscoe frequently deferred to Kevin, whose encyclopedic knowledge of the telephone company's computer-ized control switches was well beyond his own. And where Roscoe was clearly taken with the idea of a reporter trailing after him for weeks on end, Kevin was altogether uninterested in Eddie. In fact, Roscoe too seemed temporarily to lose interest in the reporter when Kevin was around, so deep was his concentration on the business of phreaking with Kevin.

Eddie was struck by the patience and perseverance the two youths displayed when seated before a computer screen. Eddie had seen Roscoe spend an hour at a time simply scanning for dialing codes, but the display of endurance when Roscoe and Kevin joined forces was in another league. For five hours on one occasion, they sat in front of a computer terminal that was connected to a phone company computer, watching a series of numbers scroll by. Roscoe and Kevin grew increasingly excited over the hieroglyphics on the screen, but their excitement passed right over Eddie. If the reporter asked them to explain what they were seeing, he received a sidelong glance of amused condescension. If these two phreaks were breaking the law, Eddie couldn't tell. Before long he stopped trying to understand what they were doing and found himself struggling to stay awake.

Eddie had an inkling that Kevin might be even more important than Roscoe for his story. Not only did Kevin seem to know more than Roscoe about telephones, but Eddie got the impression that Kevin had taught Roscoe much of what he knew. Eddie figured he should probably take the kid out for a cheeseburger, but he decided he wanted nothing to do with him. As he felt himself pulled further into this strange world, he realized that the phreaks were beginning to make him nervous. In reality, Roscoe's life wasn't much richer than the lives of the lonely souls who hung on his conference line. Yet Roscoe regarded the others with a mixture of delight and disdain: delight at their obvious respect for him, and disdain for the emptiness in their lives. But they just depressed Eddie. Between the desperate conference-line callers and the sinkhole of technical language, clouded further by Roscoe's impossibly stilted speech, Eddie was beginning to regret that he had tackled such a difficult story. He began to yearn for an easy, more familiar assignment, perhaps a backstage interview with Joey and the Pizzas. At the least, his stint as an honorary member of the gang had come to an end. It was time to write his story.

Eddie's cover story in the *L.A. Weekly* in the summer of 1980 left a lasting impression on those who saw it. Mostly it was a profile of Roscoe, a phone phreak who could do anything with a telephone. Shortly after the story appeared, Eddie was at a party and overheard a conversation about it. When he mentioned that he had written it, people said they wanted to meet this Roscoe and learn a few of his tricks.

▲ ▼ ▲

Meanwhile, Susan was obsessed by her desire to get something on Roscoe, something she could hold over him and something other people would believe. It wouldn't be difficult, she figured, since his forays often led him into dubious territory. And she wouldn't mind getting Kevin as well. Kevin and Roscoe had a long-standing agreement to share information with each other, usually to the exclusion of Susan. No, she wouldn't mind one bit if Kevin got pulled into the undertow.

The first opportunity for revenge came with the U.S. Leasing break-in. With a little programming flourish, Roscoe had given himself system privileges on a computer at U.S. Leasing, a San Francisco company with subsidiaries specializing in leasing electronic equipment, railcars and computers. In most large multiuser computers there is a hierarchy of privilege meted out to users. The system manager has the highest level of privilege, acting as *de facto* God in the computer system, while ordinary users have their capabilities more stringently curtailed. Such a pecking order works only if the lesser users cannot find ways to masquerade as the system manager.

Because its network address circulated widely throughout the phreak community in 1980, U.S. Leasing had one of the most popular computers for phreaks to play on. Both Roscoe and Susan, in fact, enjoyed posting the computer's network address on electronic bulletin boards, offering guided tours of the system to phreaking neophytes. U.S. Leasing used nothing but Digital Equipment PDP-11 computers. All the computers ran RSTS, a notoriously insecure operating system. RSTS had been designed in the 1970s as the ultimate in user-friendliness. Ask the system for a password and it would assign you one automatically. Ask for a system status report and you were provided with the name of every user on the system. More often than not, people chose their names as their passwords. To make things easier for its customers, Digital even supplied sample passwords, such as *field* and *test* for field technicians. And the field technicians often had highly privileged accounts.

Getting to computers such as the one at U.S. Leasing was deliciously easy. The first step was to dial into the Telenet network. Telenet was the first commercial network designed solely to link together computers. Computer networks differ from telephone networks in that instead of each conversation getting its own circuit, many computer-to-computer conversations on a single network can share a single circuit. Because the information being transmitted is digitized, strings of ones and zeros, it can be broken up into small packets. Each packet contains an address that tells the network where it's going. This is known as packet-switching.

Telenet was structured in such a way that you could choose to communicate with any computer within the network simply by typing a sequence of numbers, whereupon you were automatically connected to the computer you chose. The computers, each with an assigned number sequence, formed a mesh. Any system connected to Telenet, be it a computer at Bank of America or one at General Foods, could be reached with one local phone call.

Gathering some physical evidence was Susan's first task. She knew how difficult it was to track down people who broke into computers: the only fingerprints they left were electronic, and those were nearly impossible to attach beyond a reasonable doubt to an individual. Reams of printouts logging an intruder's electronic joyride through a computer or a network of computers were worthless unless there was stronger evidence linking the specific person to the incident. Susan's first step was to get something in Roscoe's handwriting. When he jotted the number of the U.S. Leasing computer and several company passwords on a sheet of paper, Susan asked to keep it so she could learn RSTS.

▲ ▼ ▲

The computer staff at U.S. Leasing were baffled when, one day in December 1980, the computers that ran their business began acting strangely. The company's computers were behaving in an unusually sluggish manner. So it was a relief to the computer operator on duty late one afternoon when someone called to say that he was a software troubleshooter working for Digital Equipment Corporation. The slowdown problem, he told the operator, was affecting all of Digital's sites. The situation was so widespread, he said, he wouldn't be able to come in person to fix it; he would have to walk someone through the procedure over the phone. The cheerful technician asked for a phone number for the computers, a login and a password. He said he would then insert a "fix" into the system. The computer operator at U.S. Leasing was only too happy to oblige—it was a procedure he had gone through before with Digital. He thanked the Digital technician, who assured him that everything would be back to normal in the morning.

But the next morning, the computers were as phlegmatic as ever. If anything, the problem was worse. The computer operator called John Whipple, U.S. Leasing's vice-president for data processing, who called the local Digital office in San Francisco and asked for the helpful technician who had called the previous day. There was no employee by that name in San Francisco. So Whipple called Digital headquarters in Massachusetts. Not only was there no such employee on record, but Digital had no plans to make a universal repair. Whipple went straight to the computer room to find the operator. "Someone has been in," he told him.

Whipple's only choice, he decided, was to find and destroy any unauthorized accounts, to call everyone who used the U.S. Leasing computers and have them change their passwords. Later in the day, the computer operator had a second call from the "technician." He was as friendly as ever, explaining that the fix hadn't taken. "I can't seem to get into your machine," he said in a concerned voice.

This time, the operator played dumb. "Give me your number and I'll call you back."

"I'm not really reachable," came the response. "Let me call you back."

When Whipple got to work the next morning, the computer operator was beside himself. The printer connected to one of the computers had been disgorging printouts all night. The floor was papered with them. And every page of every printout was densely covered with type. Covering each printout, repeated hundreds of times was a spiteful message: "THE PHANTOM, THE SYSTEM CRACKER, STRIKES AGAIN. SOON I WILL CRASH YOUR DISKS AND BACKUPS ON SYSTEM A. I HAVE ALREADY CRASHED YOUR SYSTEM B. HAVE FUN TRYING TO RESTORE IT, YOU ASSHOLE."

Another read: "REVENGE IS OURS!"

And finally, in neat rows of type, marching across the page: "FUCK YOU! FUCK YOU! FUCK YOU!"

Interspersed among the vulgarities were names. Roscoe was one. Mitnick was another. Mitnick? MITnick? Could that mean that MIT students had done it?

People who were attracted to the computer profession in the 1960s and 1970s could hardly be considered flamboyant. By and large, they were technical loners. Those who entered the field before the computer industry mushroomed, beckoning thousands of ambitious self-starters with the promise of fortunes to be made, were mostly men like John Whipple. They went to schools like MIT and the California Institute of Technology, and they latched onto computers as an extension of an adolescent compulsion to sit in their rooms and pull radios apart. Others had a simple fascination with math, and computers were just the logical next step.

Whipple had been in the computer industry for twenty years, long enough to have witnessed many a harmless prank. MIT students were famous for harmless pranks. But this was no practical joke. These electronic vandals might as well have broken into the room itself, sprayed graffiti all over the walls and taken a hatchet to the computers. Not only had they plastered printer paper with their malicious messages, they had gone into the files in system B themselves, deleted every scrap of information on inventory and customers and billing notices and replaced everything with their invective. They had, that is, destroyed the computer's entire data base. Whipple ordered both of U.S. Leasing's computers to be cut off from any telephone lines and sealed from outsider access. His chief worry wasn't so much over the lost data, because he also had the information on backup tapes. His main fear was that somehow the intruders had devised a program that would let them back into U.S. Leasing even after all the passwords had been changed, a software contrivance called a trapdoor. That day, the company's computer operations remained shut down while the computer staff loaded the backups, reconstructed the data and restored it on the machines. The entire restoration process took twenty-four hours. After a day, the computers went back on line.

Although all the passwords had been changed, and the system seemed as secure as it was going to get, Whipple decided he wasn't going to rest until arrests had been made. His first call was to Digital headquarters. All he wanted was some idea of how the intruders might have broken in at all, and some assurance that it wouldn't happen again. Anyone at Digital, he figured, would sympathize immediately with what Whipple was going through. At the least he expected a sympathetic ear, and he wouldn't have been surprised if the company dispatched some eager young security expert to San Francisco to take care of everything. But the response he got was tepid if not cool. Within a few minutes, he was mired in a bureaucratic procedure. In order to have Digital even consider the matter, Whipple would have to fill out a purchase order, then have his supervisor authorize the expense. A purchase order? Some invisible creeps were erasing his data, writing "FUCK YOU" all over the place, threatening his other computer, and Digital wanted him to fill out a *purchase order*?

This wasn't the reaction Whipple had expected from Digital, of all companies. Digital was founded in 1957 by Ken Olsen, an individualistic and outspoken MIT engineer. Just as the International Business Machines Corporation was passing the billion-dollar mark in sales of machines that sat behind glass enclosures and processed information in huge batches, Olsen set out with $70,000 in venture capital to build a smaller computer that interacted directly with the user. The idea of an interactive computer had come from a pioneering generation of computer researchers at MIT, and one of their early machines became the model for Digital's first computer, the PDP-1. When Digital delivered one of its first computers to MIT, it was installed in a room one floor above an IBM machine. The IBM computer was locked behind two layers of glass; problems were submitted on batches of IBM cards and results didn't come until the next morning. The Digital computer was accessible to students any time of the day or night; commands were typed on the keyboard and the computer responded in a few seconds. A love affair began. Students gathered around and worked on the new computer until 3:00, 4:00 and 5:00 in the morning. The MIT administration even considered removing the Digital computer because people were so obsessed with it they stopped washing, eating and studying for their classes. And from the beginning, Digital designed its computers to be used in networks.

Digital found its first large customer base at universities and other research institutions. For the same reasons of accessibility and speed, it didn't take long for Digital computers to gain acceptance among commercial customers, too. The complexity that came with buying an IBM mainframe was often overwhelming for people outside of university computer science departments. Digital computers provided a package that was as professional as IBM's, but not nearly as complex.

Whipple had believed Digital to be a company driven by its customers' requirements. He was surprised and dismayed by the company's response to his request for help. Evidently, the bureaucratic tangle served to mask the true situation: the people at Digital headquarters in Massachusetts wanted nothing to do with the event.

So Whipple called the FBI. Three agents arrived at his office the next day. They showed far more concern than anyone at Digital had. But their questions were geared toward trying to figure out whether this was a federal case. There was no doubt that a crime had been committed. There were no federal laws governing computer crime at the time, but if this case fell under federal jurisdiction, it could be prosecuted under federal wire fraud statutes. At the same time, the state of California had a year-old law on the books prohibiting unauthorized access to computer systems. After some deliberation, the FBI decided that, since the break-in didn't appear to have involved interstate telephone calls, it should be handled by local authorities.

Whipple's next call was to the telephone company. He asked Pacific Bell to place traps on the lines. The phone company agreed to do the traces. But that presumed that whoever had committed this act would call again. Whipple had the feeling, even the hope, that these intruders would be arrogant enough to do so. And they were. The call came in the afternoon. This time, the computer operator's object was to keep the "technician" on the line while a trace was completed. "I don't know what's going on here," he moaned into the phone, "but everyone is upset." The operator kept a dialogue going for as long as he could. He cursed U.S. Leasing, saying it had taken away his password, so he couldn't log on to the computers even if he wanted to. Oddly, the friendly technician seemed in no hurry to hang up. With an eerily professional knowledge of the system, he told the operator exactly what to do to bring the computers back on line. Whipple and a phone company employee were listening on another line, and when the operator put the hacker on hold, Whipple could hear two people talking in the background. "I think they're on to us," one said to the other.

The trace went only as far back as an outbound Sprint port in San Francisco—the electronic doorway of GTE Sprint, then a fledgling long-distance telecommunications company. Using several different long-distance accounts to cover their tracks was one of the phone phreaks' favorite tricks. Someone wishing to break into a computer could make himself more difficult to trace by first dialing through the equipment of one or even several long-distance carriers before finally calling his target. In every case false credit card numbers were used. Such a ruse made the phone company's task of completing a trace considerably more difficult. Crossing company boundaries made it easy for phreaks to hide behind layers of bureaucracy that slowed law-enforcement officials.

It took Whipple several phone calls before he finally reached Sprint's security department, such as it was. The security people gave him a number he could call around the clock. If the intruder called again, they told him, he should call the number at once so that they could start their own trace. A few hours later, the intruder did call. Again the unhappy operator kept him chatting, and again the phone company traced the call to the same Sprint telephone number. With the glee of an angler whose hook is securely lodged in the mouth of a prizewinning marlin, Whipple called the number the Sprint security officials had given him. There was no answer.

Whipple had never been much of a candidate for confrontation. He was easygoing and low-key. But now he was on the warpath. To arrive at work one morning and see his printer spewing out obscenities and all of the information on one of his computers obliterated was something Whipple had never imagined. He wanted to find out if other Digital customers in the San Francisco area had been hit. Or did these miscreants have a particular reason for wanting to attack U.S. Leasing? That afternoon, he began calling other large Digital installations. At least a half-dozen others told him they too had had troublemakers in their systems. One large hospital that had just bought its Digital system didn't even have it fully installed, yet had already been getting repeated phone calls from a "Digital technician" demanding the computer dial-up number and passwords. When the hospital computer manager explained to the caller that the computer wasn't even up and running, the indignant caller demanded to speak with the hospital's chief administrator. Whipple asked the hospital administrator if he would be willing to join him in pressing charges. No, came the answer. After all, the hospital had been spared any damage, and the negative publicity, he told Whipple, would far outweigh the satisfaction of seeing justice done.

Whipple realized that if anyone was going to take up a crusade against these electronic foes, it would have to be U.S. Leasing alone. He took the systems down for another day to change everyone's password a second time. He set up the computers to monitor all unsuccessful log-in attempts. The following day, the computers were back on line. By four o'clock that afternoon the intruders were back, too, trying without success to break in again. Like insects smacking themselves senseless against a screen at night to get to the light inside, they kept flying at the system, trying password after password. Once again, U.S. Leasing immediately shut down all its outside lines.

By this time, the company was buzzing with talk of the intrusions. Whipple was stunned when he met a senior executive in the hallway who chuckled and said, "Maybe we should hire this kid." Whoever was doing this certainly seemed to have more than a passing familiarity with the computer's operating system. In fact, from the commands he was typing, he seemed to know more about exploring its nooks and crannies than Whipple and his staff. But hire such a meanspirited person? That would be like giving the Boston Strangler a maintenance job in a nursing-school dormitory. Whipple's reply to the executive was curt: "How many people will we have to hire to watch him?"

As incredulous as he was disappointed, Whipple began to resign himself to writing this off as a colossal, embarrassing waste of energy and time. There wasn't much left to do but count up the losses and try to forget about it. In the end, he figured, the break-in had cost the company a quarter of a million dollars in lost time and business. The chances of ever catching the creeps who had caused all the headaches were slim at best. A few weeks later, Whipple dropped the matter entirely.

Years later, Roscoe and Kevin would claim they had been framed by Susan Thunder, that she had plastered their names all over the U.S. Leasing computer in order to pin the break-in on them, while Susan would hold firm to her claim that Kevin and Roscoe had done it.

▲ ▼ ▲

Continuing her quest for revenge, Susan decided on a frontal assault. She started out slowly, even harmlessly. She began compiling a fact book on Kevin, Roscoe and Steve Rhoades. She talked her way into the telephone billing office, got copies of their phone bills and devised complex frequency charts on the numbers they called. She was also able to get a customer name and address report on each number called. By keeping close tabs on the phone bill of Jo Marie, Roscoe's fiancée, she could track him closely.

It didn't take Roscoe long to figure out that Susan had turned on him. She called him at strange hours and left spiteful messages on his answering machine. One day she called Ernst & Whinney, where he worked in the data processing department, to inform the personnel office that one of the staff members was using the company's computer terminals after hours. As a result of Susan's call, Roscoe was fired. To elude her, he began changing more than just his phone number; he also changed his "cable and pair," a specific set of wires assigned to his apartment by the phone company switching office that ordinarily would have stayed the same when his number changed. But Susan's solution to that was simply to get Jo Marie's bill.

There were other ways of getting phone numbers. When Kevin Mitnick changed his number, as he often did, Susan would drive out to his apartment in Panorama City, clip a test set to the phone line and call a standard number the maintenance people used to find out which number belonged to which cable and pair.

It wasn't long before she realized that Kevin was someone to reckon with. He was even more familiar with the ins and outs of the phone company than Roscoe was. In fact, Kevin had outsmarted Susan by logging in to a phone company computer and disabling the automatic number identification test function for his telephone. If she stood outside his house and entered the code used to read back his number, all she got was a failure.

By then Kevin had begun listening to Susan's private telephone conversations, by attaching a hand-held ham radio and clipping it to Susan's phone lines in a phone terminal box under a carport a few yards from her apartment. Anyone else who wanted to listen just needed to tune in to a little-used frequency on an FM radio. Kevin and Roscoe began taping her calls, most of which were long late-night dialogues with another phone phreak she had started to date. Kevin and Roscoe tittered in the background as a languorous Susan described the tricks of the prostitute trade to her boyfriend in full detail. And they taped her own taped greeting from the Leather Castle, the bordello where she plied her trade. "Jeanine," unmistakably Susan but uncharacteristically sweet, gave out the current rates for each service: $45 for a half hour "if you're dominant," $40 "if you're submissive" and $60 "if you'd like to wrestle." For a while, Kevin and Roscoe followed her around in separate cars, communicating with each other over their radios like two cops on a trail.

When she discovered them, Susan escalated the war by bringing it into the public domain.

▲ ▼ ▲

Bernard Klatt was one of the tens of thousands of technical workers in Silicon Valley, 350 miles north of Los Angeles. These intense engineers, predominantly men, populated the twenty square miles south of Stanford University in increasing numbers after World War II. They created a technological paradise that spawned first the semiconductor, then the microprocessor and finally the personal computer.

An electrical engineer, Klatt worked as a technician at Digital Equipment Santa Clara, servicing and troubleshooting the company's computers. For Klatt, a tall, dark-complexioned Canadian who was aloof and formal with strangers, as for many others in Silicon Valley, computing was part vocation and part passion.

He lived with his wife in a two-story apartment building in Santa Clara, several blocks off El Camino Real, Silicon Valley's main commercial thoroughfare. The building was one of thousands built as Silicon Valley exploded with new industry in the 1960s: a contemporary prefab structure with thin walls and a kitchen separated from the living room by a belly-high divider that served simultaneously as wall, counter and breakfast table.

Like many in Silicon Valley, Klatt had given over his spare bedroom to his computers. But while others tinkered with inexpensive personal computers, Klatt was busying himself with a surplus Digital Equipment minicomputer. One of the company's most successful computers in the 1960s, the machine, known as a PDP-8, was awkward to program by today's standards, but it was one of the first computers that didn't require an industrial-strength life-support system. Klatt was able to power it by simply plugging it into his apartment's wall outlet, an unusual feature for large computers of its time.

Klatt decided to make his computer useful to a wider audience. Working with a friend, he wrote a program in the version of the BASIC computer language used on the PDP-8 that would allow a caller with a modem to dial in and transmit and receive messages. The PDP-8 stored the messages and then waited to display them for the next caller. This turned it into a bulletin board system, or BBS for short.

Bulletin boards like Klatt's grew in popularity in the early 1980s, spreading around the United States as part of the personal computer explosion. They were a little like neighborhood bulletin boards found in laundromats or community centers, offering lawn cutting services and free kittens, but with more vitality. Since messages could be appended to other messages and categories could be organized to be scanned quickly by computer, the systems subdivided into special-interest categories. Most of the discussions were about computers, but subjects ran the gamut from science fiction to odd sexual practices. Private messages could be left for a particular user, while public messages were accessible to anyone who logged in. Callers could check in every day to see what had been added and to append their own comments. A kind of electronic stream of consciousness emerged as BBSs became the digital substitute for a neighborhood chat on the front stoop. By 1980 there were well over a thousand BBSs around the country. Today there are at least ten times as many.

They covered a remarkable variety of features. Some permitted callers to store and retrieve software. This in turn gave rise to the world of free software and "shareware." Some programmers made substantial sums of money by giving their software away on approval, requesting that the users send a payment only if they found it valuable.

Inevitably, by the early 1980s, pirate BBSs had emerged as well. Legendary boards like Pirate's Cove in Boston permitted callers to download (using modems to transfer it to their own computers) commercial software without paying for it, and shared advice on how to break software copy-protection schemes. Some of the pirate boards were well known, their telephone numbers posted widely. Other boards were more secretive, accessible only to a limited membership.

Klatt's bulletin board, called 8BBS, was one of the nation's first bulletin boards for phone phreaks. Bernard Klatt had a fanatical commitment to freedom of speech. After putting his system on line in March of 1980 and publishing its number on several other systems, he insisted that his board be a free haven for its users. The ground rules, which were set out for new callers in an introductory computer message, were simple:

Notice:Uncontrolled message content.
    Proceed at your own risk.

8BBS management specifically disclaims any
responsibility or liability for the contents
of any message on this system. No
representations are made concerning accuracy
or appropriateness of message content. No
responsibility is assumed in conjunction
with message 'privacy'. 8BBS acts in the
capacity of a 'common' carrier and cannot
and does not control the content of messages
entered.

As a result, within months after 8BBS first began operation, a ragtag collection of phone phreaks and computer aficionados had discovered it and established it as one of the nation's premier clandestine electronic meeting places. Anyone dialing in was immediately struck by the dedication that the community of several hundred regular callers had to the subculture of phone phreaking. They were attracted by the notion that they were participating in some kind of high-technology avant-garde. Others called just to browse or "lurk," reading posted comments without making their presence known. 8BBS was frequently busy around the clock.

Soon the board had national scope. Callers dialed in from as far away as Philadelphia. Some would place their calls using purloined telephone credit card numbers to avoid long-distance charges. Others would exchange illegal information. Credit card numbers, computer passwords and technical information on telephone networks and computers were all stored and read on 8BBS. It was impossible to tell whether the person sitting at the other end of a message was a skilled professional systems programmer or a teenager in the family den perched in front of a Commodore 64.

It was in December of 1980 that both Roscoe and Susan, already locked in a bitter war, found their way to 8BBS. For several months they became regular callers, leaving general tips and trading information, while at the same time, as many phreaks are prone to do, flaunting their particular skills.

Even their first messages were revealing. Susan was ever the seducer:

```
Message number 4375 is 14 lines from Susan
Thunder
To ALL at 04:38:02 on 04-Dec-80
Subject: COMPUTER PHREAKING

I am new in computer phreaking and don't
know that much about systems and access. I
have however, been a phone phreak for quite
awhile and know alot about the subject of
telephones . . . (I think) . . . anyway, I
would very much like to chat with anyone
interested in sharing information,
especially about computers. By the way, I am
a 6 ft. 2 inch blonde female with hazel
eyes, weight 140, and I enjoy travelling
alot. If anyone can suggest any neat places
to go on weekends, let me know . . .
```

Roscoe was the cocky self-styled techno-wizard, breathless in his first private message to Bernard Klatt:

```
Message number 4480 is 20 lines from ROSCOE
[RP] to SYSOP at 18:38:27 on 06-Dec-80.
Subject: ROSCOE

MUST CONSERVE SPACE. I AM ROSCOE, FAMOUS IN
L.A., CA AND IN THE PAPERS OF L.A. FOR MY
PHONE-COMPTR PHREAKING, FREE ACCESS TO ALL,
AIRLINE TICKTS, ETC. VER IMPRSED WITH YR
SYSTM. NOT HAD CHANCE TO REVIEW ALL MSGS
YET. I WILL BE LEAVING YOU MORE DETAIL MSG
WITHIN 7 DAYS BECAUSE I CAN BE OF GREAT
ASSIST TO YOU AND ALL USRS. BASICALLY: TELL
ME WHAT INFO TO WHAT YOU NEED, I CAN GET
. . . ANYTHING. FREE AIRLINE. FREE HOTEL,
FREE CALLS . . . BEST PHONE PHREAK IN L.A.!
(I HAVE REPUTATION) AND ANM KNOWN BY MANY,
WAS ON THAT'S INCREDIBLE, T.V. SHOW IN L.A.,
AND AM MOST POWERFUL IN LA.A. HAVE DEC
SYSTEM 1,2 PASWR & DIAL-UPS IN ALL 48
```

STATES, HAVE MANY PRIVED ACCTS, CAN CRASH
SYSTEMS WITHIN 20 DAYS OF REQUEST TO DO SO,
CAN GET ANY PHONE NO TO ANYTHING!!! (MORE
THAN JUST NON-PUB . . . CAN RUN DMV VIA
TERMINAL . . . CAN RUN INTERPOL VIA TERMINAL
. . . CAN RUN AIRLINE TICKET VIA TERMINAL
. . . CAN RUN PHONE COMPANY VIA TERMINAL
. . . FREE PHONE SERVICE, FRE CUSTOM
CALLING, ETC. CAN DO MUCH MORE . . . I HAVE
SO MUCH TO SAY THAT CAN'T FIGURE OUT WHAT TO
SAY FIRST . . . EXUSE POOR REAIBILITY OF
THIS LETTER . . . PLEASE CALL ME AT
(213)469-. . . . VOICE AND LEAVE MSG ON HOW
TO CALL YOU, OF IF YOU DON'T MIND, I WILL
GET YOUR NUMBER, BUT ONLY IF YOU DON'T MIND.
I ALSO WIT WANT TO KNOW ON HOW TO GET THAT
PHONE-PHUN BOOK?
THANK YOU FOR THE XCLNT SYSTEM ROSCOE.

With the arrival of the Southern California phone phreaks on 8BBS in late 1980, the board took a disturbing twist. The anonymity that once was a source of comfort now led to rising paranoia. Susan and Roscoe's war spilled out in bitter public messages, and the ominous possibility that not only the computer underground but law-enforcement agencies were reading 8BBS became a concern.

Susan, who was growing more and more computer literate, was often antagonistic and haughty in her public postings. In February 1981, she claimed that Roscoe was collaborating with the enemy:

Message number 6706 is 16 lines from Susan
Thunder
To **>ALL<** at 00:44:18 on 27-Feb-81
Subject: GOING ON VACATION . . .

I AM GOING ON A CRUISE OF THE CARIBBEAN AND
BAHAMAS AND WILL BE BACK IN ABOUT A WEEK OR
SO . . . DON'T EXPECT ANY REPLIES FROM ME
DURING THAT TIME.

THERE IS A VERY GOOD CHANCE THAT ROSCOE IS
COLLABORATING WITH THE FBI TO TRAP PHREAKS:

```
GIVEN THE TROUBLE HE HAS RECENTLY HAD, I
KNOW THAT SOMETHING'S AMISS WHEN HE STARTS
LEAVING PUBLIC MESSAGES, ESPECIALLY TO
ANTON,* SEEING AS HOW HE AND ANTON TALK ON
THE PHONE FOR SEVERAL HOURS EACH DAY . . .
WHY WOULD HE LEAVE THE MESSAGES IN A PUBLIC
SYSTEM ANYHOW? DON'T CALL THE SYSTEMS HE
LEAVES: IT'S EITHER A TRAP, OR HE WANTS YOU
ALL TO DO HIS REVENGE WORK FOR HIM!!
HAVE FUN!

THUNDR@MIT-DM,MIT-AI
```

Outside of his spat with Susan, Roscoe was a good citizen on the BBS. Out of concern for the well-being of the board, he told other callers he believed the phone company had installed a line monitor on 8BBS. If telephone numbers were being recorded then everyone was potentially in trouble, regardless of their aliases. The news cast a pall over the BBS. The rising paranoia also affected Roscoe. He began to contemplate getting out of the underground.

```
WELL IT SEEMS THAT I, ROSCOE, AM GETTING
TOO OLD TO KEEP UP WITH SO MUCH PHREAKING.
MY INFORMATION GATHERING IS TAKING UP ABOUT
4 TO 5 HOURS OF MY TIME EVERY DAY, AND
THAT'S TOO MUCH CONSIDERING THE FACT THAT I
WORK FULL TIME AND ATTEND SCHOOL FULL TIME.
SOON TO RETIRE . . . ROSCOE.
```

Then, as if to help Roscoe out the door, Susan attacked him more viciously than ever, adding statutory rape to the list of crimes she wanted the others to believe Roscoe had committed:

```
YOU ARE GOING TO JAIL FOR A LONG TIME FOR
CRASHING AND ZEROING THE U.S. LEASING RSTS/
E. THEY WILL GET A CONVICTION. DID YOUR
CHILD-MOLESTING CASE EVER GET RESOLVED? 16
COMPLAINTS FROM AS MANY DIFFERENT 11-15 YEAR
OLD GIRLS IS PRETTY SICK. THEY SHOULD HAVE
PUT YOU AWAY A LONG TIME AGO.
```

* ["Anton Chernoff," an early computer programmer, was Kevin's nickname of choice on 8BBS.]

In mid-1981 the private war that Susan had taken public began to wane—at least on 8BBS. After Roscoe and Susan were gone, 8BBS continued to operate for almost another year. The board had lost some of its computer underground edge and shifted back in the direction of less controversial computer hobbyist concerns.

But its phone phreak roots brought 8BBS crashing down. In early 1982 Bernard Klatt received a faster modem in the mail, a gift from some Philadelphia phone phreaks wishing to improve the speed of 8BBS. He didn't know that it was stolen, purchased by mail using a false credit card. In April of 1982, while he was on vacation in Canada, a team of law-enforcement officials from the Santa Clara County sheriff's office and telephone company security agents came to his apartment with a search warrant. A fire ax shattered his front door and the agents confiscated all of the 8BBS disks and backup tapes.

When he returned, Bernard Klatt wasn't prosecuted, as he hadn't known that the gift modem was stolen property. But when his employer found out about the search of his home and the nature of the computer hobby, he lost his job. The 8BBS era had ended.

▲ ▼ ▲

Perhaps, Susan said years later, if Roscoe hadn't filed a civil harassment complaint against her, dragging her into the ugly legal arena first, she wouldn't have used the evidence. It was early 1981 and the warring had been going on for several months when Susan was summoned to the Los Angeles city attorney's office for a hearing regarding Roscoe's complaint that she was disrupting his life with obscene and threatening telephone calls to him and to his mother. Susan nodded when the hearing officer admonished her to cease this behavior. Her opportunity for decisive revenge finally came later that year, after the COSMOS incident.

It was at the Shakey's Pizza Parlor during Memorial Day weekend in 1981 that Kevin and Roscoe decided to break into Pacific Bell's COSMOS center in downtown L.A. Understanding COSMOS was essential for efficient phreaking. COSMOS (the acronym stands for "Computer System for Mainframe Operations") is a large data-base program used by local telephone companies throughout the nation for everything from maintaining telephone cables to keeping records and carrying out service orders. In 1981, most of the hundreds of COSMOS systems installed around the country were running on Digital Equipment computers. In order to manipulate COSMOS, it was necessary to know a dozen or so routine commands, all of which were outlined in telephone company manuals. A thirty-minute foraging session in the trash bins behind the COSMOS center could yield considerable booty: discarded printouts, notes passed between employees and scraps of paper containing passwords.

Though most trashing was done late at night, even those who preferred to trash at midday were seldom spotted. If they were, they usually had a plausible enough reason for being found knee-deep in the garbage. Susan's favorite method of sneaking manuals and printouts out of dumpsters was to bring along a bag of aluminum cans and hide the booty among the cans. Roscoe claimed he was hunting for recyclable material.

Roscoe, Kevin and Mark Ross, a friend and occasional phreak, set out from Shakey's late one night in a caravan of three cars. When they arrived at the phone company parking lot, it was a short walk through an open chain-link fence—why, after all, should the phone company bother locking up its trash?—into an area containing several Dumpsters. They clambered in and started wading through old coffee grounds, discarded Styrofoam cups and other garbage in search of the occasional gem.

Kevin and Roscoe already knew quite a bit about COSMOS, but they wanted to get a firm grip on the password setup. Certain passwords were needed for certain levels of privilege in the COSMOS system. The password to the most privileged account of all, called the root account, still eluded them. The root account was the master key to all the other accounts on the computer system; its password would give them the power to do anything in COSMOS.

The trash cans' offerings were paltry that night. Perhaps another phreak had been there earlier in the evening. Kevin found a handbook to a computer called the "frame" that would probably be useful, but otherwise the pickings were thin. After taking what they had found in the trash to their respective cars, the three phreaks huddled briefly to discuss whether to go into the COSMOS center itself. Kevin wanted to find room 108, the office that he knew contained the COSMOS computer itself, a guaranteed lode of information.

At first, Roscoe wanted Kevin to approach the guard by himself. Kevin would impersonate a Pacific Bell employee, tell the guard that two friends would be arriving in a few minutes for a tour of the facility and, if the guard bought it, Kevin would use his hand-held walkie-talkie to transmit a Touch-Tone signal to Roscoe's walkie-talkie telling him that the plan had worked.

But Kevin was nervous; he didn't want to go in by himself. They would all go together or not at all, he said. The three argued about the matter for several minutes until finally the other two agreed to accompany Kevin to the back entrance, where a guard sat at a reception desk. They piled into Kevin's car and backed out of the trash area. In order to be conspicuous in arriving this time, Kevin drove around to the main parking lot in the back of the building and parked with his headlights beaming in through the glass door.

If the guard thought there was anything peculiar about a Pacific Bell employee arriving at 1:00 A.M. on a Sunday to give a guided tour to two friends, he didn't show it. Kevin appeared far older than his seventeen years, and he talked a convincing line. As Roscoe and Mark stood by, Kevin chatted up the guard. First came idle chitchat: Kevin had a report due on Monday, he said, and was upset that he had to come in to work on Memorial Day weekend. Apparently welcoming the distraction from his nighttime vigil, the guard didn't ask to see Kevin's company ID, nor did he ask which department the pudgy young man worked for. In a friendly and inquisitive manner, Kevin strolled over to a television monitor that wasn't working and asked the guard about it. They both shrugged. "Just on the blink, I guess," Kevin suggested. Then, as if he had done it a thousand times, he signed the name Fred Weiner in the logbook, and Sam Holliday for Roscoe. His imagination apparently spent, he entered M. Ross for Mark Ross.

Once inside, after a false start that led them into the mail room, the visitors dispersed through the halls. It didn't take long for them to find the COSMOS computer center in room 108. A metal basket attached to the wall held a list of dial-up numbers for calling the COSMOS computer from outside. Kevin removed the list and put it in his pocket. And in a paper tray on a desk they found a list of codes to the digital door locks at nine telephone central offices.

In the Rolodex files on several desks, the three young men planted cards printed with the names John Draper (the real name of the infamous Captain Crunch) and John Harris. Next to John Harris's name was the telephone number of the Prestige Coffee Shop in Van Nuys—a number that phreaks routinely intercepted and rerouted to a loop line; next to Draper's name was the number of a nearby pay phone. The idea was to create instant credibility: if one of them called claiming to be John Draper, a Pacific Bell employee, and if as a security measure the recipient of the phone call consulted his Rolodex file to verify the name and number, he would find the information.

Next to room 108 was the COSMOS manager's office. The walls were lined with shelves holding six large manuals. A quick look through the books was all they needed to see that this was perhaps the most important thing they would find that night; they were the COSMOS manuals, which contained information on everything that could be done with the COSMOS computer. The group perused the shelves and set aside the books they wanted to take. Once they had gathered what they thought they could use, they piled it all on one desk and sorted it so as to reduce the bulk and appear less suspicious to the guard. Roscoe spotted a brief-case next to a desk; he picked it up and put the manuals inside. Then, each carrying as much as he could, they walked back out to the guard station, where Kevin signed out and exchanged hearty farewells with the guard. Just as casual as he was about letting them in, the guard seemed thoroughly unconcerned that these three young men had arrived empty-handed and were leaving laden with a briefcase and armloads of books. Kevin drove the others to their cars and the three drove in single file to the nearest Winchell's Donuts to divide their plunder. It had all taken less than two hours.

Unfortunately, they had been too greedy. When the COSMOS manager arrived on Monday morning, he couldn't help but notice that his shelves were nearly bare. He notified security immediately, and questioned the guard who had been there over the weekend. The guard mentioned the threesome from Sunday morning, and said he was sure he could recognize them again. Then the manager found two cards in his Rolodex with unfamiliar handwriting, which belonged to no one in the office. It was time to call corporate security.

▲ ▼ ▲

Meanwhile, Susan was sitting on the evidence she had been gathering for nearly a year. She had intimations that the L.A. district attorney was preparing to indict her on a dozen or so felony charges, including unlawful entry into Pacific Bell headquarters and conspiring to commit computer fraud. So she decided it was time to visit Bob Ewen, an investigator in the district attorney's office, to blow the whistle on her erstwhile buddies. She told Ewen that she had come to him because she was concerned for her nation's security: Kevin knew she had some extremely sensitive information, and Susan was certain that he wanted to get his hands on it.

By way of example, Susan told Ewen that she had once spent several days locked in her apartment, and that when she emerged she told Roscoe and Kevin that she had been downloading and printing out missile firing parameters and maintenance schedules for intercontinental ballistic missiles. She boasted that she knew the schedules of the men who worked in the hole (the men authorized to turn the keys), and that she knew when the maintenance was done and what the backup systems were. Susan claimed that, given the right information, one pimply adolescent scrunched into a phone booth with a terminal and a modem— and *certainly* Roscoe and Kevin—could set off the necessary chain of commands to release hundreds of missiles from their silos and send them hurtling across the globe. She hadn't tested her theory, of course, but she claimed to have developed a method for fooling the computer, leading it to believe that the necessary sequence of actions preceding a first or retaliatory strike had already occurred. The real vulnerability, Susan believed, lay in the nation's communications system. The Pentagon's heavy dependence on the Bell System made it a sitting duck for the likes of Roscoe and Kevin. Susan asked for immunity from prosecution in exchange for doing her civic duty and testifying against Kevin and Roscoe.

Ewen knew enough to be leery of many of the "theories" that issued forth from this rara avis. For one thing, he recognized her as the same tall, buck-toothed blond whose face was familiar to guards at Pacific Bell offices throughout the county. She had been spotted in Pasadena with Mitnick and Rhoades, and in the San Fernando Valley with the one who called himself Roscoe. But Ewen wanted to see Roscoe and Mitnick convicted. He sent her to see the district attorney. As it happened, the district attorney felt exactly the same way. He offered to give her immunity in exchange for her full cooperation. Susan, of course, was fully prepared to take the witness stand and tell her version of everything.

When Bob Ewen came onto the COSMOS case in 1981, he had already been dealing with phone phreaks for some time. For years, he kept a collection of confiscated blue boxes and other electronic fraud gear in a box under his desk. Usually there wasn't much to fear from phone phreaks, but when he went out to serve a warrant on Mitnick, Ewen wasn't sure what he was dealing with. As Susan Thunder had described the suspect, he was potentially dangerous. Ewen went first to Mitnick's residence to carry out a search. Mitnick wasn't home, but his mother was. Reed-thin and wearing a skirt that struck Ewen as too short for a woman approaching forty, Shelly Jaffe, who bore some resemblance to Popeye's girl, Olive Oyl, seemed confused and flustered. As Ewen picked his way through her son's room, Shelly stood at the doorway and popped her gum nervously. She reminded him of a teenager who hadn't outgrown the habit of reading movie magazines and still harbored a dream of becoming a Hollywood discovery.

The search wasn't easy. Mitnick appeared to be something of a pack rat. He had apparently saved every scrap of paper and every printout he had ever obtained, and Ewen had to scrutinize all of it. He unearthed printouts filled with telephone company information, computer passwords and material from one of the computer centers at the University of California at Los Angeles. Shelly became increasingly flustered. She did not, she said, know anything of her son's illegal doings. She assured Ewen that Kevin could not be responsible for the things the police suspected. Kevin, she insisted, was an angelic son.

"Then what's this?" Ewen asked, holding up a printout of telephone company information.

"I've never seen that before," she responded.

"When is the last time you were in your son's room?"

"I don't remember."

Ewen got the impression that Shelly, who displayed no hint of malice herself, might be intimidated by her son. It was evident that she tried to maintain a decent home. Aside from Kevin's cluttered room, which was actually tidy compared with some Ewen had seen, Shelly kept a neat household. But she appeared to have no control over the boy she had raised or the situation at hand. She seldom saw Kevin. Her early hours as a deli waitress rarely coincided with her son's more nocturnal schedule. Clearly, she knew nothing about computers and had no desire to try to learn anything about them. It was Ewen's hunch that even telephones might overwhelm her.

Among all the evidence Ewen gathered in Kevin's bedroom, he found nothing linking the suspect to the COSMOS break-in. There were no manuals, directories or other signs that Kevin had been there. Still, on the strength of the guard's identification of Mitnick, Ewen had obtained an arrest warrant. Although the Mitnick family hadn't been particularly observant Jews, Kevin often went after school to the Stephen Wise Temple in the private community of Bel Air, where he worked part-time. Ewen drove to the synagogue with three other officers; the four men divided into two cars and waited in the parking lot for Mitnick to show up. When Mitnick began to pull into the lot, he apparently saw the men sitting in their unmarked cars, because he hesitated, then drove by. As soon as he took off, the police peeled after him, sirens ablare. Mitnick's car accelerated. Just as Mitnick pulled onto the 405 freeway headed north, his pursuers overtook him and forced him onto the shoulder.

At first, Ewen wasn't sure what he was up against. The suspect was very large, at least two hundred pounds, and possibly armed. Ewen feared Kevin and his buddies might have modified the COSMOS computers and planted a "logic bomb"—a hidden computer program designed to destroy data at a given time. This made Ewen worry that he might even have a terrorist on his hands. So they handled him as roughly as they would a murder suspect: guns drawn, they pushed him onto the hood of his car and handcuffed his hands behind his back. It wasn't until the officers felt not muscle under his shirt but soft and pliable flab that they realized this young man was no physical threat and loosened their grip. Mitnick complained that he had to go to the bathroom immediately. Then he began to cry. "You scared me. I thought I was going to die."

"Why did you think you were going to die?" Ewen asked.

"I thought *they* were after me."

"Who?" asked Ewen, bewildered.

"You know, there are a lot of people that don't like me." Kevin didn't mention it to Ewen, but a few months earlier he had been hunted down by an irate ham operator who was fed up with Kevin's high jinks. Ewen told Kevin he was sorry to hear that people didn't like him. He was just doing his job, he said, and would have to take the teenager to jail in handcuffs. In the car, Kevin began to talk. He admitted to knowing Roscoe and Susan and Steve Rhoades. Then, as if he suddenly realized he was talking to an officer, he stopped talking. When Ewen asked him if he had put a logic bomb in the phone company computers, Kevin looked as if he might start to cry again. No, he insisted, he would never do anything to hurt a computer.

Recovering the COSMOS manuals proved more difficult than finding the suspects. Once Kevin had been taken into custody and his mother and grandmother had hired a lawyer for him, Ewen began to press the lawyer for the missing manuals. Ewen was particularly worried that the manuals would be copied and distributed throughout the phreaking community. He shuddered at the possibility. With passwords and dial-up numbers, phreaks could shut down much of the phone service in the greater Los Angeles area. Kevin's attorney insisted that Kevin didn't have them, but that he might know where to find them. A few days later, the lawyer delivered all of the stolen manuals.

The prosecutor decided to fold the cases against Kevin and Roscoe together. Both were charged with burglary and grand theft, and with conspiring to commit computer fraud—felonies under California law. The charges would be for the U.S. Leasing incident, already more than a year old, and the COSMOS break-in. U.S. Leasing's John Whipple, who had long since dismissed the entire incident as a bad dream, was surprised to hear that the vandals who had crippled and defiled his company's computer had been apprehended and charged with having "maliciously accessed, altered, deleted, damaged and destroyed the U.S. Leasing system." Mark Ross was charged with burglary and conspiring to commit computer fraud in connection with the COSMOS break-in.

After Roscoe was arrested, his mother gave him a choice. He could either use her money to hire her lawyer, an old friend from Argentina who knew nothing about computers or computer crime, or draw on his own meager resources and hire whomever he pleased. Although he had found a good criminal lawyer who had experience in cases such as his and would charge less than his mother's friend, Roscoe could do nothing to persuade her to pay. He ended up with the computer-illiterate Argentine.

While the defendants were out on bail and the prosecution was preparing its case, Ewen decided it might be a good idea to eavesdrop on one of the Shakey's meetings. So one night before the trial, Ewen, the juvenile prosecutor, and Pacific Bell and GTE investigators dressed in the most casual clothing their closets could produce and convened at the Hollywood Shakey's. They seated themselves at a table about thirty feet away from the group. Neither Roscoe nor Mitnick was in attendance. Rhoades was also absent. Instead, seated at the phreaks' table, the center of attention, was Susan. The group of law enforcers sat in disbelief as their star witness took over the meeting, trading phone information with the others, oblivious to the presence of the men she was working for. When she and another phreak got up to leave, Ewen followed her to the parking lot and watched as she retrieved a printout from the other phreak's car. "Susan," Ewen called to her as she was returning to the restaurant, "what do you think you're doing?" It took Susan a few seconds to focus on Ewen; indulging her vanity, she seldom wore her glasses. When she recognized Ewen, she grew defensive. "Hey, I thought I could get some stuff you guys could use."

Ewen looked at her sternly.

"Okay," she conceded, "so I still like to do this stuff."

▲ ▼ ▲

Shortly before the hearing, Kevin Mitnick pled guilty to one count of computer fraud and one burglary count. In exchange for the dismissal of two additional charges and in the hope of being sentenced to probation instead of a term in the California Youth Authority prison system, Kevin agreed to testify for the prosecution, even if it meant betraying Roscoe.

On December 16, 1981, a pretrial hearing was held. The prosecution's first witness was Susan, who cut an imposing figure as she strode to the witness stand and swore to truthfulness. Susan had been given immunity. Under assistant district attorney Clifton Garrott's gentle questioning, she recounted Roscoe's trespass into the U.S. Leasing system. He had arrived one evening the previous December to take her out on a date, and as she was getting ready he sat down at her computer terminal and dialed into U.S. Leasing. When she asked him what he was doing, Susan told the court, he said he was using a program called *god* to get into a privileged account, in order to run a program called MONEY to print out the passwords on the system. Susan went on to explain in detail how the MONEY program could only be called up using a certain account.

Unable to follow the exchange between Garrott and his witness, the judge interrupted. "Slow down a minute," he demanded. "This is a whole new education for me over here."

The proceeding digressed for a few minutes while Susan presented an introductory course in computer terminology for the uninitiated in the room. She explained passwords, logins and account numbers. When Roscoe got the MONEY program to run, she continued, she saw him start to write down passwords and account numbers. And when she looked at the screen a few minutes later, she recalled, the words *delete* and *zeroed* had appeared.

At that point, Garrott produced a ragged eight-and-a-half-by-eleven-inch sheet torn from a three-ring binder and held it up for the judge to see. It was the very piece of paper on which Roscoe had jotted down numbers and passwords on the evening in question. He was shocked to see this incriminating evidence in the prosecutor's hands.

Garrott continued his questioning. He asked Susan if she had asked Roscoe what he was doing at her terminal. "Yes," she answered. He said he was "taking care of business . . . getting even." Then, she said, he began, "laughing hysterically, crazylike." At the prosecutor's gentle prodding, she said she was upset by Roscoe's behavior that night. "The vibes weren't right. There was a lot of tension between us," she told the court. "I said, 'Why did you do that? You know it was destructive.' "

"What did he say?" asked Garrott.

"I got no response."

Jose Mariano Castillo, Roscoe's attorney, did his best to discredit Susan as a witness.

"When in December of 1980 did this incident take place?" he asked her.

"I believe it was the twelfth or thirteenth."

"What day of the week was that?"

"Thursday or Friday."

"Were you employed?"

"Yes."

"Where were you employed?"

"The Leather Castle."

"What was your occupation at the Leather Castle?"

Garrott leapt to his feet to object. The question was irrelevant and immaterial. "What is the relevance of the type of work that she was doing there?" the judge asked Castillo.

"Well, it goes to credibility, Your Honor."

"How?"

"Well, let me ask her specifically," and he turned back to Susan. "Were you a prostitute?"

"No," Susan replied hotly, expressing indignation at the very suggestion.

Garrott objected to the question and the judge sustained his objection.

Castillo continued. "Have you ever been prosecuted for prostitution?"

This time, Garrott objected before Susan could answer.

Try as he might, Castillo could pursue no line of questioning that might help to impugn her testimony. Each time Castillo asked a question of Susan, Garrott objected to it as irrelevant. The judge sustained all but a few of the objections. When the prosecutor didn't object, Susan was coy with Castillo, often craning forward and asking him to repeat the question, as if she were hard of hearing, or confused.

Castillo asked her if she had been in love with Roscoe.

"Infatuated, perhaps," Susan responded. "In love? No."

Castillo asked if Roscoe had been rejecting her advances. "Yes" was Susan's reply.

"And you felt upset because of the rejection?"

"No. He wasn't the only man I was going out with."

"How many people were you going out with, dating, at that time?"

This time, the judge overruled Garrott's objection and instructed Susan to answer the question.

"Three or four." ·

When Castillo began to ask her about her drug problems, Garrott was quick to object again. Castillo persisted. Susan's drug problems, he claimed, had some effect on her emotional state. Moreover, he said, she had sexual problems. The judge rebuffed him and sustained Garrott's objection.

The hearing dragged on this way into the afternoon. Every time Castillo tried to impute vengeful motives to Susan, to sully her credibility or to introduce other names—Kevin Mitnick, Steven Rhoades—as possible suspects, Garrott leapt to his feet and the judge ruled in his favor. All told, Roscoe's prospects were beginning to look bleak. If this pretrial hearing was the legal equivalent of a movie's rough cut, Roscoe could easily imagine how the finished film would turn out.

At the end of the afternoon Kevin was called to testify. He began by explaining the term *phone phreak* to the court. While refraining from placing himself in any category in particular, Kevin described different "types" of phreaks. "One type is a person that likes to manipulate telephone lines and computers," he lectured. "Another one is the type that just likes talking with other people on conference lines. And stuff like that."

Kevin recounted the visit to the COSMOS center in precise detail. The Shakey's meeting. The trashing. The unsuspecting guard. The foraging through room 108. The stealing of the manuals. Garrott showed particular interest in what transpired later at Winchell's Donuts. Kevin told the prosecutor that the manuals were left in the cars but the door code list was brought inside, and that he later made a copy of the door codes for himself. Afterward, Kevin said, he and Roscoe divided up the manuals from behind Kevin's car.

Despite his best efforts, Roscoe's attorney wasn't making much headway in building up a convincing case for his client's innocence. Rather than take chances with a trial and risk fumbling the defense in court, Roscoe's attorney decided to work out a plea bargain with the prosecutor. On April 2, 1982, a day before his twenty-second birthday, Roscoe pled no contest to the conspiracy charge and to the charge of computer fraud against U.S. Leasing. Two months later, he was sentenced to 150 days of work furlough. Ross got thirty days. Kevin had considerably more luck than his confederates: after a ninety-day diagnostic study mandated by the juvenile court system, he got a year's probation.

▲ ▼ ▲

While Roscoe served out his sentence, Susan dropped her alias and refined her new entrepreneurial persona. She became a security consultant. In April of 1982 she appeared on the television news show "20/20." Susan wore a purple blouse beneath a wide brown jumper. The outfit was set off by a pair of oversize star-shaped earrings that swung conspicuously amid limp, straight blond hair. Why, asked an inquisitive Geraldo Rivera, had she and her friends wrought such havoc with the phone company? "I was getting off on the power trip," she replied. "It was neat to think that I could screw up the Bell System." Hassling someone—disconnecting his phone or canceling his car insurance—was all standard practice. "Is there any system that can't be gotten into?" she queried herself. Then she answered her own question. "God, I live by an old saying: if there's a will, there's a way. There is always a way."

In the next two years, Susan became a peripatetic Cassandra, traveling around Los Angeles, painting doomsday scenarios for her clients. She developed a standard set of these grisly prognoses and trotted them out for reporters and lawmakers. A year after her "20/20" appearance, under the auspices of an FBI agent, she flew to Washington to expound her theories before the U.S. Senate. During her testimony, Susan explained about phreaking and trashing, or "garbology," as she put it. She recalled with some nostalgia the talents and feats of the L.A. gang. She talked about posing as repair personnel and altering credit ratings. She described the practice of using the U.S. Leasing computers.

William Cohen, the chairman of the Senate subcommittee holding the hearings, then asked, "There was no indication they felt there was anything wrong about going in and using U.S. Leasing's program? In other words, if you walked into somebody else's house and decided you wanted to watch television for a couple of weeks or a year and just walk in and turn on somebody's set, there would be nothing wrong with that?"

"I guess if people were stupid enough to leave their door unlocked," Susan replied. "Hey, if people are stupid enough to leave the system passwords around that way—that's how the group felt."

Upon learning that Susan's formal education had stopped at the eighth grade, the incredulous senator remarked, "It seems it would be so easy for one of our major adversaries to secure the services of people like yourself who have no advanced degree or training."

"I agree," Susan responded. "It poses a very serious national concern, a very serious threat to national security. I wanted to find out how everything worked. I have got to admit to a certain interest in something that was supposedly classified as top secret."

"Journalists have the same fascination with classified information," Cohen remarked.

"I studied to be a journalist when I was in junior high school."

"I was afraid you were going to say that."

▲ ▼ ▲

While in Washington, Susan got a chance to demonstrate her "social engineering" skills. As Susan later told the story, a team of military brass —colonels and generals from three service branches—sat at a long conference table with a computer terminal, a modem and a telephone. When Susan entered the room, they handed her a sealed envelope containing the name of a computer system and told her to use any abilities or resources that she had to get into that system. Without missing a beat, she logged on to an easily accessible military computer directory to find out where the computer system was. Once she found the system in the directory, she could see what operating system it ran and the name of the officer in charge of that machine. Next, she called the base and put her knowledge of military terminology to work to find out who the commanding officer was at the SCIF, a secret compartmentalized information facility. Oh, yes, Major Hastings. She was chatty, even kittenish. Casually, she told the person she was talking to that she couldn't think of Major Hastings's secretary's name. "Oh," came the reply. "You mean Specialist Buchanan." With that, she called the data center and, switching from nonchalant to authoritative, said, "This is Specialist Buchanan calling on behalf of Major Hastings. He's been trying to access his account on this system and hasn't been able to get through and he'd like to know *why*." When the data center operator balked and started reciting from the procedures manual, her temper flared and her voice dropped in pitch. "Okay, look, I'm not going to screw around here. What is your name, rank and serial number?" Within twenty minutes she had what she later claimed was classified data up on the screen of the computer on the table. A colonel rose from his seat, said, "That will be enough, thank you very much," and pulled the plug.

Computer security experts had been suspecting for years what Susan was proving time and again: the weakest link in any system is the human one. Susan liked to illustrate her belief with the following scenario: Take a computer and put it in a bank vault with ten-foot-thick walls. Power it up with an independent source, with a second independent source for backup. Install a combination lock on the door, along with an electronic beam security system. Give one person access to the vault. Then give one more person access to that system and security is cut in half. With a second person in the picture, Susan said, she could play the two against each other. She could call posing as the secretary of one person, or as a technician in for repair at the request of the other. She could conjure dozens of ruses for using one set of human foibles against another. And the more people with access the better. In the military, hundreds of people have access. At corporations, thousands do. "I don't care how many millions of dollars you spend on hardware," Susan would say. "If you don't have the people trained properly I'm going to get in if I want to get in."

Kevin and Lenny

The list of cohorts with whom Kevin was ordered not to associate was comprehensive. It included Roscoe, Susan, Steven Rhoades and Mark Ross. Kevin also had strict orders to stay away from all phone phreaks in general. In effect, his circle of acquaintances was pared considerably. But there was always Lenny.

Lenny DiCicco had never been attracted to phone phreaking, or trashing, or even social engineering. His was a fascination with buttons. As a little boy in Oak Park, outside of Chicago, Lenny wouldn't bother with a toy unless it had a knob to turn or a lever to push. Even as a toddler, he had a knack for finding the switch to stop an elevator between floors, or the button that would halt an escalator. When he was five, he managed to activate the fire alarm of one of Chicago's largest hospitals, sending nurses and physicians scurrying through the halls. But to his parents, Lenny's compulsion to play with gadgets was an enigma. Gilbert DiCicco (pronounced "duh-SEE-ko") was an illustrator at the *Chicago Tribune*; Vera DiCicco was an artist, too. Their only child eschewed crayons in favor of toys with moving parts and motors.

When Gil DiCicco took an illustrating job with a San Fernando Valley newspaper in 1977, the family moved to Los Angeles. In his very first minutes in Los Angeles, twelve-year-old Lenny exhibited surprising enterprise. As Gil and Lenny waited for Gil's brother to pick them up at the airport, Lenny discovered that a quarter was refunded for each rental cart returned to the automated dispatcher. For half an hour, Lenny ran furiously around inside the terminal, rounding up stray carts and collecting the refunds. An auspicious omen, thought Gil, for starting their new life.

Lenny enrolled at Sepulveda Junior High School in Mission Hills. His formal introduction to computers came at Monroe High School, in John Christ's introductory computer class. Lenny demonstrated an intuitive understanding of the labyrinthine integrated circuits around which computers were built and an aptitude for programming. It was the start of the personal computer era and he developed a passion for computing. And like many his age, the gangly teenager also displayed a special interest in communicating with remote systems over telephone lines. When Lenny managed to log on one day to the school district's central computer, Christ could only chuckle. Two years earlier, a student of Christ's had demonstrated the same talent by using the simple classroom terminal to access the school district's Digital Equipment computer system. His name was Kevin Mitnick. He had dropped out of high school and taken an equivalency test for his diploma. "Oh, no," Christ groaned and smiled at Lenny, "not another Kevin!"

Nothing could have pleased Lenny more.

Lenny knew of Kevin Mitnick, who was something of a legend in Los Angeles high school computing circles. As a student at Monroe, Mitnick had absorbed information like no student before him. His telephone company exploits had been described at length in the *Los Angeles Times*. By Mitnick's own account to the FBI, the paper reported, he and his friends had gained unauthorized access to computers all over the United States—systems supposedly impenetrable to all but selected people who know the passwords. Mitnick also told the FBI that he had obtained sensitive data from "the Ark," one of the main systems in Digital's software development group. In suburban Southern California, where high school heroes traditionally emerged on the gridiron, Kevin had been the perfect antihero: a nerd who used technical wizardry to befuddle the authorities. There seemed to be no end to his clever tricks. While at Monroe, Kevin had reconfigured a modem line so that he could use it to dial out of the school's main computer and into others. The modification gave him a perfect cover for his activities. Anybody who attempted to trace the phone calls would inevitably run into a cold trail at the computer center.

Lenny first met Mitnick in 1980 when a mutual friend introduced them. Lenny and his friend were on their way to a conference of an organization for Digital Equipment customers called DECUS. They had stopped by Kevin's house to invite him to come along. He declined, but asked them to keep their eyes out for interesting literature. National DECUS meetings usually attract as many as twenty thousand Digital users from around the country. For high school students to be intrigued by Digital computers was unusual enough, but for them to go out of their way to attend a DECUS convention was a sign of extreme dedication. Mingling with the attendees, they could be as anonymous as they pleased. The convention hall was a sea of terminals clustered within exhibition booths that demonstrated the latest in Digital computers, software and product literature.

The day after his trip to the DECUS meeting, Lenny got a call from Kevin, who asked if he could get copies of the manuals Lenny had picked up. Lenny had finally met someone whose unusual passion matched his own. Like Lenny, Kevin Mitnick displayed little respect for his elders and even less for the institutions over which they presided. Someone had told Lenny that when Kevin was an adolescent, his mother had so much trouble controlling him she sent him to a disciplinary camp for incorrigible children. On top of the much-publicized COSMOS incident, Mitnick had gained notoriety for breaking into computers at college campuses around Los Angeles. Lenny knew he was dealing with a veteran, and although Lenny was learning his way around the computer underground quickly, he knew he still had a long way to go.

Lessons from Kevin were far more interesting than anything Monroe High School had to offer. Lenny had always found school a useless bore. His attendance record at school was already spotty, and once he met Kevin it got worse. Kevin had a lot of time on his hands. He was working as a delivery boy at Fromin's, a delicatessen in the heart of the San Fernando Valley owned by Arnold Fromin, who was living with Kevin's mother in Panorama City. Hungry for a terminal to use, the pair discovered the San Fernando Valley's Radio Shack stores. Each store had a demonstration model of a personal computer called the TRS-80. Because the computers were also used to send inventory updates each evening to Radio Shack's Fort Worth headquarters, they were equipped with modems as well. Lenny and Kevin merely had to supply a communications software program to transform one of these computers into a full-fledged terminal. Using stolen MCI codes, they could dial long-distance into any computer they could find. Kevin's initial method was to talk someone out of a password. Once he was in, he no longer needed to employ his verbal skills.

What were supposed to be five-minute demonstrations became marathon six-hour reconnaissance sessions. Kevin sometimes pushed his luck, imploring a manager to keep his store open well past closing time. They could last for about a month at each store. When a manager got irked enough by their constant presence to ask them to take their hobby elsewhere, they simply combed the Yellow Pages for new possibilities. They became techno-nomads.

The two teenagers turned their attention to the University of Southern California. Kevin had already been caught fiddling with computers on the USC campus a couple of years earlier, but, as usually happened on the college campuses where he was caught, he was "counseled and released." When he returned with Lenny at his side in 1982, no one recognized him. Kevin already knew the locations of campus terminal rooms that stayed open twenty-four hours a day, available to anyone who could pass as a student.

Both Lenny and Kevin were largely self-taught and they were still learning. Most of what they picked up about computers came not from time spent in classrooms or computer labs but from hours spent with whatever system they could steal some time on. Most of the time, the computers were Digital systems, the most popular computers on college campuses. So it wasn't surprising that the computers they came to know best were Digital.

By the early 1980s, Digital had augmented its PDP series of minicomputers with the VAX line (VAX stands for "Virtual Address Extension," which refers to a way to expand memory that speeds a computer's performance). These computers were designed in such a way that all models could use the same software and share data over computer networks. With the VAX, Digital strengthened its position in the commercial market and started competing successfully in the banking, insurance and accounting markets, traditional IBM strongholds. But Digital still had its roots in the technical and university community. And as its loyal customers at universities such as USC migrated to the VAX, so did Kevin and Lenny.

Phreaking as an end in itself had never done much for Lenny. His affinity for the telephone began and ended with his frequent lengthy conversations with friends. He felt uneasy about misrepresenting himself in order to get information or coerce a password out of someone. And Lenny didn't have Kevin's appetite for electronic revenge. Kevin used phone company computers to wage his private wars against whoever he thought had crossed him. Lenny watched one evening as Kevin attached a local hospital's $30,000 telephone bill to the home phone of a fellow ham radio buff whom Kevin disliked. But there was something that drew him to Kevin. His own reluctance notwithstanding, Lenny was intrigued by Kevin's ability to feign and cajole his way around almost any phone company office, or even a police department. Kevin could call someone at a switching office, for example, and convince him to drive fifteen miles just to turn a computer on. The secret lay in convincing the person on the other end of the line that he was the supervisor of that person's supervisor. He would gain that person's trust by speaking the same language. It was often remarkably simple. Kevin's use of telephone company vernacular snowed most customer service operators, whose full cooperation he needed to, say, pull a victim's toll records.

And some of Kevin's phreaking tools came in handy for Lenny. The telephone company's loop lines—the special numbers it used for testing customer service—were ideal for certain tasks. For example, when Lenny and Kevin applied for jobs and needed to supply a reference, they could give out a test line number, sit on the line waiting for the call, and give the reference themselves. And they made frequent use of phone company test numbers that provided a constant busy signal, or simply rang and rang. To them, it seemed more like harmless hoodwinking than fraud. Free long-distance calls using credit card numbers were perhaps the best part of Kevin's phreak repertoire. But when Lenny tried to show his father how to cash in on the free service, he got a stiff reprimand.

Computers were Lenny's abiding passion. At fifteen, he was still too young for his driver's license, so on weekends and school holidays his father drove him to the nearby Northridge campus of California State University to use the school's computers. Lenny usually asked to be picked up six or seven hours later. The senior DiCicco understood nothing of Lenny's fascination with computers and things technical. Occasionally Lenny tried to explain things to him, but his words sailed over his father's head. Gil DiCicco had grown up in the 1940s and 1950s, when relatives would shout into the telephone to make themselves heard on long-distance calls. Gil didn't see much to celebrate in this so-called computer revolution of the 1980s. If anything, he lamented the loss of communities as he understood them, where people met in person and not electronically, and where teenagers did more than sit in front of computer screens. He was sad to see that young computer enthusiasts felt more comfortable interacting with machines than with people. He was a humanist who prided himself on having nothing to do with these machines. Still, he knew enough to guess that Lenny's computer fixation would land him a well-paying job some day, so he was glad for the time Lenny spent inside the Northridge computer center. Lenny had told him that he had made many friends there, and although he spent a large portion of the day playing computer games, it was keeping him off the streets.

But when Gil got a call from the campus security office one Sunday afternoon, saying he should come retrieve his son, who had just been caught illegally logging on to the highly restricted administrative computer used by the school, among other things, for recording grades, he was sure Kevin Mitnick had something to do with it.

Kevin Mitnick was in fact Gil DiCicco's worst nightmare. Lenny had never shown much interest in school, but he usually managed to bring home fair grades. But when Lenny started hanging around with Kevin, his grades took a dramatic plunge. Gil had only met him once, but there was something unsettling about the overweight teenager. He was always whispering things to Lenny in front of others. This guy wasn't just a bad influence. He was hurting Lenny's chances of ever making good on the future Gil was trying to give his son.

First, Gil admonished Lenny, trying to impress on him how lucky he was that Cal State-Northridge hadn't pressed charges. Furthermore, he told him, he didn't want to see him spending time with Mitnick. The senior DiCicco then decided to take a rash step to see to it that his son's unwholesome friendship with Kevin Mitnick ended: he went to Fromin's and confronted Kevin. "Look," he said, "just stay away from Lenny. It's doing both of you no good." Gil didn't want to see Kevin's face around the DiCicco house again, he told him flatly. From the dull expression on Kevin's face, Gil figured he might as well have delivered his speech to the nearby bowl of half-sour pickles. Kevin shrugged and smiled, avoiding eye contact with Gil. "Uh, okay, I've gotta get back to work now," he replied, and turned away. Gil had no illusions that this little chat had made any impression.

Now that Northridge was off limits, Kevin and Lenny spent weeks at USC, building up their own small empire of purloined accounts. They had managed to get accounts on all the university's computers. When they arrived one evening as usual, they saw that of their six accounts, all but one had been disabled. It was obvious that someone in the computer lab was on to them. Lenny was worried, and he warned Kevin that it was probably a setup. He wanted to leave, but Kevin wouldn't budge, insisting that they restore their lost accounts and that they stay on campus to use the university's own high-speed links. In the end, Lenny shared Kevin's delusion that what they were doing was undercover work worthy of a Hollywood spy thriller. Brazenly walking onto the campus was the kind of thing Robert Redford would do as the hunted CIA researcher in the movie *Three Days of the Condor*. Kevin's favorite movie, it told the story of a former English literature graduate student hired by the CIA to extract plots from novels, who stumbles onto a conspiracy within the agency. In one dramatic scene Redford masquerades as a telephone lineman and throws the agency's high-tech surveillance gear off the track by crossing some wires. And Lenny's own code name was Falcon, after Christopher Boyce, the young TRW employee and protagonist of the book *The Falcon and the Snowman* who in the mid-seventies fed the Soviets secret military technical data that traveled over TRW's satellite communications system.

▲ ▼ ▲

For all Mark Brown knew, it could have been a faculty member who was monkeying around with the system. He wouldn't have been surprised if it turned out to be a student looking for a way to make trouble. Four years earlier, in 1978, two USC students, unhappy with their grades and financial aid, had made news when they were arrested for trying to tap into the university's computer to improve their standing. Those two failed, but years later, in 1985, a year-long investigation would uncover a ring of some thirty USC students in league with someone in the school's records and registration office, who routinely changed grades and created fraudulent degrees. They charged handsomely for the service, selling doctoral degrees for as much as $25,000.

Brown, a young assistant in the computer lab, hadn't encountered any malicious computer attacks. In 1982, USC's attitude toward computing was typical of most universities': it wanted to keep its computers open and accessible to its users. In the late 1970s and early 1980s, security had yet to become a serious issue on college campuses. Setting up a reliable password system was just about the only security barrier at places like USC. Because so many of Digital's customers had told the company that they were more interested in convenience than in security, the operating system software was shipped with many of the built-in security features disabled. While customers could, by reading the instruction manuals, activate those security features, most customers chose not to, since increased security meant increased inconvenience.

In 1982, computers and computer terminals were proliferating on college campuses so quickly that it was almost impossible for computer managers with their limited resources to pay much attention to security. Besides, placing undue restrictions on computer usage would have been like sectioning off the university library for selective access. Librarians weren't there to keep people from the books, but to lead them as easily as possible to the books they were seeking. So it was with computers as well.

Still, a modest hierarchy had to be imposed on the USC computer systems; the people who watched over the system needed a master key to everyone's account, just in case they needed to retrieve a lost file or fix a student's account. Like any other computing center, USC had system managers with full access to everyone's account, and Mark Brown was one of them. Part of his job was to monitor the system's activity, to make sure that the machines were running correctly and to provide assistance where needed. One day, he noticed that some strange things were happening. Someone was logging in to privileged accounts normally used only by system managers. So, taking the role of detective, he started trying to track the trespasser down. His first step was to place some secret traps in the operating system to catch anomalous behavior. Next, he wrote a program that watched for unusual activity on the accounts the trespasser used most frequently. Finally, he began printing out the records of the break-in attempts.

After a few days, it was clear that the USC interloper was exploiting a loophole in the campus computers' operating systems to get access to the most privileged accounts on the system and then create his own account names and passwords. Alternatively, he used the accounts of others to break into privileged accounts. The printouts showed that he was coming in over off-campus phone lines. Brown's first move was to call the phone company and have the line traced back.

To do that, he learned, he would need to notify the police and obtain a court order. So he decided to take it on himself. Rather than shut the intruder out of the system entirely and give up the chance to track him down, Brown limited his access and minimized the amount of damage he could do. Judging from the intruder's actions, his motives seemed relatively benign—that is, until the day one of the log scripts showed him downloading university accounting files. These are files that keep records of the establishment and deletion of accounts, and they often include password entries. Not only did the intruder have the accounting files, he also had managed to use a privileged account to steal personal mail files from systems people like Brown, presumably in order to glean further information about the computer system. That was stepping over the line. Gentlemen do not read other people's mail. Brown began taking countermeasures.

Brown considered himself a hacker in the hallowed, traditional sense of the word. To him, hacking was the honorable pursuit of perfection in computer programming. It did not mean breaking into someone else's system. Brown decided it would take someone who thought like an intruder to catch one. He called for help from a couple of his fellow programmers, including Jon Solomon, a former phone phreak, and started the chase.

After a week of intensive monitoring, they realized that on certain days the phantom visitor was entering the school's computers at much higher speeds than usual. He could do that, they reasoned, only from a specific terminal room, where the only high-speed links could be found. Brown had been around the computer community long enough to see a lot of cockiness, but this was incredible!

They knew that the next time the high-speed connection came, it would take a bit of maneuvering to isolate where the intruder was in the computer lab, but it could be done. To make things more efficient, they decided to risk scaring him off with their obvious tampering: they limited his access to one machine.

One night when Brown went out to dinner, he left Solomon to watch for the trespasser. No sooner had he left than the intruder was on line— and his connection was coming from inside the campus. It took Solomon only a few minutes to trace the terminal. It was spooky: the interloper was at work in the very lab where Solomon was sitting. Solomon walked across the room and stood a few feet behind an overweight young man with a familiar face who was sifting through a stack of papers he seemed to be using as a reference. He was oblivious to anything else in the room. Solomon stepped a couple of feet closer to look over the intruder's shoulder. The papers were copies of the stolen accounting files. Next to the intruder sat another man who seemed much younger than the first. They appeared to be together. Suddenly Solomon placed the familiar face: he had met the chunky hacker weeks earlier at a DECUS conference. The hacker had struck Solomon as pretty cocky, boasting about his exploits. Solomon knew that his name was Mitnick, and that he had been arrested in the past for breaking into computers. Neither Mitnick nor his friend was enrolled at USC. Hurriedly, Solomon stepped away and called the campus police.

In the few minutes it took for the police to arrive, Solomon watched as the pair kept typing, logging in and trying out different passwords. Solomon was beginning to get the feeling that he could have watched them for an hour without their noticing him, so engrossed were they in the task at hand. When the two officers appeared in the doorway, Brown's assistant pointed to the older of the two strangers and they flanked his chair. When the plump young man looked up, his face expressed none of the guilt or remorse that Solomon expected to see. Here was a known computer criminal who wasn't merely breaking into the computers of a private university with which he had no legitimate connection—he strolled onto campus to do it. Solomon expected to see some sign of self-rebuke—at least some surprise—that he had been caught in the act. Instead, when the police questioned him, Mitnick took offense. "I'm not doing anything wrong," he said. The papers he was working with, he claimed, belonged to him. His friend sat nearby, discernibly more nervous. When the officers asked Mitnick to step outside, he picked up his thick stack of computer printouts, tucked the sheaf under his arm and obliged. The officers took both of them to the campus police station.

At the security office, the police took their names: Kevin David Mitnick and Leonard Mitchell DiCicco. When Brown and Solomon questioned Mitnick, his attitude was supercilious, as if under other circumstances he wouldn't deign to answer. He taunted Brown for what he claimed were unsophisticated sleuthing tactics. They never would have caught him, he claimed, if he hadn't made himself such an easy target by coming onto campus.

In a records search, the campus police discovered that Mitnick was on juvenile probation for breaking into other computers and that DiCicco, barely seventeen, had been caught six months earlier doing the same thing at Cal State-Northridge. They handcuffed Mitnick and DiCicco to a bench and called the L.A. police to come fetch them. In spite of Kevin's cool reaction to Brown, his fear of getting caught was so profound that it gave him heart palpitations so severe that he would soon have to depend on heart medication usually taken by people three times his age. Nonetheless, he seemed to realize the inevitability of arrest. He had once told Lenny that he knew he would be caught again after the COSMOS stunt. He had been lucky that time, getting probation instead of jail. Nevertheless, his compulsive side had won out over his fear of the consequences, and now here he sat, handcuffed to a bench. It was a conflict that would play itself out for years to come as Kevin's obsession intensified. And, at his side, not so much because he sought it but rather because he couldn't avoid it, would be Lenny, a faithful but not always willing companion.

The bench was in a narrow hall next to the door, twenty feet away from a security officer working at a desk. By now Lenny was familiar with the deep fear that came with getting busted, but as a juvenile he knew he was better off than Kevin. When he had been caught at Cal State-Northridge, he had been detained and released after a few hours. His parents hadn't even punished him. Kevin, on the other hand, was now nineteen and could get this put on his adult record. Lenny prodded Kevin. "Hey, Kev," he said, reaching into his back pocket for his wallet. He grinned as he produced a handcuff key. Lenny unlocked his own handcuffs, then Kevin's. The two sat speechless for a moment. Lenny broke the silence. "You go ahead. You've got a lot more to risk. Take off." Both looked over at the officer, who appeared to be paying no attention to his wards. The door next to the bench was unlocked; Kevin's car was in a parking lot just a few yards away.

Lenny's mind was perhaps too full of comic strips and spy thrillers. He had a madcap side to him that tended to surface just as a situation was spinning out of control. His escape plan for Kevin came straight from fiction. Lenny's plan would have brought Kevin closer to being a true criminal than Kevin could have imagined for himself. Kevin rejected Lenny's plan not out of reason but out of fear. He was too scared to try to flee by himself. Kevin would try it only if Lenny came too.

Lenny sighed and snapped his own cuffs back into place. He dropped the key onto the floor and pushed it under the bench with his foot. When Kevin tried to lock his own cuffs again, he couldn't. The noise Kevin was making in his efforts to engage the hardware caught the attention of their guard. He got up from his seat and checked Lenny's cuffs, then turned to Kevin. As he pulled on Kevin's arm, it swung free. "So we've got a Houdini over here," the officer remarked. He searched both of them thoroughly and shackled Kevin back to the bench. "Do this again and I'm going to handcuff you to the toilet," he warned.

When Solomon and Brown returned to the computer room to examine the terminals where Mitnick and DiCicco had been seated, they saw that Mitnick's terminal had just logged out of a computer called Elex-

Wash. Solomon recognized it as a Defense Department computer, but he couldn't tell what Mitnick had been doing with the account. Brown looked over the stack of printouts that had been confiscated from Mitnick, and he saw that it had a list of passwords to all the local accounts that had been created in the last two months. The stack also contained account names and passwords of companies Brown knew to be affiliated with the Defense Department, and what appeared to be secret information about genetic research by a company called Intelligenetics. All of the computers were connected to the Arpanet, the computer network funded by the military that connected nonclassified military installations, military contractors, universities and research centers around the world.

At first, Lenny thought he was going to get out of this one without a scratch. For some reason, USC decided to drop the charges, and as punishment at home, his parents grounded him for a week. Then, in a puzzling reversal, six months later USC refiled the charges and Lenny was summoned to appear in court. He pled guilty to a felony as a juvenile and got a year's probation.

Dominick Domino, the detective in charge of the Los Angeles Police Department's young computer crime unit, wanted to see Kevin somehow rehabilitated. In his police report, Domino wrote a brief summary of the case, ending it with an ominous observation: "This defendant is expert at computers and apparently enjoys the challenge of breaking computer codes. He will undoubtedly continue to be a police problem in this area unless maturity rechannels his energy and ambition."

So Kevin went to jail: six months at the California Youth Authority's Karl Holton Training School, a juvenile prison in Stockton, California, with about 450 inmates. Karl Holton was one of the more secure prisons in the state; violence-prone teenagers who were difficult to control were sent there for crimes ranging from armed robbery to murder. Kevin was doubtless the only inmate convicted of breaking into computer systems. Living conditions at Holton were harsh: the prison was overcrowded, there was minimal personal privacy and there was a great deal of violence. But Kevin tried to make productive use of his time there. He became something of a jailhouse lawyer, and he developed a computer program for tracking fellow wards of the court. He also worked with the Los Angeles police to put together an instructional videotape on computer security. He was released in late 1983.

▲ ▼ ▲

Richard Cooper wanted to know why Kevin Mitnick was using the computer so much. Why was he helping himself to Cooper's phone line all the time? And how was it that he could magically reconnect lines that had been disconnected?

When Cooper presented himself at the L.A. district attorney's major fraud section in October 1984 for a talk with investigator Bob Ewen, he described himself as a sales consultant at Video Therapy, a curiously named enterprise on Ventura Boulevard in Sherman Oaks. Cooper said he was a partner with Donald Wilson, the owner of National GSC, a merchandising company with a mystifying product line consisting of gourmet desserts and "solar barbeque" franchises. Donald Wilson, a friend of the Mitnick family, had agreed to hire Kevin soon after his release by the California Youth Authority to work for Great American Merchandising, one of several National GSC subsidiaries. These grandly named subsidiaries, it seemed, sprang full grown from Wilson's head and usually fizzled into bankruptcy shortly thereafter. Cooper ran one of the ventures, called Video Therapy, but he had had about enough of Wilson's bizarre businesses and was unnerved by the presence of the young man Wilson hired to handle clerical duties and work with the computer.

Before Mitnick arrived, the only person who touched the office computer was a secretary, who used it for three to four hours a day to write letters. Mitnick was now working on the computer all day. When Cooper came at 9:00 each morning, he usually saw Mitnick's black Nissan, conspicuous for its vanity plate that read "X HACKER" and its mobile radio antenna, already in the lot. And Mitnick often stayed at least until Cooper left at 6:00 each night. When Cooper asked Wilson why Mitnick was glued to the computer so assiduously, Wilson replied that he was working on a number of special projects.

But the answer didn't satisfy Cooper. Every time he passed Mitnick's desk and looked at the screen, he saw that Mitnick was in the middle of patching in to the computer of TRW's credit bureau to inquire into credit standings. Cooper knew that neither Great American Merchandising nor its parent company would be making such inquiries. And there was no reason to run credit checks through TRW, as Wilson's franchisees generally did not apply for credit.

Then the telephone nonsense started. In early October, Cooper told Ewen, he moved his office into the "client relations" room, where Mitnick and the computer sat. Every time Cooper saw the phone line for his Video Therapy light up, Mitnick was using it.

All day long, Cooper told Ewen, he heard Mitnick on the telephone.

He appeared to be talking with Pacific Bell employees, because he heard Mitnick refer to such things as COSMOS, satellite operators and work orders. He introduced himself variously as Gillie, Paul, Peter and Steve to whomever he called. And there was one apparently Scottish person Mitnick often made passing reference to during these telephone calls: R. C. Mac. Ewen had to smile at that. Cooper had no idea that RC MAC was the Recent Change Machine Administration Center, an internal telephone company department that processed orders and changes in service via computer.

Coincidentally, Bob Ewen had already opened an investigation two weeks before Cooper's visit, on an allegation that Steve Rhoades and Kevin Mitnick were illegally getting into telephone repair and billing computers, and fraudulently using an access code from Satellite Business Systems, a long-distance company, to make toll calls. Mitnick and Rhoades also appeared to be harassing people on MIT computers, where they had guest accounts. And six months earlier, a rash of complaints had come from the ham radio community, claiming that Mitnick had been provoking hams over the air once again. To avoid getting sent back to prison, Mitnick had surrendered his ham radio license to his parole officer. Based on what Cooper was now telling him, Ewen decided there was enough cause to issue a search warrant for the offices of National GSC and an arrest warrant for Mitnick.

Ewen took with him three other investigators and Terry Atchley, a Pacific Bell security official. From the Erector-set feel of the tiny office, Ewen got the impression that this was a fly-by-night operation at best. Donald Wilson told the group that the reason he had hired Kevin Mitnick was that he felt sorry for him and he wanted to give him a break. Wilson said he wasn't aware of any calls made to TRW, but when Ewen questioned the secretary, she said she had received two calls for Mitnick from a woman saying she was from TRW. When Ewen asked where Kevin might be found, Wilson said he didn't know. Kevin had left for lunch just fifteen minutes earlier. He added as an afterthought that he had once overheard Kevin say he would flee to Israel before going back to jail.

When Ewen showed up at Shelly Jaffe's Panorama City condominium an hour later, Shelly seemed calmer but even less cooperative than the first time Ewen had paid her a visit three years earlier. It was as if Kevin had coached her on what to say in case the police arrived at the door, or so it seemed to Ewen. Kevin was living with her, she told the investigator, but she hadn't seen him for a few days and refused even to venture a guess as to his whereabouts. So Ewen checked out some of Kevin's usual haunts including Warner-Elektra-Atlantic Records in Burbank, where Lenny DiCicco was working as a computer operator. Did Lenny have any idea where Mitnick might be? Israel, perhaps? No, Lenny told Ewen, he had heard that Mitnick might be in Las Vegas.

Ewen's next call was to TRW Information Services. The company's security auditor checked the credit bureau's records and confirmed that dozens of inquiries on Mitnick and Rhoades had been made in recent months. They had come from William Pitt Jewelers and Security Pacific Bank. The inquiries appeared to be something of a prank. The jewelry store alone had made 350 of them into such alleged credit applicants as Steven Rhoades, Steve's grandmother Juanita, Kevin Mitnick, Lenny DiCicco and Gretchen Dog—the name of Juanita Rhoades's Doberman pinscher. Mitnick and Rhoades were merely playing around with a purloined TRW account number. It wasn't surprising that Mitnick and Rhoades hadn't altered any of the credit ratings, as that would have been far more difficult to do.

During the search at Donald Wilson's office, Ewen and his men seized a Xerox personal computer, a printer, a disk drive, a modem, a monitor and various floppy disks. But he missed Kevin. Somehow, Kevin must have found out he was in trouble. He was well-known for running frequent warrant searches on himself.

That might explain the strange phone call that came in to the warrant section of the Los Angeles Police Department on October 24, the day before Ewen served the search warrant. At 6:00 that evening, a man who identified himself as Detective Jim Schaffer from the LAPD's West Valley office asked if there was a probation violation warrant out on one Kevin Mitnick. The computer operator who took the call said yes, there was a fresh warrant in the computer. The detective thanked him and gave him a number where he could be reached. About an hour later, a second call came from a male with a different voice who also identified himself as Detective Schaffer. He said he had Mitnick in custody and wanted to confirm a parole warrant. Suddenly suspicious, the operator put the second caller on hold and dialed the number he had been given by the first caller. "West Valley detectives," came the salutation from a woman, or at least a female-sounding voice, who confirmed Detective Schaffer's existence. When the operator came back to the caller, there was no one on the line. When he called the L.A. number again, an answering machine picked up the line. "Hello, you have reached Roto-Rooter. We're closed . . ."

Kevin may have fancied himself an aspiring Condor, but when things got rough, the bravado shattered. It turned out that while walking to his car at lunchtime, Kevin saw Bob Ewen, the security investigator for Pacific Bell and three other men headed for the elevator of his building. Kevin called Lenny from a pay phone across the street from his office. He was in a panic. He told Lenny he knew there was a warrant out for him, and he was going to leave Los Angeles.

Lenny was sure Kevin would be in touch at least once a day, as always. But he didn't hear from his friend after he went into hiding.

Instead, once a week or so, Kevin put in a call to Roscoe. Without disclosing his whereabouts, he asked for gossip about the amateur radio scene. Roscoe had long since given up his HOBO-UFO conference line. He was now married to Jo Marie, the woman whose presence in Roscoe's life had crushed Susan's pride four years earlier. Jo Marie had completed law school and Roscoe had gone relatively straight. He had dropped out of USC after only a year, opting instead for a quick diploma from a local vocational school called the Computer Learning Center of Los Angeles. Rather than parlay his weeks of fame from the cover of the *L.A. Weekly* into a security consulting job, as he had hoped, he settled for a position as data processing manager at an auto parts importer north of Long Beach. It wasn't exactly rarefied work, but it suited Roscoe's orderly style. Now that he had settled down, Roscoe enlisted his wife's help and had his 1982 conviction reversed, thus putting the whole unpleasant mess behind him. A weekly check-in call from a fugitive was the last thing he needed.

Kevin wouldn't tell Roscoe where he was, and with each telephone call, Kevin seemed more paranoid. Fleeing the warrant and probation was, perhaps, Kevin's own point of no return; he knew that what he was doing was wrong as defined by the law. This was by no means the first time he had been in a legal scrape, but it is unlikely that he had ever previously imagined himself a criminal.

Once, Kevin called Roscoe and told him he needed legal advice from Jo Marie. He had been attending a small two-year college in Northern California, and the credits were under an assumed name. He wanted to change his name back to Kevin Mitnick so that he could get the credits transferred to an L.A. school.

During his year-long absence, Kevin did leave a hint or two that he was still in business. One day in early 1985, Roscoe came across the phone number of Ronnie Schnell, an old bulletin board buddy from his 8BBS days, and decided to call. When Roscoe reminded him of who he was, Ronnie sounded surprised. "Oh, I have Kevin Mitnick on the other line. He wants me to get him an Arpanet account." Roscoe was amused, and he counted the seconds before his phone rang. "How did you know I was calling Ronnie?" Kevin demanded when he called Roscoe ten seconds later. It was a mite odd that four years had passed since either Kevin or Roscoe had talked to Ronnie, Kevin was on the lam, determined to keep his whereabouts secret, and suddenly Roscoe happened to be on the line at the same time. However much Roscoe tried to tell him it was a coincidence, Kevin remained unconvinced.

In another strange incident that seemed to suggest Kevin was still playing around with phones, Steve Rhoades was fooling around one day with the toll-free number for lost telephone calling cards, and when he called the number, the person who answered with "Pacific Bell, may I help you?" was unmistakably Mitnick. Rhoades was so amused he recorded the greeting and used it as the outgoing message on his answering machine.

▲ ▼ ▲

In the summer of 1985, Kevin resurfaced, nearly a year after his disappearance and just a few weeks after his arrest warrant expired. Bob Ewen was shocked to learn that the juvenile probation department had dropped the warrant from its books with no explanation. If Ewen had known that the warrant was going to dissolve, he would have insisted on an extension. The investigator knew that if he had been working the case steadily instead of spottily, he would have found Kevin no matter where he had gone.

Lenny didn't find out Kevin was back in L.A. until he got a call one day at work. When he picked up the receiver, all he heard was someone pushing Touch-Tone keys. The caller was spelling out his own name using a private code that Lenny immediately recognized—the kid who had escaped the law was back in town, possibly up to his old tricks. Kevin finally spoke up and the second chapter of the DiCicco-Mitnick partnership began.

Kevin said very little to Lenny about his time spent underground. From what he did say, Lenny figured that Kevin had cobbled together a false ID or two, gathered his bar mitzvah money, and caught the first plane out of L.A. But even when Lenny tried to goad him years later into telling him where he had been, Kevin wouldn't say.

When Kevin returned, Lenny was working on the four-to-midnight shift as a computer operator for Hughes Aircraft's Radar Systems Group in El Segundo. It was his fifth job in two years. Well-spoken and relaxed during interviews, Lenny was a master at talking himself into any position. His habit of abusing his employee privileges, however, would surface soon enough, and what his family referred to as "Lenny's computer shenanigans" usually drove his supervisors to fire him after only a few months.

Whenever Lenny got a new job, Kevin wanted to know what computers were there. A Digital computer was a good start; a dial-up line that allowed access to the computer from the outside was even better. And if Lenny was working a job with night hours, and could get Kevin in when everyone else had left for the day, then Lenny heard from him just about every day.

Kevin had a new project in mind: establishing accounts on the score of minicomputers that Pacific Bell operated in Los Angeles for order entry. Looking back several years later, Lenny decided that when they were dabbling in the "minis," as telephone company personnel referred to them, they gathered more power than they had ever had or would ever see again. Someone with access to these computers, which connect to switches controlling all Los Angeles telephone service, could enter commands and see them take effect immediately. The switches accessible through the minis did everything from discontinuing service on a line to initiating a trace. It was Lenny's job to put computer equipment at Kevin's disposal. Without a badge, Kevin couldn't come onto the premises at Hughes. Once, however, in April of 1986, he managed to talk the guard into letting him in. During this, Kevin's first and last visit to Hughes, he logged on to Dockmaster, a computer run by the National Computer Security Center, a division of the National Security Agency, the nation's highly secretive intelligence agency. Despite its obvious appeal to the adventuresome pair, Dockmaster didn't harbor any deep national security secrets. It was simply NSA's public bridge to the outside world.

To get onto Dockmaster, Kevin had found the name of someone outside of the NSA with a guest account. Posing as a technician at an NSA computer center, Kevin had telephoned the legitimate user and said he was issuing new passwords and needed some information: name, telephone and current password. It was an old trick that Kevin and Roscoe had refined together, and it usually worked like a charm.

But Dockmaster was incidental, a diversion from the Pacific Bell "minis" project. Since he couldn't come to Hughes, Kevin telephoned his instructions to Lenny. He called constantly. "It's the wife," Lenny's partner in the computer room would joke when he handed Lenny the phone. And when they needed to devote twenty-four hours at a stretch to their work, they checked in to one of the dozens of cut-rate motels along Sepulveda Boulevard in the heart of the San Fernando Valley. It was pure cyberpunk: the mile-long strip of seedy shelters was a magnet for local prostitution and drug rings, but neither Kevin nor Lenny cared, as long as the motel room had a telephone line that could be transformed into a data communications center. Once that criterion was met, Lenny preferred motels with pools in case he got the urge for exercise, and both were partial to places next to a 7-Eleven or other suitable convenience store for frequent junk-food runs. Kevin usually plunked down the $19.95 for the night onto the registration desk in coins filched from Shelly's tip money from waitressing. Once inside the room, they went straight to work: Kevin unpacked the terminal while Lenny popped open his tool set to rig the telephone jack so that it would accommodate their modem. They worked through the night, taking little note of some of the seamier activity around them. They often pushed the checkout time to the limit, until the manager came around to throw them out, making them *personae non gratae* at that establishment.

It took six months for them to get fully privileged accounts on every phone company mini in the Los Angeles area. And with each computer they conquered, their potential power increased. It would not have been difficult for Kevin and Lenny to take down the phone service for the entire metropolitan area; but neither of them was interested in that. The idea of power was more seductive than actually wielding it.

When Hughes got word from the NSA that someone from the El Segundo facility had been into the Dockmaster computer, Hughes management went on a witch-hunt. The night of the penetration was linked to Kevin's visit, Mitnick was linked to Lenny and Lenny was summarily dismissed. The company's corporate security department spent the days after Lenny's departure behind closed doors, conducting intensive interviews with anyone Lenny had associated with.

After Lenny was fired from Hughes, he got a job as a delivery man for a flower shop. The only vital credential was a perfect driving record. Though only twenty, Lenny already had a formidable record of outstanding traffic warrants. But that was Leonard Mitchell DiCicco's problem; the florist hired Robert Andrew Bollinger, a model motorist. As Lenny would later tell the story, with Kevin's help he had created a new identity for himself. He even went so far as to rent an apartment in the San Fernando Valley as Mr. Bollinger.

But a florist doesn't have much need for a VAX minicomputer, the mainstay of Digital's product lines and Kevin's target of choice. Now Lenny hardly heard from Kevin at all. The intensity of this friendship, Lenny realized, waxed and waned depending on his job. The degree of Kevin's loyalty was a function of the kind of equipment Lenny could provide.

Bollinger quit the flower delivery business after one month. Lenny DiCicco resurfaced as computer operator for a shipping company. For the time being Kevin kept a low profile. He seemed to have become more paranoid since his return. Kevin had always been secretive about disclosing personal data—it wasn't until months after Lenny met him that he was even willing to give Lenny his home telephone number—but now he was acting strangely. He was taken with the idea that people should be able to contact him at any time, but without his having to reveal his whereabouts, so he started carrying a pocket pager everywhere he went. And both Lenny and Roscoe noticed that the more paranoid Kevin became, the more he ate.

Perhaps in an effort to follow Roscoe's path into the world of conventional computing, in September of 1985 Kevin enrolled at the Computer Learning Center of Los Angeles. Roscoe had graduated a few years earlier and spoke well of it. Kevin's previous experience in higher education had been dismal. In 1982 he was expelled from Pierce College, a two-year community college in the heart of the San Fernando Valley, for tampering with the school's computers. Maybe this time he could keep his mind on the coursework.

The Computer Learning Center was started in the 1960s by a group of computer executives who, astutely enough, predicted that the nascent computer industry would see explosive growth in years to come and would need tens of thousands of people qualified to fill the jobs that were emerging. By the late 1970s the computer industry was a major source of white-collar employment, not unlike the consumer electronics industry when it mushroomed in the 1950s and 1960s. And just like any other technical school advertised on matchbook covers and city buses, the Computer Learning Center wanted to attract a student body hungry for work.

To gain admission to CLC, a prospective student needed a high school diploma or equivalency degree. The applicant then took a basic aptitude test designed to measure ability in symbolic logic and mathematical reasoning, the foundations of computer science. The $3,000 tuition for the nine-month program was paid in advance and the student was on the way to a fresh career. CLC graduates, many of whom were retooled English majors or auto mechanics, joined the data processing ranks inside Security Pacific Bank and the dozens of giant aerospace corporations that dotted the Los Angeles basin. Polaroid snapshots of these successful graduates were enshrined in a glass display case upstairs at CLC. And next to the first-floor admissions office, letters of glowing commendation from such satisfied employers as First Interstate Bank, Continental Airlines and Agfa-Gevaert lined the walls.

Although the school failed to offer its graduates a formal degree, as a technical school it could certify would-be technicians and programmers as proficient in using the center's computers. For the most part corporate America didn't need computer science Ph.D.'s fresh out of UCLA, Stanford or Cal Tech as much as it needed young men and women willing to work a lobster shift, changing tapes to back up the day's business for a starting salary of $20,000. The Computer Learning Center was a steady source of entry-level programmers, technicians, computer operators and data-entry clerks for the region. If not exactly the glamour posts of the computer industry, these jobs at least paid better than other clerical positions that required less training.

The Computer Learning Center was a resolutely straight, no-nonsense place. Its dress code forbade all but the most businesslike attire. Like so many scrub-faced Mormon missionaries, the men were required to wear jackets and ties, the women skirts or dress pants.

Larry Gehr, who taught an introductory class at CLC in the COBOL computer language, was vaguely aware of Kevin Mitnick's past problems, but he tried to treat him like any other student. Kevin was inquisitive, and his mind always seemed to be working at least a lecture's worth ahead of the class.

Kevin had an unusual mix of computer expertise: there were some fundamentals he didn't know, but in other areas he was well ahead of his fellow students. He displayed so much ability, in fact, that Gehr was concerned that an entry-level job wouldn't challenge him. Gehr adopted the role of teacher-counselor to this erratic student. He tried to encourage Kevin, telling him that if he could get a good entry-level job and show what he could do, he would rise swiftly.

▲ ▼ ▲

Even in the days when Susan was around, dating her way through the phreaks, Kevin had seemed entirely oblivious to women. So it was a bit of a surprise to friends when, in the summer of 1987, Kevin casually mentioned that he had just gotten married. Everyone developed his own theory about Kevin's nuptials. That the bride in question worked at GTE, one of the two telephone companies servicing the L.A. area, seemed no coincidence. At first, word went out that she was a programmer there. For someone like Kevin, his friends agreed, there was nothing more tantalizing than the promise of an inside source. And by the time Susan Thunder, now in the Downey area in south Los Angeles training for a career in professional poker, found out about it, Mrs. Mitnick was being described as a senior executive at the telephone company.

Bonnie Vitello was slight and dark, her round face and brown eyes framed by a thick mass of long chestnut hair. Bonnie's expression in repose approached a frown, but the suggestion of a sour temper would suddenly disappear with an unexpected explosion of glistening white teeth that cast a disarming spell over strangers. Bonnie's smile was her built-in edge.

Born in New Jersey, Bonnie was six when her parents moved to Monrovia, about twenty miles northeast of downtown Los Angeles. When the Vitello family moved there in the mid-1960s, the predominantly white, mostly conservative city was on the verge of change: gang fights between the recently arrived Hispanics and the entrenched whites were starting to plague the community. Nine years after their arrival, unhappy with the growing racial tension, the Vitello family of six moved south to Rowland Heights, a quiet, arid strip of a town on the eastern fringes of Los Angeles County. Twenty minutes north of Disneyland, Rowland Heights is a community of new condominium subdivisions interrupted occasionally by clusters of single-family homes—an ideal place to start over. A listless student back in Monrovia, Bonnie suddenly became interested in school in Rowland Heights. At her new high school, the teachers weren't busy breaking up gang fights and she no longer had to guard her back. She graduated early and enrolled at a small junior college not far from home. Then interruptions set in, not the least of which was her first marriage, at age eighteen, which lasted just six months.

Bonnie's impulsive jump into marriage exacted its price. She dropped out of college and put her academic pursuits on hold indefinitely. At twenty-one she was doing temporary office work when she got an offer to work full-time as a secretary for GTE at its Monrovia headquarters. Reluctantly, she went back to the place she had left so gladly. When the phone company moved its headquarters from Santa Monica to Thousand Oaks, a suburb on the northwestern edge of the Los Angeles area, Bonnie transferred there.

Bonnie hated the monotony of clerical work. After several years of filing and typing, with no real prospects for advancement, she put her mind to learning about computers. It was an easy transition for her. Personal computers had made their way onto desks throughout the company. When Bonnie was given an IBM Personal Computer, it was with the expectation that she would use it for nothing more than word processing. But the little machine captivated her. She pored over the manual and learned the machine's every feature. Word of Bonnie's computer prowess spread and she became the department's PC troubleshooter. With GTE paying the way, Bonnie enrolled as an evening student at Computer Learning Center.

One night, Bonnie was at the main console learning how to control one of the central training computers. The student operators sat at computer consoles, simulating the activity of a large corporate data processing operation: computer jobs were running, the system was monitored and data files were regularly backed up. Suddenly Bonnie saw messages flashing on her screen from someone sitting across the room in a spot reserved for people with full privileges on the system. "Don't flush my file!" came the message, then, "Let my job go first!"

"Who is that guy?" Bonnie asked one of the instructors.

"Oh, that's Kevin Mitnick," he replied. "Don't piss him off. If he doesn't want you to flush his job, then don't." Bonnie looked across the room at the overweight stranger and smiled.

She sent him a reply. "What else do you do besides tell people not to flush jobs?"

"I like to go out to eat," came the response. "Would you like to go out?"

"I can't. I'm engaged," Bonnie replied.

"That's too bad. You have a nice smile," he wrote back.

Bonnie had been engaged for six months to an engineer who was a great deal older than she, but the excitement of the relationship's early days had gradually diminished. She considered this for some minutes, then wrote back to him: "Well, I'm not so happy with the relationship."

He answered within seconds: "Then maybe you'll go out with me."

She was tempted. "Maybe next week," she typed back.

Kevin Mitnick's electronic entreaties persisted. For several weeks after that, each time he flashed her a message asking her to have dinner with him, she politely declined. Late one evening, he abandoned the computer messages and approached her desk, eating something. "Well, obviously you're not seeing your fiancé tonight," he said midchew. "Do you like Thai food?" At that point, although Bonnie had never considered dating anyone quite so bulky, she finally said yes.

Over dinner, Kevin asked her where she worked. When she replied that she worked for a phone company, he began to laugh so hard he nearly choked. But he wouldn't explain why. Bonnie found him charming and interesting. It surprised her to discover that he was just twenty-three, six years her junior; and he professed to be equally surprised that she wasn't his age, as he had at first assumed. He seemed more mature, particularly in comparison with some of the other, younger CLC students who were just out of high school. What was more, he could explain difficult computer and mathematical concepts as no one else seemed able to do. He wasn't boastful, and yet it was clear that his knowledge of computers ran deep.

They began to see each other often, sometimes taking a bottle of wine to the beach in Santa Monica on a Friday evening. Not much of a drinker, Kevin gamely swirled his wine in his glass as they talked. And although Kevin had never done much dancing, Bonnie managed to coax him into local discos. When Kevin was around Bonnie, in the early stages of their relationship at least, another side of him emerged. In contrast to the young man who reacted against a difficult and isolated childhood by lashing out at hams, monkeying with people's phone service or reaching for a computer—the one thing that gave him a feeling of control and power—around Bonnie he was easygoing and engaging. Love had its salutary effects as well: Kevin began to drop his excess pounds like so much ballast, and Bonnie watched his body transform into a more appealing shape.

Within a few weeks, Bonnie officially dissolved her engagement, and Kevin all but moved into her apartment in Thousand Oaks. It was a small one-bedroom place, but there was enough space for the two of them. Before long, marriage became Kevin's obsession. Every week or so, he asked her to marry him. Something about it felt right to Bonnie. She was ready to accept. When Bonnie joked to Kevin that she was his shiksa—the Yiddish word for "non-Jewish woman"—he asked her what the word meant.

▲ ▼ ▲

Steph Marr had been around computers long enough to know that a good system administrator knows how to read his machine. And something was wrong with the system at Santa Cruz Operation.

A blind church organist can tell how many people are in the church by the way his music sounds, but he probably can't explain how he does it. A good jockey can tell what kind of mood his horse is in, and can often tell how well the horse is going to run by the way it walks to the starting line. A good computer system administrator develops a feel for the patterns and sounds of his computer. When they are aberrant he always investigates. Computers are supposed to do the same thing over and over again, and when they do something different there is always a reason. The internal rhythm of a computer is seen in the delays in getting a response to typed keys, the staccato sound of a disk arm rattling, the flickering lights of modems and disk controllers, or the daily routine of log and journal file entries.

Named for the Northern California coastal town where the company had its headquarters, Santa Cruz Operation got its start selling a version of the UNIX operating system to run on personal computers. In eight years, Santa Cruz Operation had grown from a father-son consulting start-up to a multimillion-dollar company.

Marr was one of the people who worked to keep Santa Cruz Operation's network of computers up and running. He had been there for a year, long enough to know that certain users not only had certain privileges on the system but also had individual habits. Engineers logged on from their homes late at night; secretaries logged on only from work and only during working hours.

Steph was sufficiently tuned in to his computer's own circadian rhythms—the times of heavy use and the periodic lulls—to notice an aberration in late May of 1987. Not only did he feel the trouble, but the system was telling him in no uncertain terms that something was going awry: one of the secretaries who used the computer was acting out of character. She was logging in after hours, cruising the system and trying to peek into other people's directories. She was accessing files that had been dormant for many months, including an out-of-date "help" system. When Steph asked her about it, she professed ignorance. Apparently someone had ferreted out her password and was blundering around inside SCO's computers.

When Steph realized that an intruder had entered the system, instead of immediately trying to throw him out, he set up an alarm system, and put limits on what the stranger could do. The trespasser must have felt the presence of someone electronically peering over his shoulder, watching his every move, because a few days after the monitoring started, he typed onto the screen, "Why are you watching me?"

"Because it's my job," Steph typed back.

Once he had engaged the company in an electronic conversation, the hacker made a request that signified so much audacity that the system administrator knew this was no ordinary intruder. The uninvited guest told Steph that he wanted an account that gave him unlimited privileges throughout the system. By this time, a group of onlookers had formed a half-moon around Steph's desk, and a lively debate ensued over whether or not Steph should give him an account on the company computer.

Steph believed his own experience as a hacker on the edge of the law gave him a better understanding of just what motivated this person. In his younger days, Steph had done his share of system cracking. As a college student, his favorite pastime was to slide past the security on university computers, then call the system administrators to inform them of what he had done before they discovered it. He, too, believed in free access to information. But Steph also knew when to stop. He believed in exercising restraint and respecting someone else's right to run a computer system without having to build elaborate safeguards against outsiders. There was something about this person's attitude that made Steph sense that he knew he was breaking the law, and was perhaps even proud of his trespass.

A high-level account was out of the question. If Steph gave him such power over other users, and if he was clever enough, the intruder would be able to change the machine's operating system, or even do something as patently malicious as depriving others of the use of the computer. If he wanted to, he could become the electronic equivalent of a mad gunman in a bank, holding the computer hostage. Finally, Steph decided to give him an account that masqueraded as something more powerful than it actually was. The account was called Hacker, a label of the intruder's own choosing. In setting up an account for him, to which Steph also had full access, Steph figured he was erecting an aquarium, in which he could watch every move of this felonious fish. Moreover, letting him lounge around on the computers would make it easier to trace his calls. Within forty-eight hours, Pacific Bell security was doing just that. But when Pacific Bell security investigators started to trace one call, they discovered that somehow the trespasser was wily enough to get access to the phone company computers themselves and block certain commands requesting information about the line he was using.

Despite the long hours the intruder was dedicating to wandering through the SCO computer, he appeared to have no discernible quest. He seemed merely to like to prowl through the system, checking out directories but seldom opening files themselves.

But after a week or so, the hacker's probing seemed more directed. He fastened onto programs that would allow someone to modify the operating system, but he didn't have the necessary privileges to do so. It was clear after a while that he seemed to have a peculiar goal of modifying XENIX, a derivative of the UNIX operating system and the crux of the company's business. More disturbing was his apparent goal of transmitting a copy of XENIX to his own computer.

But the owner of the Hacker account gave himself away. The simple oversight that exposed him lay in his use of MCI to gain access. A special MCI feature automatically identified the telephone from which the trespasser's call to the local MCI access number was placed. It was a telephone in Thousand Oaks, California.

▲ ▼ ▲

When investigators from the Santa Cruz Police Department flew to Los Angeles on the morning of June 1, 1987, to conduct a search at apartment number 404, 1387 East Hillcrest Drive in Thousand Oaks, years of experience investigating fraud couldn't have prepared them for the odd collection of evidence they encountered. The officers had been told what to look for—computer printouts, a computer, a modem and notes of phone numbers and access codes—but they weren't sure why. When the officers knocked on the front door there was no response, so they let themselves in with a pass key borrowed from the manager of the complex. Detective Patricia Reedy of the Santa Cruz Police dutifully noted in her report: "On the dining room table was a computer with a black box with multiple small red lights on it. Detective Nagel advised the black box was a modum [sic]. The 'MC' light on the modum was lit. Lying next to the computer was a beige touchtone phone with the front plate removed. The computer, the modum and the phone were all hooked together." To prevent anyone from calling the apartment and erasing possible evidence from the computer, Detective Reedy followed what she must have thought was the proper procedure: she removed the receiver from the phone. As she noted in her report, the receiver then "made a loud screeching noise." Any computer crime investigator would have warned her to keep the phone in place, as the suspect might well have been in the process of transmitting data.

As they made their way through the small dwelling into the bedroom, they first encountered large piles of clothing on the floor. On a bedside table they counted fifty-five computer disks, in boxes and scattered on the tabletop. From underneath the bed they retrieved "a book marked 'OS Utilities,' numerous loose sheets of printed materials on computers and a plastic bag containing a large quantity of hand-written notes and computer printouts." They gathered all of it as evidence. Also under the bed was a loaded Charter Arms .38 special two-inch revolver. Inside the bedroom closet the detective found a second, much larger firearm—a Remington .87 shotgun. On the bedroom floor they found two small plastic bags containing what appeared to be marijuana, and a glass bong. They took the items as evidence. When Detective Reedy checked the clothing in the closet, she found $3,000 in $100 bills in the pocket of a man's suit jacket. She initialed each of the bills and left them on the kitchen table.

The detectives knew they didn't understand enough about computers to analyze what they had found and decided to call in a computer specialist from the Ventura County sheriff's office. "The specialist advised us that what we had in front of us was a computer terminal that had no way of storing information. He explained the black box on top of this terminal was in fact a modum. He was able to put the receiver back on the phone . . . and access into the terminal. He was able to bring up on the screen information that said 'abort.' He advised we probably interrupted the computer accessing into somewhere else."

The search ended at 4:00 P.M.

When Bonnie came home that afternoon, she thought they had been burglarized. The computer terminal was gone, as well as the modem, the floppy disks and all of their computer books. But when she saw a stack of $100 bills on the table, money Kevin had been saving for their wedding celebration, next to a document bearing an official seal, she knew they had been searched. She packed a bag and went out to find Kevin. When she told him who had just been to the apartment, he flew into a panic.

One of Kevin's first calls was to Roscoe, demanding to know whether Roscoe had informed on him. Roscoe tried to calm Kevin down and ask him questions. But Kevin was too excited to listen; he began to sound as if he were talking to himself. Should he get out of town? Without waiting for an answer, he kept talking. Then again, he rambled on, he didn't want to leave Bonnie. Should she go too? Roscoe cut him off and told him to get in touch with an attorney.

The officers returned to the apartment at 8:30 the next morning. They knocked on the door and got no answer, so they went to interview the manager, Alice Landry, who told them that the tenant in number 404 was Bonnie Vitello, a nice-looking woman in her twenties with a thin build and dark hair who worked at GTE. She said Vitello had had "her brother" staying with her. He was tall, heavyset, nice-looking and clean-shaven. Overall, the manager told the inquiring officers, Ms. Vitello had been a good tenant, although there had been a couple of complaints about noise from inside the apartment in the evening, the sound of people arguing. The brother, she said, was at home a lot.

Later that morning, the officers called GTE's security manager, who verified that Vitello had worked there for several years. That morning she had called in and asked for vacation time. She had said she was moving and would be back the following Monday. Detective Reedy asked the security manager if he recognized the name Kevin Mitnick. Not only did the manager recognize the name, but he rattled off a list of agencies he knew had investigated Mitnick in the past. Mitnick, it seemed, was something of a household name around GTE.

Reedy's next call was to the L.A. County district attorney's office, where she was directed to Bob Ewen. Ewen told Reedy that Mitnick had been a suspect in several major computer crimes in Southern California, and that he himself had worked a case involving Mitnick several years before. At one time, Ewen said, there had been a warrant for Mitnick in the office's computer, that Mitnick had found out about it and that he had then fled to Israel. There were no current warrants out for Mitnick, but Ewen believed the FBI was working some cases involving him. In any case, Ewen warned, Mitnick was extremely dangerous, capable of destroying computer systems remotely using "logic bombs."

Logic bombs? Israel? Southern California was bizarre enough, but this sounded absurd. That afternoon, the officers called Santa Cruz Operation to say they believed they had found the intruder. The arrest warrants that were issued from the Santa Cruz County Court charged both Kevin David Mitnick and Bonnie Lynne Vitello with unauthorized access to a computer, a felony under California law, with bail set at $5,000 each. The officers who had conducted the search described Vitello to the court as "dangerous and bright." And twenty-three-year-old Mitnick was a known criminal with a long list of previous offenses. Three days after the warrant was issued, the suspects surrendered voluntarily at the West Hollywood police station. Once it was established that Mitnick had acted alone, Santa Cruz Operation dropped the charges against Bonnie.

For all the trouble he had been in through the years, Kevin still had a clean adult record and he wasn't about to let that change; he refused to plead guilty to a felony charge. His attorney asked that, in exchange for Kevin's full cooperation in explaining how he had cracked the Santa Cruz system, the charge be reduced to a misdemeanor. So a misdemeanor it was, with a small fine, thirty-six months' probation and a three-hour meeting between the Santa Cruz Operation computer staff and Kevin Mitnick in the presence of their respective attorneys.

When Steph Marr, the system administrator at Santa Cruz Operation, met Mitnick for the first time, he thought he should offer some words of praise. After all, Mitnick had managed to muscle his way into the Santa Cruz Operation computers and, for a time at least, had eluded detection. "Well played, well met," Marr said as he greeted him. But Mitnick barely responded. When Marr asked him a technical question, he responded to the Santa Cruz Operation attorney instead. And his attitude as he described his methods was annoyingly condescending. It was hardly the hacker-to-hacker session Marr had hoped for. When Marr got a call a year later from Mitnick asking about a job, it only confirmed that this was indeed a young man with extraordinary gall.

Kevin and Bonnie were married that summer, while the Santa Cruz charge was still hanging over Kevin's head. It was a trying time, especially the frequent trips to Santa Cruz for court appearances. It wasn't exactly the way Bonnie might have chosen to spend her thirtieth birthday, which arrived just a month after their apartment had been ransacked by the police. But for all the problems Kevin was causing her, Bonnie still wanted to marry him. She had already been through one big Catholic wedding and she didn't want another extravagant affair, and Kevin didn't seem to mind one way or the other, so they went to City Hall and emerged fifteen minutes later with their vows in place. To appease Bonnie's mother, the newlyweds dressed in wedding garb and went to her house for a celebration party. The couple stood smiling for the obligatory photographs. Bonnie was happy. Kevin had given her a ring and his heartfelt word that his computer shenanigans were forever behind him.

If Kevin was planning to make a clean break with his past, it didn't help that he became the primary subject of a Pacific Bell security memo a month or so later. A ham with whom Kevin had a less than friendly relationship happened to work at the telephone company, and read the memorandum over the air. Written by a security manager, the memo detailed the contents of the computer disks found during the search after the Santa Cruz break-in. Describing the events surrounding Kevin's case as "alarming," the manager listed what had been found in the Thousand Oaks apartment: "The commands for testing and seizing trunk testing lines and channels; . . . the commands and logins for COSMOS wire centers for Northern and Southern California; . . . the commands for line monitoring and the seizure of dial tone; . . . references to the impersonation of Southern California security agents . . . to obtain information. . . ." The list went on.

The memo concluded that computer hackers were becoming more sophisticated in their attacks on phone company computers. The author suggested that it was possible that hackers could incapacitate an entire central office switch by overloading it or tampering with the computers that controlled it. The memo also voiced concerns that terrorists or organized crime groups might get their hands on "underground computer technology."

When Kevin heard this, he panicked. He had to see the memo. He called Lenny, then Roscoe, who may have stopped his extralegal activities but didn't mind being pulled in on the occasional clever hack. The three worked out a plan for getting the memo. Kevin called the secretary in the San Francisco office of the manager who had written the memo and, posing as another security manager, told her he had never received his copy of the security memo. Would she mind faxing it to him? Of course not, she replied, she would be glad to; she even had the number programmed into her fax machine's speed dialer. When she pushed the button to send the memo on its way, she had no way of knowing that Kevin had programmed the number her fax machine was calling to forward the memo to Roscoe's fax machine at work. Roscoe had reprogrammed his fax machine so that when it responded to the secretary's machine, it looked as if the proper fax machine was responding. Even after Roscoe received the memo and read it aloud to Kevin, Kevin wanted to see it immediately. He didn't want to wait and pick it up after work. So he had Roscoe fax it to him at a copy shop in the San Fernando Valley.

Roscoe later leaked a copy of the memo to reporters and a story describing its contents landed on the front page of *The New York Times*. He couldn't resist embellishing slightly the story of how the memo was obtained. He told the reporters that it had been intercepted by tapping a phone line between two Pacific Bell fax machines. Company officials confirmed the memo's authenticity, but said they were mystified by how it had landed in the hackers' hands.

▲ ▼ ▲

The first call from Pierce College came in to the Los Angeles Police Department on the afternoon of February 17, 1988. A Pierce College security officer was on the phone to report that since January 13 two young men had apparently been making illegal copies of software. On that January evening, Pete Schleppenbach, a computer science instructor, had walked into the computer science lab and seen a tall, slender man in his late teens or early twenties hovering studiously over one of the system terminals. Schleppenbach was taken aback by the sight of a complete stranger standing so authoritatively in a place clearly off limits. "No Admittance—Authorized Employees Only" warned a sign suspended directly above the stranger's head. And one would have to be unable to read English to be oblivious to the warning notes Schleppenbach had taped to the terminal, obscuring the screen: "Do not turn this terminal off. Leave it on!" and "Students: Don't use this terminal unless all others are in use!" To see what was on the screen, the young man had taken Schleppenbach's handcrafted monitions and flipped them onto the top of the monitor. Schleppenbach stood and watched the young man—a student? a Digital technician?—as he reached behind the terminal and turned it off, then on. Then the stranger sat in front of the machine as if he owned it and began to type.

Schleppenbach approached him. "What are you doing?" he asked.

The young man barely looked up. "Just looking," he mumbled. His speech had the tone of someone convinced that his actions were beyond reproach.

"What are you doing on the terminal? Didn't you see the signs? You're not a student here, are you?"

He shrugged at Schleppenbach. "No, I'm not, but she said it was all right," and, without turning around, he thrust his shoulder in the direction of a student employee working in the computer lab. Unsatisfied, Schleppenbach told the stranger to leave. The teacher then asked the student worker if she had given the stranger permission to use the terminal. She said she hadn't.

When he turned around, Schleppenbach saw that the trespasser had not left the room, but was seated at another terminal with someone who appeared to be his friend, a pudgy complement to the rangy, cocksure stranger. Might *he* be a Digital technician? He looked old enough to be out in the world of gainful employ. The friend seemed engrossed in what he was seeing on the computer screen, and he typed in short, feverish bursts. As Schleppenbach began to approach the two, he saw the slender one tap the shoulder of his friend, who turned around to face Schleppenbach. He turned back to the computer and, as if in a big rush, typed something quickly and got up to meet Schleppenbach halfway across the room. The pudgy stranger spoke first. Friendly and inquisitive, he told Schleppenbach that he and his friend were there to find out about the course Schleppenbach was teaching in office computer networks. Cautious despite the young man's friendly, almost engaging demeanor, Schleppenbach explained the course's prerequisites, and told them how to register for classes. He told them they shouldn't be using the system unless they were enrolled at the school. "Okay, we were just leaving," replied the heavy one, at once defensive and vaguely arrogant. "We just know a little about Digital computers."

When they were gone, Schleppenbach walked over to the computer where the two had been sitting. He saw that a tape was inside the tape drive, and that a light on the drive was flickering to indicate that something was being written on the tape. The only person aside from Schleppenbach who was authorized to mount tapes for backup was another instructor. Schleppenbach went straight to the classroom next door, where the other teacher was holding class, and asked his colleague if he had been doing work on the MicroVAX II, a small Digital Equipment computer. The other instructor shook his head. Schleppenbach rushed back into the lab and saw that the light next to the tape drive was still on.

Schleppenbach enlisted the help of one of his students to try to figure out what was going on. They went to a terminal and typed, "show system"; their screen displayed the name of every job running on the computer. When Schleppenbach saw a program called CP.COM, he began to get nervous. Schleppenbach knew that program hadn't been there before, and when he displayed it on the screen, he saw it was a simple seven-line program, a command procedure for making a complete tape backup of every program on the college's system. At first, Schleppenbach decided the only thing to do was to abort the work of the tape whirring inside the tape drive, but he thought better of it. The backup process lasted forty minutes. When Schleppenbach removed the reel of tape from the drive, he saw right away that it did not belong to Pierce College. When he reloaded the strange tape into the tape drive, he asked the computer for a listing of its contents. The tape contained a copy of every file in the system. If he hadn't intercepted those two young men in the middle of their task, they would have walked out of the room with a copy of software worth $20,000. It wouldn't have deprived the college of the software, but in Schleppenbach's view it was theft nonetheless. Schleppenbach figured he had seen the last of them, but for safekeeping he put the tape in a locked room accessible only to faculty. The next morning, he called the chairman of the computer science department to tell him what had happened.

When Schleppenbach walked into the lecture room at 7:00 P.M. on February 9 for the first class of Computer Science 64, he was shocked to see the same two youths seated in the back of the classroom, grinning and tapping pens against their desks. They had enrolled.

Two days later, Schleppenbach ran into Anne Delaney, the former computer science chairwoman, now a professor in the department. He had hardly finished telling her of the incident with the tape drive when she interrupted him. "It isn't Kevin Mitnick, is it?" He looked down at his class roster. Yes. Kevin Mitnick was one. The other was Lenny DiCicco. Delaney looked stricken. She told him that in 1982 Mitnick had been expelled from the school's computer science program for tampering with the school's computers. He was trouble. She told Schleppenbach to alert everyone that Kevin Mitnick was back on campus.

▲ ▼ ▲

This was not the first that Jim Black had heard of Kevin Mitnick. Tall and wiry with an appeal suggestive of Montgomery Clift, the forty-seven-year-old LAPD computer crime detective had loosely kept track of Mitnick for years. When the call came from Pierce College, Black's hunch was that this was more than a simple matter of copying some software from the computers of a junior college. He had heard enough about Mitnick through the years to suspect that he might be up to something big. This time, he wanted to see Mitnick and his friend spend some real time in prison. Black had heard that Mitnick didn't like jail one bit. The detective suspected there wasn't much that would disrupt the young man's impulse to seek control of electronic devices other than a healthy dispensation of justice. He dropped everything to work full-time on the case.

Black had begun to specialize in computer crime in 1982, when two embittered employees of Collins Food, a restaurant chain, were accused of planting two "logic bombs" in the company's computers. The insidious software was designed to destroy payroll, inventory and sales records, and it was only a lucky accident that led an employee to discover the potent little pieces of code before they did their damage. Black had begun to handle other aspects of fraud eight years earlier. He was working auto repair fraud when he was asked if he'd like to join in the Collins Food case investigation. He spent several years on the case, and although the prosecution couldn't gather enough evidence to convict the suspects, the challenge of working a case in which someone had the ability to do something so destructive without leaving a trace was such an interesting departure from routine investigative work that he moved into the department's fledgling computer crime unit. By 1988, Black's division was one of a dozen or so units around the country devoted to computer fraud.

Black saw people like Mitnick not just as a general threat to computer systems, but as a personal threat. Mitnick was known to take direct revenge on members of the law-enforcement community. Black spoke with one of Mitnick's former probation officers, who said the telephone service at her home had simply gone dead one day, but when she called Pacific Bell to report the problem, the company told her that according to its computer her service was just fine. It had taken her days to convince the phone company that her line was really dead, and still longer to get it fixed. Black didn't want to take any chances. He made special arrangements with TRW so that anyone attempting to see his credit rating would have to go through extra steps. The telephone company made similar provisions for Black's home telephone service.

The day after the initial call, Black was at Pierce, talking to the staff and administration. That afternoon, he ran a search on both Mitnick and DiCicco. DiCicco's record showed several outstanding traffic warrants. Curiously enough, Mitnick came up clean. Black put in a call to Bob Ewen in the DA's office, and Ewen told him about the Santa Cruz case. When Black called the Santa Cruz Police Department, he learned that Mitnick had pled guilty to a misdemeanor eight months earlier, and was currently on probation.

Black pulled driver's licenses for Mitnick and DiCicco, and called the letter carrier to see who received mail at 8933 Willis #13 in Panorama City. The mailman said that Mitnick got mail there. So did Bonnie Vitello, Mitnick's wife, and Shelly Jaffe, Mitnick's mother. Black's next call was to the local FBI office. An agent there said he had gotten a call from the FBI office in Baltimore a few months earlier, linking Mitnick to the penetration of a National Security Agency computer from the Hughes Radar Systems Group in El Segundo, but as far as he knew the

L.A. office had no open case on Mitnick or DiCicco. A few days later, Black and a deputy district attorney met with Schleppenbach at Pierce. They told Schleppenbach that before and after each class he should watch both suspects and jot down anything that looked suspicious. Black's next step was to call the local Digital office and describe the problems at Pierce. An engineer from Digital's Los Angeles office took on the job of analyzing the tape DiCicco had left in the college's computer and monitoring the duo's computer activity at the school.

On March 3, surveillance began. Black wanted their every movement accounted for. First, the campus police took up the watch. At 6:30 that evening, plainclothes officer Kenneth Kurtz watched Lenny DiCicco arrive at the computer science lab, sit down at a computer and begin to type. Thirty minutes later, Mitnick arrived. Once Schleppenbach arrived, the students in the lab moved to the lecture room; DiCicco spotted Mitnick and sat next to him. For the next hour, the two students watched the instructor, occasionally leaning over to whisper to one another. At 8:00 P.M., both Mitnick and DiCicco bolted from their chairs and beat the rest of the students to the adjacent lab. The officer sat nearby and watched as Mitnick began helping his stumped peers with their assignments. DiCicco got into a conversation with Schleppenbach and, by way of explaining his thorough knowledge of VMS, the operating system for Digital's VAX computers, told him that he worked at TRW on a VMS system. Kurtz left a few minutes before class was to be dismissed and waited outside the building for Mitnick and DiCicco to appear. At 9:45 P.M., Mitnick came out and began to walk the perimeter of the computer science building. Kurtz climbed onto a roof and watched as Mitnick reentered the building. About three minutes later, Mitnick and DiCicco emerged together. Kurtz hopped from roof to roof of adjoining buildings. He scrambled down in time to see DiCicco get into his small brown Toyota and drive away. Mitnick drove in his car from a separate exit.

Black and his partner picked up the surveillance from there. They followed Mitnick's black Nissan Pulsar as it traveled a seven-mile stretch of the western edge of the San Fernando Valley. By the time Mitnick reached a smaller, winding road in Calabasas, the detectives noticed that he seemed to be following another car. Both cars turned into a parking structure beneath a Home Federal Savings and Loan building. The cars parked out of sight, but Black saw Mitnick walk toward the front. He looked up and saw a light go on in a second-story office. A man with dark hair stood in front of the window. The light went out.

The officers left just before midnight, making a quick swing through the parking area. The Toyota and the Nissan were the only cars there. For the next month, Black and his partner continued the surveillance. On occasion the two detectives had help from the department's Special Investigative Section, or SIS, which carried out most of the sophisticated surveillance needed by the L.A. Police Department. Officers in SIS usually busied themselves with violent criminals, and although these two suspects posed no physical threat, SIS was needed to carry out the fine art of successful tailing. With ten to twelve men, two to each unmarked car—Camaros, Volkswagen bugs, and pickup trucks—the officers carried out a tag-team approach to its surveillance. As one car peeled off, another picked up, sometimes accelerating to one hundred miles per hour to catch up to the suspects, but always staying at least a couple of blocks behind them.

The two suspected software pirates stuck to something of a routine. After leaving the campus on Tuesday and Wednesday nights at about 10:00 P.M., they would drive to the Calabasas office building. They frequently stopped along the way at a Fatburger, an L.A. fast-food chain famous for its tacky decor and gigantic burgers. This Fatburger outlet was squeezed into an L-shaped pink-stucco minimall between a 7-Eleven and a taco restaurant. The two would emerge from Fatburger laden with bags. Judging from the sheer volume of food the two carried upstairs, Black figured they were packing in for a long night.

It would have come as a surprise to the two police officers to learn that Mitnick was supposed to be on a strict diet. As far as his wife, Bonnie, knew, her husband subsisted on a low-calorie menu consisting of oatmeal for breakfast, two ounces of turkey for lunch, and a salad for dinner. It was a sensible diet for someone with heart palpitations who frequently checked himself into nearby emergency rooms with chest pains and kept a drawer at home filled with medical insurance claims. Not surprisingly, the regular pit stops at Fatburger had been the downfall of many Mitnick diet plans. Computers and eating went together for Kevin, which was one reason he never managed to fall below 240 pounds for very long.

The two police officers got permission to use a residential driveway across the street. They tucked their car as far back in the driveway as they could while still keeping a clear view of the insurance office. Occasionally, one of the suspects would venture outside, look to his left and right as if preparing to cross the street, then slip back inside. The vigils livened up on the occasions that Mitnick came outside, crossed the street, walked a few paces to the Hotel Country Inn and began using the pay phone in the lobby. Mitnick's pay phone sessions would last about twenty minutes. Whatever DiCicco was doing inside seemed to be related to Mitnick's time on the pay phone: DiCicco would walk to the landing outside the office and peer down the street toward the hotel.

One of the more difficult things to establish was just what office door of the Calabasas building the two suspects were entering. The manager of the Home Federal Savings and Loan branch told Black that neither Mitnick nor DiCicco worked at the bank. There were at least a half dozen other businesses that rented space there. Two members of the surveillance team were dispatched to the roof of the building, where they lay until they figured out that it was the door to suite 101, the offices of a company called VPA, which stood for Voluntary Plan Administrators.

When Black did some research into VPA, he found that it was a company that administered disability programs for larger companies. The company used a MicroVAX computer made by Digital Equipment. But notifying VPA that it might have computer criminals in its midst was out of the question. Black didn't know if the company was involved with whatever Mitnick and DiCicco were doing.

On March 17, when Black's partner got to work, he received a frantic phone call from Pete Schleppenbach, the Pierce teacher, who wanted to report a bizarre and annoying incident. Five hours earlier, at 3:00 A.M., Schleppenbach had been awakened by a telephone call from a man identifying himself as Bob Bright, an officer with campus security. The officer told the bleary-eyed Schleppenbach that he had just apprehended two male suspects who had been caught wheeling heavy equipment out of the computer science lab. He described one as tall and slender, the other as shorter and fat. "That's Mitnick and DiCicco," Schleppenbach said. "You've really hit the jackpot with these guys." Schleppenbach said he would come right over. The officer told Schleppenbach that that was a good idea, and that he had just put in a similar call to two other Pierce officials who were also on their way. When Schleppenbach and two other Pierce faculty members arrived at the campus security office, there was no one in sight but one lonely security guard, who said it had been an uneventful evening. There had been no burglary and he had no colleague named Bob Bright. It was a mean and childish trick, but for Lenny and Kevin it had served a useful purpose: now they knew they were being investigated.

There wasn't much the police could do except log a mischievous phone call. It was clear that Mitnick and DiCicco weren't going to make this easy. Black and his partner gathered as much background as they could on the two pranksters. DiCicco, at least, appeared to have a full-time job at VPA, as his car would usually stay parked in the garage beneath the building all day. Mitnick, it appeared, was unemployed.

In March, Black got a disturbing tip. Someone from inside the police department had heard that Kevin Mitnick was getting ready to start a job in computer security at Security Pacific Bank.

▲ ▼ ▲

For anyone with Kevin's record, the Security Pacific job should have been out of the question. When he applied in early March for the opening as an electronic funds transfer consultant in the bank's audit department, he knew the chances of getting it were remote. His reputation as a computer criminal was more difficult to shake loose than he had thought it would be. A few months earlier, he had taken a job at GTE as a COBOL programmer. He had worked there for only a week before the security department checked his background. A security officer then approached him one day, escorted him to his car and waited to see that he left the premises. It was a humiliating episode. Most of his other jobs had opened for him through family connections. Had the bank managers at Security Pacific known that a computer criminal would be in constant contact with the computer system that executed, monitored and logged hundreds of millions of dollars of transactions every day, they would have been aghast.

Kevin diligently completed the job application. For the standard query, "Have you every been convicted of or are you pending trial for a criminal offense?" he placed a small, bold check next to "No." An unusual collection of references included Donald Wilson, his former employer at National GSC; Arnold Fromin, owner of Fromin's Delicatessen and his mother's boyfriend; and Roscoe. Then, in a surprising concession to the cloak of secrecy he had maintained for the past three years about his time as a fugitive, Kevin tipped his hand. Perhaps to boost his otherwise anemic educational background, he noted on the application that in the winter and spring of 1985, just when the police had him on a kibbutz in Israel, he was a student at Butte College, a two-year community college in Oroville, a small town in Northern California about 250 miles northeast of San Francisco. Kevin had apparently remained anonymous by enrolling under an assumed name.

Kevin told Lenny that if he got the job at Security Pacific, he would stop breaking into computers for good. And it seemed that he might get the job. With a little embellishment here and there, on paper Kevin didn't look at all bad. He listed his job at National GSC, where he had spent so much time on the telephone, as programmer/analyst. At Fromin's Delicatessen, where his primary job was that of a delivery man with a little computer work on the side, he was also a programmer/ analyst. Both businesses were owned by friends of the family who could be expected to give good references.

When he received the letter from the bank's personnel office confirming an annual salary of $34,000, Kevin was ecstatic. His first day of work, the letter informed him, would be March 25, when he was to report to the training center downtown for an orientation. The last paragraph of the letter was boilerplate: "As discussed, your employment is contingent upon satisfactory reference checks." When Bonnie got home from work, a smiling Kevin told her he had been offered the job. They went out to dinner to celebrate.

Black got the tip about Mitnick's new job on March 23. He called the bank's security department immediately. The next day, Black got a call from Peter Kiefer, a Security Pacific vice-president. Yes, he told Black, the bank had just offered Mitnick a job in its electronic funds transfer section. Mitnick was to start work the following day.

Late that afternoon, Kiefer and a colleague named Barry Himel arrived at police headquarters to talk with Black. They presented Black with Mitnick's application and his list of references. They also showed him an article that had been clipped from the *Los Angeles Times*, dating back seven years to 1981 and telling of three young men, a Kevin Mitnick among them, who had been arrested for stealing manuals from Pacific Bell's downtown computer center in order to wreak havoc on the telephone company and its computer systems. The story had been brought to their attention by a bank employee who knew of Mitnick from his involvement in amateur radio. The worried executives asked Black if this happened to be the same Kevin Mitnick who was about to start work as a security consultant at Security Pacific. By that point, Black's confirmation was hardly necessary. The two security officers had already established a strong link: as one of his references, Kevin Mitnick had listed the true name of Roscoe, one of the three arrested in 1981 for breaking into the Pacific Bell COSMOS center.

It was an awkward situation, to be sure. A decade earlier, Security Pacific had been the victim of a historic heist. The thief, Stanley Rifkin, a plump and soft-spoken thirty-two-year-old computer expert, had worked there as a computer security consultant and walked away one afternoon with the day's code for the bank's electronic funds transfer system. Later that same day, Rifkin phoned the wire transfer room and, using a fictitious name, said he was with the bank's international division. He rattled off a few security codes and his $10 million withdrawal sailed through. Rifkin was eventually caught, but the bad publicity surrounding the bank's security system had stuck with the bank over the years. No one wanted a repeat performance.

When Barry Himel called Black the following day, he said Mitnick had been informed in person that the employment offer had been withdrawn. Mitnick's response, Himel reported, was simply to smile. There was a pause, then Himel asked: Did Jim Black think Mitnick was the type of person to seek revenge in any way? If he were to seek revenge, Black answered, it would most likely be through his knowledge of computers and telecommunications.

Two weeks later, Black got another call from Himel. He told Black that one of the officers of the bank had just had a call from a news service in San Francisco seeking confirmation and more details about a press release from Security Pacific that had come over the wire earlier that day. The release stated that for the first quarter of 1988, Security Pacific was going to show an earnings loss of $400 million. The only clue Himel had that the release was a forgery was the absence of a customary sign-off identifying the source of the story. Of course, Himel told Black, the story was completely false. Apprised of the fraudulent press release, the bank's corporate officers were horrified. The potential damage to the bank if such a release got into the newspapers was incalculable. In plunging stock value and account closures alone, it could exceed the $400 million reported in the fictitious press release. Fortunately, the hoax was stopped in time. Again, Himel paused. Could this egregious act possibly have been committed by Mitnick? It was possible, Black replied, but he couldn't verify it. Nobody ever could.

Computer crime cases were notoriously difficult to investigate, and that was part of the appeal for Black. The little evidence that could be gathered was difficult to tie directly to a suspect. Companies and universities whose computers had uninvited nocturnal visitors could produce dozens of pages of computer printouts covered with blatant evidence of the intrusion, but they didn't necessarily add up to much in the way of evidence. Telephone traps were useful only as far as they went: intruders such as Mitnick who had started out as phone phreaks were expert at covering their tracks. When someone like Mitnick got into a telephone company computerized switch, he could treat the vast telephone network like a series of disappearing stepping-stones, manipulating the machines to create fictitious billing numbers and forward calls to nonexistent telephones.

Black saw computer crime as the ultimate challenge. Not only was the technology interesting, but he enjoyed thinking about a computer criminal's mind-set. Computer criminals hardly fit the common profile of an outlaw. The old investigator's axiom, "We catch only the dumb ones," seemed to break down when it came to computer crime. There were no dumb ones. More often, they were caught because a friend or an associate snitched. Black believed that Mitnick and his circle had been snitching on one another for years, stretching all the way back to the COSMOS incident, but the original gang Mitnick was involved with appeared to have long since split up. Susan Thunder hadn't surfaced for a few years. After a prostitution arrest in 1982, followed by a bizarre incident in 1984 in which she and a friend tried to spring their friend Steven Rhoades from jail by impersonating a deputy district attorney, Susan had apparently given up her computer security consulting in favor of beating the odds at professional poker. Both Roscoe and Rhoades seemed to have gone straight by this time, and as far as Black could tell, neither of them was involved in this incident.

Black had never worked directly on a case involving Mitnick before, but Mitnick appeared to be one of the few old phone phreaks, perhaps the only one, who had kept his skills honed. From experience, Black knew that this was a case that would require a lot of close cooperation from Pacific Bell and Digital Equipment. Not only was Pierce College using Digital equipment, but this VPA outfit was as well. Black had little doubt that Mitnick had devoted hundreds of hours to refining his talents on Digital computers.

Black knew that Mitnick's technical ability far exceeded that of most investigators, especially those unschooled in the subtle art of exploring computer crime. The Santa Cruz Operation incident was a case in point. Black and his colleagues had been stunned to hear of the careless way in which the Santa Cruz police detectives had conducted the search of Vitello's apartment. Why hadn't anyone told them not to take the phone off the hook?

Now that Black knew where the two suspects were working from, he asked Pacific Bell to put a dialed number recorder on the VPA phone to register outgoing calls. After eleven days of bureaucracy bashing, Black was able to get the tap. It yielded a number of surprising calls: several to nonworking, unassigned phone numbers; others made to the MCI local access using pirated authorization codes. Among the most interesting calls were those made to a New Jersey Bell COSMOS computer similar to the one Mitnick had been caught breaking into at Pacific Bell seven years earlier. There was also a call to a New York City–based subsidiary of Security Pacific. Called Precision Business Systems, this company managed data communications for the West Coast bank. The FBI's New York office had recently been investigating a possible wire theft case involving Security Pacific: in early April someone had attempted to get into the bank's secured data communications lines.

Although Black thought he had enough evidence from the Pierce incident to nail Mitnick and DiCicco on computer fraud charges, he kept holding out for more. While intriguing, the Precision Business Systems lead was still too sketchy. Black wanted enough evidence to warrant a substantial prison sentence and he hoped to get it from the VPA surveillance. But in late summer, another case forced him to put the investigation on hold and end the surveillance temporarily.

In the meantime, Pierce College had brought its own case against Mitnick and DiCicco. The college held a disciplinary hearing for which the two suspects put together their own defense, exhibiting enough legal acumen to draw the hearing out to an exhausting twenty-one hours. When Black set aside the criminal case, they were appealing their expulsion.

It had been a frustrating exercise for Black. These investigations hinged on the close cooperation of law-enforcement agencies and commercial institutions working together against a common foe. The detective had hoped for instant help from both Pacific Bell and Digital Equipment. Instead, Pacific Bell's bureaucracy had stonewalled his efforts. And although he had received immediate help from the local Digital office, Black couldn't seem to get the attention of security people at headquarters in Massachusetts. He hadn't been able to trace any of Mitnick and DiCicco's outgoing calls to a major computer system. Still, he was sure that somewhere out there a computer system administrator was having Black's problem in reverse, wondering who the invader could be.

▲ ▼ ▲

When Lenny began his job at Voluntary Plan Administrators in May of 1987, Kevin had no reason to visit: there were no modems. And even though Lenny kept the news from Kevin when the company bought modems that summer, Kevin somehow figured it out. Just as a well-fed cat learns to run to the kitchen at the sound of a can opener, Kevin started showing up regularly at VPA. The company, it turned out, was an ideal place from which to practice his craft. As one of the principal computer operators there, Lenny had the run of the place once everyone had left for the day.

But things were beginning to sour between them. Lenny knew Kevin's capacity for threatening people. He had already threatened to turn Lenny in for creating the fraudulent identity that got Lenny the flower delivery job. But there was more to it. It was also a classic form of *folie à deux*: Kevin could appeal to criminal aspects of Lenny's nature that might have remained unrealized if the two had never met. Also, in Lenny Kevin found someone with the flashes of intuition that Kevin lacked and that were so necessary for their work. Lenny too had become addicted to the excitement of breaking in to computers.

Kevin had stolen, or at least tried to steal, software in the past. Now he had a major project in mind: the acquisition of Digital Equipment's most important software, the latest version of the company's VMS operating system. Lenny became project assistant. They started out from VPA in a low-key fashion. The first thing they established was a way into the Arpanet, the vast research and military computer network. The duo found an account at Patuxent Naval Air Station in Maryland. For a few months in the summer of 1988, Kevin and Lenny used the Patuxent computer as a convenient storage locker for the software they were stealing electronically. When system managers at Patuxent noticed something was going on and closed the electronic doors, the two looked for another place to stash their data. They found it a few weeks later, when they sneaked back into USC's computers.

▲ ▼ ▲

Mark Brown noticed immediately that someone was in the system. Except for the highly publicized degrees-for-sale incident three years earlier in 1985, nothing very damaging had happened to the USC computer system in the past few years. Brown was now manager of research and development for USC's computing services, and he had all but forgotten about the two teenagers who had so boldly waltzed onto campus and broken into the system six years earlier.

But now, someone was dialing into the USC system from off campus. Having found a bug in the system program, the intruder was able to modify the VMS operating system subprogram that acted as the computer's gatekeeper. Brown had to admit the trespasser was extremely clever. Somehow, the electronic interloper had altered the program that supervised user log-ins so that every time someone logged in to the computer, a copy of the password was slightly jumbled and stored in an innocuously named place inside a file. And he had altered the program so it left open a "back door" that allowed him to return any time to harvest passwords.

But there was a flaw. In trying to cover his tracks at the same time that he was installing his rogue code in the system, the thief occasionally crashed some of the USC computers accidentally. Users around the campus, of course, simply assumed that the system had crashed. It was the sort of annoyance that users expected from time to time. But Brown could tell that these crashes were connected to the break-in. From what he could gather, he had fallen victim to the infamous West German Chaos Computer Club, which had achieved international notoriety that fall after claiming responsibility for a summer of poking around inside NASA's SPAN computer network, an international web of computers used by scientists for space and physics research. The Chaos Computer Club had attacked the NASA VAX computers using the very same software trick. They called their program "the *loginout* patch." Whoever was breaking into USC wasn't just installing the *loginout* patch, but was using the school's computer as a launchpad to break into other computers on the Arpanet. That was precisely what Chaos had done on the SPAN network. And since most of the activity seemed to occur in the late afternoon and early evening, it made sense that West German troublemakers would be at work late into the night from Hamburg or Hannover or Berlin, or wherever they lived.

A second odd occurrence left Brown even more concerned. Soon after the break-ins started, he noticed that disk space was disappearing from a USC computer with thousands of user accounts dedicated to coursework in physics and chemistry. Huge chunks of storage capacity, forty megabytes at a time—the equivalent of dozens of textbooks—were being eaten up with no corresponding files to account for the missing space.

After a few days of pulling apart the operating system in a hunt for the source of the mysterious problem, Brown finally figured out that the intruder was creating files and disguising them as system index files, which are directories that describe other files—the last place anyone would think to look. When Brown opened the files to examine them, he was amazed. Someone was salting away the source code—closely guarded original programs—for Digital Equipment's proprietary VMS operating system.

Brown didn't want simply to lock out a trespasser who had entered privileged accounts and seemed to know the operating system. There was no telling what might happen if USC locked him out. It might make him angry enough to find his way in again and start doing some real damage—deleting files or crashing systems.

Software designers write source code in what are called high-level languages. A translator, known as a compiler, then converts the high-level code to binary form—ones and zeroes that can be understood by a digital computer but are difficult for humans to decipher. Computer companies guard their "human readable" source-code files as if they were the crown jewels. Only the binary code, which cannot easily be understood or modified, is distributed to customers. In this, computer companies are like a master chef who serves a six-course meal without giving away the recipe. A competitor can easily copy a computer's hardware, but recreating the operating system that runs on that computer is a far more difficult and expensive undertaking. Companies guard their source code not only because they're worried about competition, but also because access to source code makes it easier for a saboteur to open a secret back door into the computer, known as a Trojan horse. A Trojan horse is a seemingly innocent program planted inside a computer that is designed for a special purpose, such as capturing passwords or even destroying data. It is often difficult to tell that a change has been made. Access to source code can make the planning and execution of Trojan horses easier.

The thief was stealing Digital's lifeblood, millions of lines of software that run on most of the VAX computers in the world, and electronically stashing it in the USC computers. Furthermore, whoever was doing this wasn't just taking any old VMS code; he was copying the very latest version, called VMS Version 5.0. And the only place from which he could have been lifting this software was a group of development computers at Digital's laboratory in New Hampshire. Version 5.0 was so new that Digital customers themselves didn't have it yet and would eventually get only parts of it on microfiche. Perhaps the intruder had managed to break into Easynet, Digital's internal network, which connected tens of thousands of Digital computers around the world. From Easynet, the interloper must have found a gateway to the USC machine on an academic network.

Brown guessed that this thief was keeping his loot on ice at USC because he didn't have enough storage capacity of his own or he didn't want to be caught with the stolen goods. Every time the pilferer logged out, Brown opened the files to see what was in them. He watched in amazement as Digital's trade secrets scrolled by on the screen. Within just a few weeks, many megabytes of disk space had been consumed in the process of stashing stolen material in the illicit treasure chest.

But when Brown called Digital's security department about the theft taking place seemingly under the company's nose, he was surprised again, this time by the lukewarm reception he received. He could understand that the world's second-largest computer manufacturer, with tens of thousands of customers around the world, including a growing number of banks and government agencies, wouldn't be eager to have it widely known that some thief was breaking into its computers and making off with its most prized software. After all, what did that say about the safety of information stored on the customers' computers? But Brown wanted to stop this thief and assumed Digital would rush to offer its full support. He was hoping that the company would arm him with some high-tech monitoring technology that would enable him to peek electronically over the invader's shoulder while he was in the act of burying his plunder. He had heard that Digital had special programs that allowed a system operator secretly to watch people who were on the system in real time—that is, while they typed.

When he called the local Digital office, he explained the situation. "Look, we've got this guy breaking in here," Brown said, "and it's pretty big time. He's got some of your sources." Brown asked if he could get one of the company's monitoring programs.

He was told to wait for a return call. The call came from Chuck Bushey, the chief security investigator at Digital headquarters in Massachusetts. But instead of offering to send a team out to Los Angeles to investigate the situation, Bushey asked for an account on the USC system so that his experts could look things over themselves. He promised he would send Brown the monitoring programs he was asking for. But the software never came.

From that point forward, Brown felt that the matter had drifted out of his hands. Digital didn't seem exactly to be denying that someone had penetrated the very core of the company, but the company wouldn't acknowledge, to Brown anyway, the gravity of the situation. From his conversation with Bushey, Brown got the impression that Digital was somehow trying to brush the matter off. It was like dealing with the Pentagon. Brown surmised that while the security people appeared phlegmatic, software engineers back in Massachusetts were madly scampering to plug their security holes.

The Digital security experts seemed much more concerned with the lax security all over Easynet—it wasn't called Easynet for nothing—than with the fact that the company's source code was being stolen. Brown could have built his own VMS monitor, a program for watching the intruder's keystrokes while he was on the USC system, but it would have been time-consuming and tricky. VMS was an esoteric, user-un-friendly operating system; it wasn't built for customers who liked to tinker under the hood. It was built for large commercial and scientific customers who were content never to touch it because they expected Digital to handle their problems.

So after a handful of unsatisfying encounters with the people in Massachusetts, Brown threw up his hands and decided to let the intruder keep his millions of characters of disk space and have at it. After all, he didn't seem to be hurting anything on the USC computer. And although this was a lot of space to be giving to some uninvited guest, it wasn't more than an engineering graduate student might use up in one night. All he could do, Brown decided, was to keep careful logs of what he was able to see.

The intruder became bolder. He was beginning to use the USC system as a way into other computers more frequently. So Brown shut down access to his privileged accounts on the system. And no sooner had he done that than the Brian Reid hoax happened.

Chris Ho, who worked with Mark Brown at USC, got a telephone call one afternoon in August from someone who identified himself as "Brian Reid from Stanford."

"We're having a break-in here, and it looks like he's coming from USC," said the caller. "We need a privileged account on your system so we can track him down."

"Sure," Ho replied. "Just give me a number where I can call you back and we'll set it up."

"I'm not in my office right now."

"Oh, then give me a time when I can reach you and I'll call you back then."

"I'll have to get back to you." With that, the caller hung up.

Chris Ho was more than a little skeptical of the call he had just received. He called Stanford at once to check on Brian Reid, a Stanford computer science professor. A secretary in the computer science department told him that Reid had left Stanford to work at Digital, in the company's Western Research Lab in Palo Alto. Ho's suspicion confirmed, his curiosity led him to follow through. He called the Digital office in Palo Alto, and after several days of leaving messages and imploring secretaries to get Reid to the telephone, he finally succeeded. Reid needed to utter no more than two words for Ho to detect the difference between Reid and the impostor. Where the timbre of Reid's voice was deep and resonant, the earlier caller had spoken in a much higher pitch, as if each word got trapped in his larynx before escaping from his mouth. No, Reid assured the anxious USC worker, he had not called Chris Ho, and yes, he had been aware of the break-ins at Digital for months.

After the phone hoax drew new attention to USC's plight, cooperation from Digital improved considerably. In October, Bushey flew out to Los Angeles and met with Brown and Ho. But Bushey seemed interested only in logging the intruder's every move. And Brown knew what that would mean: late-night vigils for the next two months, or until whoever was using USC as an electronic warehouse was caught. He wasn't exactly eager to lose weeks of sleep to help Digital, but he agreed to keep a watchful eye on the situation.

▲ ▼ ▲

Lenny had always enjoyed the aspect of traveling through computer systems that made him feel like a fearless explorer. He liked the idea of having computers throughout the world at his fingertips. The most exciting thing about playing around on the Arpanet military network hadn't been so much the information it contained; it was the act of roving itself. Lenny's hacking may have kept him cooped up for days at a time in the VPA office, or in the Hiway Host Motor Inn, but at the same time it broadened his world far beyond the San Fernando Valley.

The Arpanet was the granddaddy of all computer networks. Started with United States military research funding in the late 1960s, the network had served as the technology testing ground for the commercial computer networks such as Tymnet and Telenet that were to follow. The Arpanet originally linked universities with corporate and military research facilities. By 1988 the Arpanet had largely been subsumed in a growing thicket of commercial, academic, scientific, government and military networks known collectively as the Internet. The Internet joined individual networks together via computerized gateways so that it was possible to travel electronically almost anywhere in the industrialized world.

With the advent of computer networks, the traditional sense of geographic space as it was known to explorers of earlier times was becoming obsolete. It had been replaced by a different notion, the idea of cyber-space. Traveling from a computer in suburban Los Angeles to a computer in Singapore was a matter of typing one command on a keyboard. It happened instantly. In fact, distance on the Internet was so transparent that a computer located in Southeast Asia would appear no different than a computer in the next building or in the next county. They would merely be different numbers on a computer host table, a listing of computers on a network.

As it happened, Kevin and Lenny's favorite computer manufacturer was also a pioneer in computer networking. At Digital, networking really meant remote computing—the freedom to move work around the network from one computer to another. The company also tried to make it easy for an engineer in Massachusetts to work on a set of data in California. Networking among Digital computers was built around simplicity, uniformity and ease of use.

In 1984, Digital built its own internal corporate network, the Easynet. Eventually the Easynet would connect thirty-four thousand Digital computers in more than twenty-five nations, giving direct access to nearly two-thirds of the company's 120,000 employees. Easynet has made Digital the world's best-networked corporation. Engineers in Germany, Japan and the United States share design work and rally support for project proposals. Employees also send messages directly to Digital president Ken Olsen. So it wasn't surprising that Easynet became a favorite playground for Kevin and Lenny. Since it was strictly a VMS system, as opposed to the UNIX-heavy Internet, Easynet spoke their language. And for two young network explorers with an eye toward procuring Digital's proprietary software, Easynet was the perfect transportation medium. Once they were on one computer in the network, they could connect to any other.

When Kevin and Lenny penetrated Easynet, Lenny was as excited as he had ever been. Later, Lenny would say that his breaking into computers with Kevin was "like we were boldly going where no hackers had gone before." But to Kevin "it was just a task, like coming to work every day, just to get the job done." While Lenny seemed content to scamper from one computer on the Easynet to another like a puppy, sniffing at each new one, Kevin considered their computer system cracking a serious endeavor with a series of discrete goals. In fact, for as long as Lenny could remember, Kevin had always approached his illicit computing as a serious project. When he sought revenge on someone, it was as if he were taking on an assignment from some invisible employer. Once an assignment was complete, he reported back to Lenny or Roscoe with news of his triumph. As early as 1981, Kevin would call Roscoe to report the results of a frontal assault on the telephone service of someone he was out to harass.

Kevin's project for 1988 was downloading Digital's VMS source code. It wasn't so that he could make pirate copies and sell them. Rather, it was both the challenge of the hack itself and his intellectual curiosity about such a complex and advanced program.

Kevin's heavy phone use had just gotten him fired from a small firm near Thousand Oaks that made electronic testing equipment. He was sleeping until at least 11:00 every morning and would usually put in his first call to Lenny around noon. Around 3:00 in the afternoon, Kevin would instruct him to get started with a particular task. Then, at about 7:00, on the days they didn't have class at Pierce, Kevin would show up at VPA, and after dinner they would return for a night of network exploring. As the night wore on, Kevin's pocket pager sounded every hour. It was Bonnie, wondering where her husband was. He lied and said he was taking an evening class at UCLA. Kevin's lack of accountability to his wife didn't seem to matter to him.

One of the first orders of business on these evenings was to go into a large telephone wiring closet on the ground floor of the building and connect VPA's modem line to another tenant's phone. To someone checking toll records, it then seemed that calls were coming from another business. By now, Kevin had refined this technique, prompted in part by his increasing knowledge of how the police and telephone security forces worked, and by a growing paranoia. He regularly checked to make certain there were no dialed number recorders on the line he was using. And to be extra safe, Kevin and Lenny looked for an untraceable way to make long-distance calls. Like many other hackers and phone phreaks, they frequently used illegal MCI calling card numbers. Most phreaks would "scan" for them—that is, they would program their computers to have their modems keep dialing a local MCI port and enter different codes until a successful code was found. Purloined codes were traded on electronic bulletin boards or "code lines," toll-free voice-mail numbers that had been discovered by hackers and modified to dispense pirated calling card numbers. But Kevin, in his inimitable way, had perfected a method for obtaining telephone credit card numbers with great efficiency. He later told the FBI that he had obtained the password for the network security account on MCI's electronic mail system from an electronic bulletin board. But Lenny knew that Kevin had talked someone out of the password, using his excellent social engineering skills.

Once logged on to the MCI network security account, Kevin and Lenny were privy to confidential information about new accounts, stolen accounts and accounts that were in a state of limbo until MCI actually discontinued them. They could also read electronic messages concerning security breaches. It was a fertile field from which to harvest account numbers, as well as an early warning system to subvert the security types. Kevin guarded the MCI network security account from Lenny as if he were in possession of the Hope Diamond. "This is something I could make money with," he would tell him.

Slipping into Digital's Easynet was a little like discovering the mother lode. All of the company's private discussions were carried out over the worldwide network. Using relatively standard tricks, Kevin and Lenny collected passwords. For instance, when a machine is first set up, there are a few preestablished accounts with preset passwords for the convenience of Digital's field service people. Each such password gives access to that account's electronic mail. Once Kevin and Lenny had found these passwords, they used Kevin's gift for intuiting how organizations functioned and who was important in an organization. An untrained but instinctively expert social scientist, Kevin could look at patterns of communication in stored electronic mail and figure out who had power and who had valuable information. In that way, they found the mail that was worthy of their attention.

Andy Goldstein's mail was by far the most informative. Goldstein was regarded by many as the most brilliant technical expert in the VMS engineering ranks. He was also a VMS security expert. As a result, most messages about VMS security problems eventually ended up in Goldstein's mailbox. One of his correspondents was Neill Clift, a researcher at Leeds University in England. From what Lenny and Kevin could tell, VMS security was a hobby of Clift's; he seemed to spend hours plumbing the depths of operating system arcana in search of security flaws. Clift was a prolific electronic correspondent. Most of his messages to Goldstein concerned problems with VMS security. And every time he described a new flaw to Goldstein, Kevin and Lenny drank in every word.

It was while snooping through Andy Goldstein's electronic mail that Kevin and Lenny found the infamous Chaos Computer Club *loginout* patch. From what Kevin and Lenny could decipher from the mail to Goldstein, it seemed that someone in Europe had broken into a European institution using the Chaos patch. The victimized company had made a copy of its entire system and sent it to Digital to analyze. Goldstein, it appeared, had analyzed the program, extracted the portion containing the rogue code and reverse-engineered what the Chaos Club had inserted. After stealing Goldstein's analysis from his mailbox, Lenny and Kevin puzzled for days over the unexpected prize that had landed in their hands. From what they could tell, someone from Chaos had changed *loginout*, the program for logging in and out of the computer. Chaos had modified, or patched, the program so that each time a password was entered a copy of the password was sent to a spot in a remote corner of the system, where it sat unnoticed until someone came to retrieve it. Lenny and Kevin couldn't believe their luck. A program to steal any VMS password, courtesy of Digital! Not only was the *loginout* patch small and smooth, but it had an added feature that made the user's presence difficult for the system to detect.

To further confound Digital's computer security experts, the *loginout* patch had been written in such a way that it successfully eluded ordinary techniques to detect software that has been tampered with. Thus no standard computer security alarms were set off by the program when the patch modified the computer's operating system. A system operator would never realize something was amiss. In fact, the patch was written so well that Digital officials had little notion that their machines had been so thoroughly compromised. Kevin and Lenny were delighted by this bit of serendipitous international cooperation. More than that, they were in awe of the Chaos Club for its ingenious hack. They had been calling themselves "the Best," but now that they had seen this, they amended their epithet to "the Best in the West."

So USC's Mark Brown had been right after all. However, it wasn't the work of the Chaos Club itself he was seeing, but that of a pair of the club's admirers who had discovered and were using one of the club's cleverest hacks.

The Chaos Club's famous hack into NASA's worldwide SPAN computer network in the summer of 1987 had been a public-relations calamity, both for NASA and for Digital, whose computers were the foundation of the SPAN network. For months, members of the Chaos Club had foraged through hundreds of computers on SPAN. By exploiting an embarrassingly obvious hole in VMS, the cocky group of young computer anarchists had first broken in to the CERN physics laboratory in Switzerland, then electronically hopped over the Atlantic to Fermilab in Illinois, and continued on to hundreds of computers on SPAN. When the incident first surfaced after Chaos held a press conference to announce its accomplishments, it revived some concerns that were first expressed following the 1983 release of the movie *WarGames*. The film, which depicted a teenager who played havoc with a North American Air Defense Command computer, roused widespread speculation that bright kids could somehow compromise U.S. national security. In what seemed to be a similar situation, the infiltration of SPAN by Chaos implied to the public that SPAN was a sensitive military network. This proved not to be the case; nevertheless, it was more than an embarrassing incident for NASA and for Digital. The company suffered a corporate black eye. Nothing like this had ever happened to computers made by International Business Machines, Digital's principal competitor.

Digital could ill afford bad publicity. The company was under fierce competitive pressure, moving thousands of employees from assembly lines and corporate offices into the sales force in an effort to slash costs and boost sales. For nearly a decade Digital's strength had been its sturdy, steady seller, the VAX, a potent force against IBM. But by 1988, the pace of technology was forcing Digital to seek out new markets while continuing to satisfy its traditional customers. Digital was locked into a race, scrambling to bring out new products and rebuild the VAX line. The computer industry was no longer growing exponentially, particularly in the minicomputer business that Digital had long dominated. Instead, the markets for personal computers and workstations were still expanding, but the company had stumbled badly in the PC business. To get back on course, Digital was going to have to go after a more commercial market than it had served in the past. Financial institutions were a new major target and they would hardly tolerate potential flaws in the security of the machines they were buying. So the company remained as quiet as possible about the NASA incident. It patched the loopholes and tried to tighten security. As it turned out, its efforts were less than successful.

The *loginout* patch was written so cleanly that Lenny and Kevin had to modify it only slightly to install it on the newer version of VMS used in the United States. Once they had done that, they were ready to insert their new Trojan horse. In the ensuing months it was to become their most valuable electronic crowbar. Kevin and Lenny dubbed it their "super-duper password scooper."

▲ ▼ ▲

But even before Lenny and Kevin discovered the *loginout* patch, they had used Kevin's wiles to get access to the company's VMS development cluster.

Digital's largest software development facility is a complex of buildings in Nashua, New Hampshire. Three of the Nashua buildings are at 110 Spitbrook Road and are code-named ZKO1, ZKO2, and ZKO3. The ZKO complex is the software capital of the company. It employs about two thousand people—more than half of them programmers—and has more than three thousand computers on-site, thirty of them mainframes. There is a heliport outside that employees can use to travel to other major Digital facilities or to Boston's Logan Airport. ZKO is set in the middle of a stand of hardwoods, near a pond. It's at least a quarter of a mile to walk indoors all the way from one end of the complex to the other, and as you walk down the hall you look out at landscape that probably hasn't changed much since the days of the American Revolution. It's an intriguing mixture of moods, and the floor-to-ceiling windows along the hallways manage to blur the boundary between modern high-tech and hardwood forest.

Managing three thousand of anything can be tricky. People who manage large collections of computers typically organize them into groups, and into groups of groups. Fred Brooks, the chief designer of IBM's OS/360 operating system, made a shrewd observation some time ago: the structure of a computer system almost always mirrors the structure of the human organization that created it. As a result, the VMS computers in Digital are largely organized into groups that correspond to the management structure: people who work together cluster their computers together under a single administration. However, technical peers work across the grain of the organization and often give one another privileges on their own machines.

The Star development cluster of VAXes in the ZKO complex was where all of Digital's VMS development work took place. Late one night, soon after Lenny started at VPA, Kevin stationed Lenny at VPA and, from a pay phone, called a graveyard-shift operator in the Star control room. Posing as a technician in the field, Kevin had her walk over to the console and type in a single command that seemed harmless enough to someone without an intimate knowledge of VMS. Without hesitation, she logged in to the computer, thereby spawning a new process. A dollar-sign prompt appeared on Lenny's screen, the signal that he could now type in the privileged commands reserved for a Digital operator. But the full import of where Kevin and Lenny had landed eluded them at first. Kevin wasn't so interested in hanging around to check out the development cluster. It looked a lot like any other VAX system. Instead, once they were in, Kevin invoked his personal battle cry: "Get priv'ed, set up, move on." Setting up was a matter of creating an account for themselves so that they could get in at some future time. Who knows, the reasoning went, when access to the company's development computers might be useful?

Only gradually did it dawn on Kevin that they had achieved their ultimate goal. Breaking into the development cluster could give them unlimited access to the VMS source code. But even though they were inside the temple, there were still obstacles. They needed to transport the software to some safer haven, some secret location.

In order to acquire the source code in such volume, one of the things the two data hijackers needed was a high-speed gateway through which they could transfer the programs. Relying on their relatively low-speed modems at VPA would have been like siphoning the ocean through a straw. Just as important, they needed to find a link from Easynet to the outside world. Digital's computer network was not, as a rule, connected to the outside world. But there were some exceptions. Digital's Western Research Lab in Palo Alto, it turned out, had just the gateway they were looking for. Established in 1982 by a group of defectors from Xerox's Palo Alto Research Center, DECWRL, as it was called (pronounced DECK-whirl), represented Digital's future, a computer science laboratory where researchers experimented with the most advanced ideas in computing.

The Palo Alto lab was composed of about twenty-five people in a four-story streamlined brick building that also served as Digital's western headquarters in the downtown section of Palo Alto, an upper-middle-class community next door to Stanford University. The Western Research Lab seemed more like a Silicon Valley start-up than a corporate outpost. The lab was also one of the few UNIX havens in Digital's otherwise VMS-dominated universe. It was logical that the researchers at DECWRL would want to work with UNIX, specifically Digital's version called Ultrix; not only had they grown up with UNIX, but in their frequent cooperative research with university scientists, UNIX was the common denominator. And there are other things that make UNIX attractive: it is highly "portable," which means it can run on almost any type of computer, from an IBM PC to a Cray supercomputer. VMS, on the other hand, is confined to VAX computers. UNIX, which was invented by two computer researchers at AT&T's Bell Laboratories, has gradually become a standard for scientists, engineers and universities. Scientists at the Western Research Lab were experts in the UNIX operating system, though they held a grudging respect for VMS.

Also at Palo Alto was the Digital Workstation Systems Engineering Group, a small group of designers inside Digital who were working around the clock on the next-generation workstation, Digital's first machine ever to be designed solely for the UNIX operating system. It was being built outside Digital's traditional development process and the company wanted to get the machine out as soon as possible in order to remain competitive with smaller, younger companies, such as Sun Microsystems. The new workstation's code name was PMAX.

High-speed network gateways may have been common at large universities and research laboratories, but they were found less frequently at large corporations. Most IBM sites have no direct connections to the outside world, mainly for reasons of corporate concern about security, and those with connections can only dial out to other computers—they cannot accept incoming calls. The Digital gateway in Palo Alto, with links throughout the world, was extraordinary in its very existence. It could crank data in and out at 56,000 bits per second, which is the equivalent of transmitting all of *Moby-Dick* in less than two minutes. Its purpose was to put the research of Digital's computer scientists closer to research that was going on at places like Stanford and Berkeley. Having good network connections mattered to a computer scientist in 1988 the way a properly equipped kitchen would matter to Julia Child. As part of the international research community, sharing ideas and papers and ongoing research with counterparts around the world, the Palo Alto computer scientists believed it was necessary to have direct connections to the global Internet. And in recognition of the open-mindedness back at corporate headquarters, the computer scientists in Palo Alto took great care to operate their precious gateway responsibly. To give the best possible oversight both for maintenance and security, Ph.D.'s in computer science took turns poring over daily log files, a job usually performed by administrative staff.

So it was only a matter of hours after the intrusions into the Palo Alto computers began that the gateway watchers there noticed something amiss. The Palo Alto scientists who monitored the gateway—Brian Reid and Paul Vixie—decided they were dealing with experts. Sometimes the intruders came into Digital over a phone line, as an employee might dial into the company's computers from home. From there, they knew how to get access to Easynet. From Easynet, they could connect to any of the small VMS computers in the Palo Alto building, often a desktop workstation, where they experimented with passwords until they managed to log on. Once they had logged on, they exploited weaknesses in VMS to gain privileges on the small system. And once they had done that, they would commandeer the small computer, forcing it to masquerade as a larger system on the network, some users of which would have expanded access rights. This is a method known as "network spoofing." At the same time, the intruders would plant their Trojan horse, a modification of the *loginout* program, and capture user passwords, and then take advantage of the natural tendency for people to use the same password on all the systems on which they worked—whether VMS or UNIX.

▲ ▼ ▲

The intruders' objective was to log on to the gateway computer—an Ultrix machine called Gatekeeper, named for a character in the film *Ghostbusters*. Gatekeeper was the only means of contact with the Arpanet. Getting into Ultrix, then, was necessary if they wanted to take software out of the company. Once the intruders had gathered their crop of passwords from a VMS machine, they could try each one of them on an Ultrix machine. And that is how they got into Gatekeeper. Long after midnight one night, Paul Vixie noticed activity in the account of one of the secretaries, who appeared to be using Gatekeeper in a very unusual and sophisticated manner. The next day he had her change her password, and disabled her account on the Gatekeeper computer.

It was a daily battle, and even the best computer scientists at the research laboratory began to feel beleaguered. The intruders were creating too much extra work. Reid and Vixie were copying all recently changed files to special backup disks in case something went wrong. The intruders' clandestine copying activities were creating an unusually large number of recently changed files, which occasionally overflowed the backup disks.

The Palo Alto scientists felt that the intruders were as clever as any they had ever seen. They were experts who didn't make mistakes. They broke into almost any computer in Digital at will. It was evident from the way they were exploiting bugs that they must have source code, some pieces of which were so proprietary that even Reid and Vixie didn't have access to them. Frustrated by his own lack of access to the source code, Reid wrote in a message to Digital's security administrators: "Clearly the intruders have the sources. When sources are outlawed, only outlaws will have sources." One thing they did to limit the loss of information was to alter the data just before the intruders transmitted it out of the Palo Alto computer, coding it in such a way that it was useless after it was stolen.

By August, at least three people in Palo Alto were taking turns monitoring the laboratory network. It usually made for an interesting evening. One thing they noticed was that unlike a typical intruder in a computer, whoever this was had a fairly sophisticated method of sifting through directories for what he wanted. Ordinarily, a meddler entered a directory that appeared to contain something worth looking at and read the first twenty lines or so of several files. But this intruder was considerably more efficient. He had an uncanny way of finding a computer with vast amounts of data and within half an hour he would have the one thing up on the screen that was worth stealing. He did it by executing a sophisticated sweep of the entire system, looking at access times and looking to see which people had been using which directories. His search strategy, curiously people-oriented, revealed an ingenious insight into corporate hierarchies. That is, he seemed to be able to figure out which people were important, then he would go look at what they had been doing recently. The kicker was that he kept opening the safe and looking inside, but he never took anything other than source code. Perhaps he was just casing the joint.

Among those involved in chasing down the interlopers, frustration grew. Sometimes, just to get some work done, or to get a decent night's sleep, the Palo Alto crew would break the network links between Palo Alto computers and the outside world, disconnecting from both Easynet and Internet by pulling the plug on the modems. If the computer scientists at DECWRL felt that they were under a sustained attack, some system managers at Digital's commercial offices were terrified, convinced that someone had taken the control of their machines out of their hands.

In an electronic mail message to his managers, one irate Digital employee wrote:

```
We seem to be totally defenseless against
these people. We have repeatedly rebuilt
system after system and finally management
has told the system support group to ignore
the problem. As a good network citizen, I
want to make sure someone at network
security knows that we are being raped in
broad daylight. These people freely walk
into our systems and are taking restricted,
confidential and proprietary information.
```

The best that could be done was to take the VAX development cluster in New Hampshire off the Easynet, which meant that engineers were unable to work on those computers from home, or from remote sites. To many of them, it was an irksome and unacceptable solution.

In the meantime, those in the Workstation Systems Engineering Group in Palo Alto who were working on the new PMAX workstation had some real trade secrets to protect inside their computers. The Palo Alto group was toiling secretly, trying to bring its workstation to market ahead of Sun Microsystems, which was working on a similar computer. The principal objective behind the development of the new workstation was to take Digital's competitors completely by surprise. Losing that element of surprise could significantly hurt the computer's chances for success. Although the intruders logged on to machines belonging to the workstation group, they didn't appear to be interested in taking anything there. But as an extra precaution, the group kept the specifications to the new machine off line, and gave them out in person to a select group of people.

▲ ▼ ▲

It might have had something to do with residual instincts left over from his days in the intelligence community, but Chuck Bushey was convinced that the break-ins were the work of an international espionage ring, with malicious West German computer hackers feeding valuable Digital programs to the East Bloc. Bushey was Digital's top cop and he had a cop's way of thinking. Once he had gone to a morning appointment at a remote Digital office two hours early to test security by confronting guards at all the doors.

But Bushey was unschooled when it came to technical matters, and the subtle points of computer penetration that a computer security expert might find significant tended to mystify him. His suspicions arose in part from the fact that he had recently received intimations of shady doings on the part of the Chaos Computer Club, and partly because some of the phone traces went back to network nodes close to East Bloc borders. Another piece of evidence led to Karlsruhe, West Germany, which Digital officials knew to be a hotbed of Chaos Computer Club activity, and to a university student who had spent the previous summer working at Digital in Palo Alto. While at Digital, the student had spent much of his time in front of the photocopier. Moreover, the German student was the only person who knew the location of Brian Reid's password-cracking software, which the intruders stole from a Palo Alto computer. By this time, it seemed that the Digital security forces were building up a strong case for the existence of an international conspiracy.

In early November, Bushey met with Mark Rasch, a young attorney from the Justice Department who had developed an expertise in computer crime. Bushey told Rasch about the months of fruitless investigation and false starts with no suspect. When Bushey posited his international conspiracy theory, Rasch found it very plausible. A tough-minded thirty-year-old New Yorker who had gone straight to the Justice Department from law school, Rasch had become proficient at investigating high-tech crimes. His first position was in the department's internal security division and his early cases had included an investigation into the unlawful export of VAX computers to Eastern Bloc countries. His practice was to assume the worst. Presented with a set of brazen, systematic electronic break-ins aimed at the heart of Digital's software development efforts, his inclination was to assume it was espionage until proven otherwise.

It was a busy week for Rasch. While meeting with Bushey he received word that a computer virus had infected the Internet the previous evening, crippling thousands of computers around the nation. Mike Gibbons, an FBI special agent in the Washington field office, tracked Rasch down at the Digital offices that morning to tell him he had already opened an investigation into the Internet virus. As Rasch was hearing the reports, there were still no leads on who had written the virus. Rasch saw no apparent link between Digital's troubles and this new computer virus, but he couldn't rule one out entirely. One of the virus's targets was VAX computers.

The Palo Alto scientists were beginning to disagree with Bushey. By November, they had been monitoring the intruders' activities for months. This was not international espionage at work, they decided. First of all, the passwords that the intruders made up when they created accounts for themselves—words like *Spymaster* and *Spoofmaster*—were in consistently flawless American vernacular. Second, it was widely known that the Soviets already had both VMS and UNIX and wouldn't need to steal them. Moreover, if they did need to steal an updated version, it was more than likely that among Digital's 120,000 employees, or among those few customers with source code, there was at least one Russian operative who had already taken home everything he needed. It just wasn't necessary to go through all the effort that the intruders undertook. Brian Reid, the Palo Alto office's resident computer security expert, believed that in all likelihood this was a game being played by some ne'er-do-well somewhere who slept into the afternoon, got up, ate some breakfast and sat down at his home computer at three in the afternoon for a long day of hacking. After a hard day at the office breaking into computers, he (experience had taught Reid that it was inevitably a male) would knock off around midnight, meet some fellow hackers for pizza and go to bed around three in the morning.

Not only did those hours correspond directly to the hours during which the break-ins took place, but Reid had spent enough time around the computer underground to know its mind-set. In the ten years Reid had taught at Stanford, he had been involved in network security, had met several such "cyberpunks" and had sparred with dozens of them inside the Stanford computers. Both Reid and Vixie were impressed with the technical skill displayed by this one, but they argued that anyone with a modicum of talent and ten hours a day to devote to the problem could probably do the same thing. The one piece of evidence in favor of Bushey's international conspiracy theory was that the intrusions were so energetic they had to be the work of at least two people, a tag team perhaps. If these were truly bored teenagers, on the other hand, then there was also in them something of the network cowboy, out on the nets having a good time roping computers like so many steers, just out to see how many network nodes they could lasso in a night. Witnessing the frenetic activity every day was a little like watching a movie with the sound turned off; Reid didn't get to hear the whooping with each vanquished system, but he could imagine it well.

▲ ▼ ▲

Retrieving the data from USC was a three-man operation, so Kevin and Lenny pulled a reluctant but curious Roscoe in on the project. Fearing that he would be recognized, Kevin stayed off campus; Lenny and Roscoe went to USC together. Lenny stationed himself at a pay phone outside the university's main computer center and called Kevin, who was stationed at a terminal at the offices of a friend. Kevin then instructed Lenny to tell Roscoe to mount a magnetic tape. Roscoe walked in and, dusting off his old social engineering skills, posed as a student working on a project, handed a tape to a computer operator and asked him to mount it. Once that was accomplished, Roscoe returned to the pay phone where Lenny had Kevin on the line, and Kevin typed in the appropriate commands to transfer the files from the USC hard disk to the magnetic tape. Then Roscoe could retrieve the tape and hand it to

Lenny. This complicated procedure had to be done a few times in order to build the collection of tapes, which were gradually stacking up at All-Purpose Storage in Mission Hills, where Lenny had rented a locker. Roscoe occasionally provided a blank tape from his workplace, but most of the tapes came from VPA.

Kevin and Lenny both considered what they had accomplished a tremendous feat. By having their own copy of the VMS source code, they could pick the operating system apart line by line, understand it, find its numerous flaws and even, if they wanted, create a modified version—say, with something like the Chaos patch included. If Kevin and Lenny really wanted to be malicious, they could send contaminated software back to the Star cluster, or even to the software distribution center in Massachusetts, which generates exact copies of what it has received. People who work at a software distribution center are typically not engineers and would have no way of checking to see if their copy of the information had been secretely altered. That meant that until the next release came from the developers, every copy of VMS that left the distribution center for a customer site would contain the corrupted software.

When the engineers back at the Star cluster discovered that someone had broken in, they had no way of knowing whether or not their master source code had been corrupted. So they were forced to go through a laborious week-long exercise of installing all the magnetic tape backups of their master files and comparing them with what was on the disks, to make sure that every change was authorized.

But Kevin and Lenny had had no plans to visit havoc on the software. Their electronic joyride through Digital was harmless by their standards. They were copying the source code, but in their minds they weren't taking anything. As they saw it, they weren't even stealing, because they left the original software intact, just where it was and just how it was.

By the same idiosyncratic view of the law, when Kevin and Lenny encountered a security program called XSAFE, which they learned about while reading the confidential electronic mail of an engineer named Henry Teng, they had no qualms about taking it. Teng appeared to be the program's chief architect and "Digital Confidential" was stamped all over any memos concerning XSAFE. The program wasn't due to be released for another year, and Kevin wanted the source code.

It didn't take Teng long to notice they had been in, so he changed his password. Getting Teng's new password was a particular challenge, as he was choosing from a list of randomly generated passwords. Although the password Teng picked did not echo back on his screen, the list from which he chose went to a log file he kept, which Lenny and Kevin could see. They went to the log file, found the list of choices, and eliminated their way to Teng's new password.

When Teng discovered they were back in, he moved to a machine that was on the Easynet but had disabled inbound connections. Through reading mail, Kevin and Lenny found the name of the machine Teng was now on. When they tried the "set host" command, a way to link one computer to another, they got a response that all inbound connections would be refused. Figuring that Teng himself had to have a way of connecting to the computer, they started looking at terminal servers (computers that act as switchboards, making connections between computers on a network) in the development center and finally found a hidden avenue that gave them a way to reach to the machine. They continued the task of transferring out the XSAFE source code.

Perusing the personal mail of network security experts like Brian Reid, and corporate security people like Chuck Bushey and Kent Anderson, who was in charge of European security, was a good way for Kevin and Lenny to track the progress of the hunt. They read mail in which network security people were told to drop what they were doing to work on the case. At one point, it appeared to Kevin and Lenny that at least three people in corporate security were working on the problem full-time.

Even if they were caught, Lenny and Kevin had read enough about computer crime to know that Digital might be reluctant to press charges against them. They were aware that few of the computer crimes detected were ever reported to the police and still fewer were made public through criminal charges. Lenny and Kevin knew that companies worried about having their vulnerabilities publicized.

▲ ▼ ▲

But in June 1988 there was a close call. Lenny got a telephone call one day from a particularly overwrought Kevin.

"Get a copy of today's *L.A. Times!*" he cried.

When Lenny looked at the newspaper, he couldn't believe it. "JPL Computer Penetrated by a 'Hacker' " was the front-page headline. The duo's traversing of the Internet had taken them briefly into a computer at NASA's Jet Propulsion Laboratory in Pasadena a few months earlier. Kevin was interested in the JPL computer because it was one of the few

VMS machines on the Internet and he had considered using it as a storage locker for the VMS source code, but then decided against it. But they didn't think they had been detected. Luckily, the laboratory appeared to suspect the Chaos Club, because of evidence of the club's trademark "insidious Trojan horse," the *loginout* patch. The newspaper article said that although no classified material was compromised by the hackers, one JPL engineer said officials there "worried that an intruder could learn how to send bogus commands" to the eight unmanned interplanetary Explorer spacecraft the laboratory controlled. Kevin and Lenny, of course, had no such intentions. They had just been poking around.

By early fall of 1988, Kevin was acting particularly paranoid, eating more than usual and making unreasonable demands on Lenny. He had no regular job, leaving him with plenty of time to be obsessed with his hacking projects. He insisted that Lenny be ready to drop what he was doing in his legitimate job and start work for Kevin promptly at three o'clock every afternoon. Or he would call Lenny at five in the morning and instruct him to go to VPA to insert the Chaos patch just as everyone in Massachusetts and New Hampshire was logging on. Kevin had changed his telephone billing name to James Bond; the last three digits of his new number were 007. He had changed Lenny's billing name, too —to Oliver North.

Kevin made it clear to Lenny that he wanted to stay several steps ahead of the growing pack of hounds Digital had deployed for the chase. By this time, Kevin had improved his phone work so much that he made it almost impossible for USC to track him down. His most effective red herring was his use of the call-forwarding feature on the telephone company's computers. This masking technique meant that when the phone company initiated a trace, it would lead to a blind alley. Kevin was ruthlessly arbitrary in forwarding the telephone numbers. And until he and Lenny read Chuck Bushey's mail one day, they had no idea that Digital had gone so far as to trace the calls back to the random numbers they had chosen to forward to. One of these random numbers led back to the apartment of a Middle Eastern immigrant in Santa Monica, a middle-aged man who was watching television when federal agents barged in looking for computer equipment. They found none.

When in the throes of the hacking, Lenny was having a good time. The idea of having control over many computers was incredibly seductive. The challenge of figuring out where to go to look for the information they wanted was more stimulating than any college programming course. And all the while they watched a twelve-billion-dollar company flounder helplessly as it headed down one cul-de-sac after the other.

Yet something in Kevin was changing. He had become so secretive that when he told Lenny he had lost the Security Pacific job, the greater surprise to Lenny was that he had gotten it in the first place. The same thing had happened when Kevin got and quickly lost the job at GTE. For Lenny, too, the thrill was beginning to disappear. He became suspicious that Kevin was storing information to use against him someday, while withholding anything about himself that Lenny could use. They began to argue frequently over trifling things. And they were constantly betting against each other—twice Lenny lost a $120 bet on whether Kevin could break into the VPA computer from outside. Kevin did it through sheer doggedness. When it came to perseverance, Lenny was a distant second. He lost most of the bets.

And since Kevin had come back from his months "underground" a few years earlier, he seemed to take his electronic penetration more seriously than ever. Perhaps he was thinking that if he was going to lie to Bonnie every night about where he was, it should be for a reason he considered defensible: he had goals to meet. On the other hand, something seemed to be beyond his control: he kept telling Lenny that he wanted to stop the break-ins as soon as they finished the project they were working on; but once they had finished one project, Kevin always seemed to want to start another.

By November, Kevin was insisting on coming to VPA every night to break into Digital. Whenever Lenny tried to beg off so that he could get some sleep, see other friends or spend time with his girlfriend, Kevin would badger him until he agreed to spend the evening with him. Sometimes he even called Lenny at two in the morning and rousted him from bed, insisting that he meet him at VPA. Kevin was obsessed and he had drawn Lenny into his mania.

Bonnie, who was paying most of the bills, had started losing patience with her errant husband. She paged him constantly, and Lenny sat in disgust as he listened to Kevin lie. "I'm at the UCLA law library working on the Pierce case," he would tell her. Or: "I'm studying for my extension class." After he hung up, Kevin would grumble about Bonnie. She was a nuisance. Getting in the way. He was thinking of divorcing her.

Lenny could feel that things were heating up. He turned to Roscoe. Kevin was getting out of hand, he told him. He felt as if he couldn't pee without checking in with Kevin first. Roscoe, too, was beginning to get fed up with this roller-coaster relationship between Lenny and Kevin. It was true that Kevin was a nuisance. He was calling Roscoe all the time, too. Mostly, Roscoe felt terrible for Bonnie, who knew of none of it. He resisted a temptation to tell her to go to the UCLA law library one night and look for her husband. Now Lenny was sounding panicked. Roscoe didn't want to get involved, but he did offer him some advice: get a lawyer.

Even now that the VMS software was encrypted and safely locked away in a storage locker, Kevin still wouldn't quit. His next project, he decided, was to steal a game called Doom from Digital's Star cluster. He was quite content with the setup at VPA. Judging from the unsuccessful traces, there was no one even close to catching them. In the face of Lenny's protests, Kevin was insistent. What had once been a friendship between like-minded teenagers had become a series of demands and contests and long nights in the stuffy offices at VPA. If this was big-time electronic crime, then Lenny wanted nothing more to do with it.

▲ ▼ ▲

It was the stunt with the paycheck that unhinged Lenny completely.

"Hello, Ms. Sandivill," came the chatty, genial voice on the other end of the line. "This is Carl Halliday with the Internal Revenue Service. I understand you're in charge of payroll at VPA."

"Who did you say you were?"

"Well, ma'am, we're doing an investigation into an employee of yours. A Mr. Leonard DiCicco. Seems he owes Uncle Sam some money. So we need your cooperation."

"Yes?"

"We need to ask you to withhold Mr. DiCicco's paycheck until we clear this matter up."

Livia Sandivill, VPA's bookkeeper, shook her head. There was something about the call that made her uneasy. She asked him to hold on and went into Ralph Hurley's office.

"The IRS is on the line and they're asking about Lenny," she told him. "They want us to hold his paycheck."

Hurley, VPA's vice-president, took the call. At first, the IRS agent seemed credible enough, but it didn't take long for Hurley to become skeptical. When he asked the agent to fax him a letter with the request to hold the paycheck, the agent said his fax machine was broken. Then, when Hurley asked for a number to call him back, the caller gave him Sacramento's area code but a number with a prefix Hurley recognized as one for California state agencies, not federal agencies. Moreover, there was something very familiar about the voice. Of course! He sounded exactly like Kevin Mitnick, that overweight, Eddie Haskellish friend of Lenny's. Kevin was a frequent caller. He chatted up the secretaries in the same friendly tone. "How was Palm Springs?" he would inquire of one. "Sorry to hear about your appendicitis" were the kind words for another. He called so often, in fact, that it had become something of a joke among the women who answered the phone at VPA. Whenever Lenny was on the phone to Kevin Mitnick, he sounded as if he were fighting with a girlfriend. When Hurley tried calling the number back, he got a recording that said there was no such number.

So just after lunch, Hurley called Lenny into his office. "Lenny, are you in trouble with the IRS?" he asked.

Lenny shrugged it off. "Of course not."

"Then are you in trouble with Kevin?"

That was all it took. Before he even knew what he was doing, Lenny found himself telling Hurley about Kevin. In a rather disjointed confession, he told Hurley that Kevin had been operating from VPA for the past year, that he had been letting Kevin into the building after hours to sit in front of the VPA computer until early morning. He told him that he and Kevin had been secretly using a computer·twenty miles away on the USC campus as their personal data storage locker. And he told him about all the break-ins at Digital. Hurley, who had suspected nothing, was stunned to hear this.

Lenny knew he was well regarded at VPA. He was the general troubleshooter who maintained the company's modest computer system. The last thing he wanted to do was lose yet another job because of Kevin.

Lenny told Hurley he wouldn't have been doing all this illicit computing if it hadn't been for Kevin. In fact, he claimed, Kevin had forced him into it. Kevin Mitnick was cannibalizing his life. So he kept letting Kevin in to use the VPA computers, night after night. If he decided he didn't want to work with Kevin one night, Kevin always managed to track him down. He was a master at that. And Lenny wasn't the only one to have been terrorized by Kevin Mitnick through the years, Lenny said. He had seen Kevin cut off a person's phone service on a whim.

Hurley leaned back. "Well, Lenny, what are we going to do? I'll tell you right now that I want to get the authorities in here. Either you can do it or I will."

"Well, geez," Lenny said, looking alarmed. "I'll just get in trouble."

Hurley tried to be understanding. He told Lenny that if he was the victim he claimed to be, he would be likely to get sympathetic treatment from the authorities. "Why don't you think it over," he said.

At 6:00 that evening, Lenny came back into Hurley's office.

"Well, Kevin's really done it this time."

"What do you mean?"

"He just forwarded all our lines to my house and busied out the whole switchboard." When Hurley picked up his phone and tried to call another line in the office, he just got a busy signal.

With this came another round of confessions. Now Lenny told Hurley that Kevin wasn't just a computer expert, he was adept with telephones as well. He told him about rerouting the wires in the telephone closet and about the MCI access codes they used to call places as far away as England. The phone forwarding stunt he had just pulled was something Kevin had been threatening Lenny with for some time, Lenny said.

"Something has to be done about this by tomorrow," Hurley said. "I want to see you in here at eight-thirty tomorrow morning." Later that evening, both telephone lines at the Hurley residence went dead.

The next morning, Lenny told Hurley he knew just whom to call at Digital, and he placed a call to the headquarters in Maynard, Massachusetts, from Hurley's office. He told the man who answered in Digital's corporate security office that he needed to speak with someone about the computer hacker who had been plaguing Digital for the past year. To make his words sink in, he started listing things: the VMS source code, the XSAFE program. Lenny said he would like to be put in touch with the FBI. He was told to sit tight for a few minutes.

Five minutes later, Chuck Bushey called him back. Lenny knew exactly who that was because he and Kevin had been monitoring Bushey's personal electronic mailbox for months. He knew Bushey was angry and fed up. And he knew that the security department at Digital would give anything at this point to catch those who had privately embarrassed the company so thoroughly. "Uh, hello," said Lenny. "I just wanted to let you know that I know who's been breaking into Digital's computers for the last year."

There was silence on the other end. "Who is this?"

Lenny stumbled. "My name, uh, my name is Leonard DiCicco and I'm calling from Los Angeles and I want to tell you about Kevin Mitnick. I know that he's been breaking into Digital's computers all this time. I know you're looking for him." Bushey told him to go find a "secure" phone and call him from there.

Lenny and Hurley went together to a pay phone at a nearby super-market, and Lenny called Bushey back using Hurley's credit card. "Well, I'm, uh . . . I'm a friend of his, and I'm calling you because I want to turn him in," Lenny said, and he launched into a condensed version of what he had told Hurley the previous day. He made certain to mention all of the key incidents. He told Bushey that Kevin had taken a copy of a highly secret security program that only a handful of people within Digital were even aware of. To prove it, he named the program. He also said that a copy of the latest version of Digital's operating system for the VAX had been taken as well.

There was silence on the other end. Lenny kept talking. He said he wanted to be as much help as possible in apprehending Kevin Mitnick, who was, Lenny intoned, a menace to society. Silence. Digital had better move fast, Lenny added, because Kevin had been talking lately about giving up his break-ins, and the company might not have many more opportunities to catch him in the act.

Bushey was speechless at the sudden lucky break. This breathless young man on the telephone was describing the break-ins that had been plaguing the company for months. That night, Bushey and Kent Ander-son, the company's head of European security, were on a plane to Los Angeles.

While the two Digital people were on their way, an FBI special agent named Chris Headrick summoned Lenny to a meeting at a hotel in the San Fernando Valley. Lenny was impressed by the very ordinariness of the agent. Slight and bespectacled, the low-key Headrick looked as if he would be more at home in a classroom in front of a chalkboard, scrib-bling out differential equations. In fact, some of Lenny's high school teachers were flashier than this. When the Digital security men arrived at the hotel, they turned one guest room into a makeshift command center. Lenny drove out to All-Purpose Storage, retrieved the three dozen tapes filled with the fruits of his and Kevin's electronic adventures and surrendered them to his interrogators.

Lenny stayed at the hotel, talking until 4:00 A.M. Feeling rather like a hero, he was brazen enough to mention to Bushey that Digital could have solved this frustrating case much sooner. If Bushey had pressed people at USC, university records would have revealed a history of problems with Kevin Mitnick. Also, Lenny told him, if Bushey had consulted Digital's old records, he surely would have seen that Mitnick had gotten into trouble as early as 1980 for breaking into a Digital computer. Then there was the Pacific Bell security memo that followed the Santa Cruz Operation incident and circulated around Digital several months later, at the same time that Mitnick's activity inside Digital was at its peak. The names had been deleted from the memo but a few phone calls to Pacific Bell security would have made the Mitnick connection.

The next morning, together with Lenny and Ralph Hurley, the authorities drew up a simple plan to verify Lenny's claims. As usual, Kevin was planning to show up at VPA at 7:00 that evening, after the regular employees had all gone home. Headrick's plan was to use a computer room on the first floor of the building to monitor everything Mitnick did from the main computer room, which was one flight up on the opposite side of the building. Hurley stayed around that night too. His presence downstairs would prevent Kevin from using the first-floor computer room. An FBI agent named Gerald Harman arrived at 5:00 P.M. and began to set up a surveillance. He wired Lenny with a tape recorder, fastening an elastic belt with two pockets—one for the tape recorder and one for the microphone—to his chest. A Digital engineer installed a monitoring device into the VPA computer that would let them watch and record everything that happened from the system upstairs.

When he arrived that evening, Kevin seemed to be in good cheer. In fact, he was on a roll. He had recently gotten a more extensive list of users on the computer at the university in Leeds, England, and wanted to try to get their passwords. He was also hoping to break in to the machine of Neill Clift, the Leeds researcher whose hobby was spotting security holes in VMS.

The agents planted themselves downstairs, and Hurley stayed in his office. Lenny greeted Kevin outside and lowered his voice. "Kev, we can't use both computer rooms tonight. Hurley's working late." The two went straight upstairs and Kevin began to dial directly into Digital. His first order of business was the Digital engineering cluster. Lenny was wearing a loose-fitting T-shirt, but he was still worried that Kevin would pat him on the back, thwacking his hand against the recording equipment. Kevin sensed that Lenny was jittery. "What's going on with you tonight?" he asked. "You're acting strange."

Downstairs, in the hastily appointed command center, the Digital people must have been surprised to learn that all their theories had been wrong. Not only was this not the work of an international spy ring, but the notorious West German Chaos group didn't seem to be involved at all. Instead, the culprit appeared to be an obese, nearsighted twenty-five-year-old high school dropout from L.A. whose diet, according to his friend, consisted of greasy cheeseburgers and Big Gulp colas from the nearby 7-Eleven—to say nothing of Lenny, a whiner who had spent months letting himself be bedeviled and blackmailed by the fat one. This must have been a far less glamorous denouement than any of them had imagined, inside a stuffy room the size of a walk-in closet, listening to two grown men taunt one another like children.

As the evening wore on, it became clear that this Kevin Mitnick was indeed their man. Mitnick seemed to have a special fascination with the Leeds computer that night. He talked someone at Leeds into giving him a password, then logged on to the British system and perused lists containing the names and access codes of everyone with legitimate access to the Leeds computer. The FBI agent seemed befuddled by what he was witnessing. Anderson explained to him that once Mitnick had obtained the list of authorized users and their passwords, he could come back later and log on as any of those people.

This young man's facility with the telephone network in general, and MCI codes in particular, was impressive enough. As soon as he finished his session on the Leeds computer, Mitnick dialed his way into an MCI Mail account, from which he was able to harvest MCI credit card numbers, so that the session would never show up on VPA's telephone bill. He seemed to know exactly what commands to type, as if he had done it hundreds of times before. But even more remarkable was Kevin's ability to monitor Pacific Bell's telephone traces. And that was precisely what had kept Digital at bay for so long. His knowledge of the phone system was so extensive that he could tap straight into Pacific Bell's monitoring equipment and keep a careful watch over anyone trying to trace his call. For that purpose, he always had two active terminals in front of him: one for his breaking and entering and a second to show what was happening on Pacific Bell's computer. If a trace to his phone line started, he could log off at once.

Anderson had been in the computer business for nearly a decade and he was aware that hackers were known for their patience and stamina. This one seemed to be among the most indefatigable. At 1:00 in the morning, long after the FBI agent and the jet-lagged engineers had grown bleary-eyed, Mitnick seemed to be just waking up. If it was true that he was recently married, didn't his wife wonder where he was?

Lenny's role seemed to be a small one, though the surveillance team suspected he might be lying low for its benefit. Lenny's principal job that night was to keep an eye on the second computer terminal that the pair used to maintain their electronic lookout for any traces.

Lenny had done his part. He had led them straight to the person who had eluded them all these months. Kevin was doing so many illegal things that Lenny was surprised the agents didn't storm upstairs and arrest him on the spot. So he stole downstairs and rapped lightly on the door of the small computer room. The door opened a crack and Harman stuck his head out. "Well?" Lenny asked. "Aren't you going to arrest him now?"

Harman shot back a stern look. "Not just yet." Harman said they had seen quite enough for one evening, and they were going to pack up for the night. He ushered Lenny inside and removed the taping equipment.

Lenny grew impatient. "What do you want? A signed confession?"

"That would be nice." Harman smiled and gently pressed him out the door.

Lenny climbed the stairs again, shaking his head.

"What's going on? Where've you been?" asked Kevin, speaking, it appeared, straight to the computer screen. He was back in Leeds.

"I was just checking to see if Hurley's still here. Listen, Kev, I'm tired and I have to get up early in the morning. Can't we call it a night?"

"No, I'm onto something great here."

"Kev, come on. Let's go." Lenny was becoming exasperated.

Kevin turned and glared at him.

It wasn't until 3:00 A.M. that Kevin was finally ready to call it quits.

When the two went outside, they saw that Kevin's car had a flat tire. So Lenny offered to drive him home—after a pit stop at the Fatburger. In a fit of generosity, Lenny told Kevin he would pay for Kevin's meal. Surprised but not about to object, Kevin ordered two King Fatburgers with everything on them, large fries and a large Diet Coke. Lenny ordered fries and watched as Kevin, out of habit, systematically emptied half a dozen ketchup packets onto his paper place mat then jabbed the mound of sauce with his french fries. His meal dispensed with, Kevin asked Lenny what time he should arrive at VPA the following night. Lenny said he was busy. As usual, an argument ensued. But rather than retread old ground, Lenny told Kevin to find his own way home and stalked out the door.

▲ ▼ ▲

The FBI agents and Digital security experts gathered at VPA the following morning to examine what they had collected. That morning, Harman and Headrick decided they had enough to make an arrest. In an unusual move, Digital, normally intent on keeping the lowest profile possible when it came to security matters, agreed to press charges and risk public exposure. It was a daring move, for once Digital customers learned that someone had been roaming freely throughout a network they believed was secure, and that he not only had helped himself to company software but, by reading people's electronic mail, had discovered security weaknesses, Digital would have a lot of explaining to do. But as far as some Digital officials were concerned, seeing this criminal behind bars might be worth a little public disgrace. Once they had Mitnick taken care of, they would handle the more nettlesome problem of Lenny. Headrick went to find Lenny. "How do we find your friend to pick him up?"

Lenny was overjoyed. He told Headrick he knew exactly where to find him. Kevin had just come by to have his car towed to a nearby garage. Lenny called the garage.

Lenny knew Kevin well enough to know that money ranked only behind food and computers on Kevin's priority list. Lenny had lost a $100 bet a few weeks earlier that Kevin wouldn't be able to crack one of the electronic door codes at VPA. When Kevin punched the correct code after just a couple of tries, Lenny accused him of cheating. He must have seen the code scribbled on a scrap of paper inside someone's desk. But Kevin not only demanded his winnings, he charged Lenny 10 percent interest for each day that Lenny failed to pay him. So Lenny called him at the garage to tell him he had just gotten some vacation pay and he had the money for him. "I'll be there in twenty minutes," Kevin said, and hung up.

Headrick told Lenny he didn't want to arrest Kevin unless he had his blue canvas duffel bag with him. Lenny had explained that the bag was Kevin's tabernacle. It contained all his diskettes, papers and other computing paraphernalia, as well as the computer-age equivalent of a lock picker's kit—handwritten access codes and passwords Kevin used to steal his way into heavily guarded computers from the safe remove of thousands of miles. Lenny had told Headrick that the bag contained documents linking him to the stolen Digital security programs, as well as to the abuse of the USC computer. In fact, Lenny had said, if they were going to get any hard evidence, it would come from that bag. Kevin's house had been searched so many times that he no longer kept anything incriminating there.

The material gathered from the eight hours of surveillance was helpful, but if the bag contained as much incriminating evidence as Lenny claimed, it would be indispensable in building a case against Mitnick. Lenny cheerfully suggested that he steer Kevin to the trunk of the car, where Kevin always kept the blue bag. As soon as Lenny saw it, he would scratch his head.

The signal was carefully arranged. When Lenny reached up and gave the back of his head an absentminded scratch, they would make the arrest. It was just after sunset, and in the gathering darkness half a dozen FBI agents slouched under the steering wheels of their unmarked government cars inside the small open parking garage beneath VPA.

Just after 5:00 P.M., Kevin's black Nissan Pulsar rolled into the garage. Luring Kevin there had been easy enough. His next challenge was to get Kevin to open the trunk. Lenny told Kevin he needed an empty disk from him to copy a piece of software from upstairs in VPA.

But snaring the quarry wasn't so easy. When Kevin pulled into the garage, he said nothing about the money. He wasn't even eager to get out of his car. "Let's go eat," he said, apparently forgetting the argument they had had the previous night. "I'm hungry."

Lenny had to think fast. "But Kev, I really want to make a copy of this great terminal emulator. Let me just run up and do it. Then we can eat."

"No, I wanna go eat now. Copy it later." Kevin's desire for nourishment didn't surprise Lenny, but his insistence seemed unusual. This was one test of wills Lenny planned to win. He stood his ground.

Kevin finally agreed to get a disk for him. Grumbling, he went to the trunk, opened it and lifted out the large blue canvas duffel bag. He put one foot on the bumper and propped the bag on his knee. Lenny reached up behind his head and gave his hair a tousle. Suddenly, the garage filled with the sound of engines starting and tires screeching.

Kevin looked around, surprised and confused. Lenny grinned broadly: "Kev, you know that queasy feeling you get in your stomach when you know you've been caught and you're about to get arrested?"

Kevin looked at his friend. "Yeah?"

"Well, get ready."

Kevin was taken completely by surprise. The broad grin on Lenny's face left him confounded. The FBI agents jumped out of their cars and shouted to Kevin that he was under arrest. They demanded that Kevin put his hands up and lean against the car. Kevin laughed a tight little laugh. "You guys aren't from the FBI. Show me your folds." Six large FBI identification folds emerged.

Kevin looked at Lenny, who was dancing in little circles and laughing. "Len, why'd you do this to me?"

"Because you fucked me over" came Lenny's reply.

The agents hustled Kevin into one of the cars.

"Lenny!" Kevin cried out. "Could you call my mom and tell her I've been arrested?"

Ignoring the plea, Lenny turned to Chris Headrick and smiled. Headrick nodded approvingly. "You did so well you should be in my business."

PART TWO

Pengo and Project Equalizer

In the late fall of 1986, two young men took the subway across the border and got off at Friedrichstrasse in East Berlin. When they arrived at passport control, Peter Carl, a dark and slightly gnomish man in his early thirties, took over. With a businesslike flick of his wrist, he slapped his passport down in front of the guard and said he had an appointment. His companion, a tall and slender teenager with a pale complexion who called himself Pengo, sat to one side and waited to be cleared. As Pengo understood it, Carl had made the initial contact a few weeks earlier by slipping a note containing secret code inside his passport. From then on, he could enter East Berlin any time he pleased, without exchanging the requisite 25 marks. A West Berliner usually had to apply a day in advance to travel the few miles across the border. The guard waved the two men through.

The corner where they emerged from the U-Bahn, Berlin's subway network, was bustling by East German standards. But the elegance of the cafés that once defined this part of Berlin had long since been replaced by tall public buildings, their dull finish suggestive of a more proletarian aesthetic. Carl and Pengo made their way to Alexanderplatz to kill some time. If not exactly friends, the two were caught up in a mutual adventure that bound them. They sat down on a bench. Peter Carl produced a joint from his pocket, lit it and offered it to Pengo. Eager as Pengo might otherwise be to do something so daring—smoke a joint in the middle of Alexanderplatz, East Berlin, in broad daylight— he declined. He was nervous.

Shortly before noon, they walked the ten minutes to Leipzigerstrasse, a wide boulevard that began at Checkpoint Charlie and flowed eastward. The building they were headed toward—number 60 Leipzigerstrasse— didn't differ in appearance from the dozens of other postwar apartment buildings that lined the wide boulevard. While West Berlin had em- braced a haphazard, extroverted approach to reconstruction, a Soviet- inspired blueprint informed East Berlin. The dreary designs of the fourteen-story buildings along Leipzigerstrasse were first cousins to the high-rise apartments ringing Moscow's outskirts. But its proximity to the West always gave East Berlin a bit of an edge over its East Bloc brethren. Technology had been creeping into East Berlin, and the change was especially evident on Leipzigerstrasse. There was a peculiarly capitalist innovation—an automated teller machine—just outside num- ber 60.

From the names listed next to the buzzers in the entryway, the build- ing appeared to be chiefly residential. They took an elevator the size of a refrigerator, which jerked its way up to the fifth floor. They were greeted at the door by a bearded, slightly stocky, well-tailored man in his forties. He and Carl shook hands like old friends. Then Carl gestured to his associate. "Hier ist mein Hacker. Pengo," he announced with a note of triumph in his voice. Sergei turned to Pengo—his real name was Hans Hübner, but he had been better known as Pengo for years—and extended his hand.

The meeting was relaxed enough at the start. A secretary served coffee. Sergei, cordial and businesslike, inquired about Pengo in accom- plished German somewhat overwhelmed by a Russian accent. What was the young West Berliner's background? What were his views, his poli- tics? Pengo responded proudly that he was a product of West Berlin's leftist scene—the sixties movement—and that he was sympathetic with what Mikhail Gorbachev was trying to achieve in the Soviet Union. Sergei seemed unfazed, perhaps even slightly amused. Then he got down to business. "Do you have something for me?" he asked, turning to Peter Carl. Carl produced a magnetic tape and diskettes from his briefcase, handed them to Sergei, and the discussion turned serious. On the disks, Pengo explained, was a security program for Digital Equipment Corporation computers, the computers of choice in the Soviet Union, and on the tape Sergei would find some other interesting software.

Barely acknowledging what he had just heard, Sergei told Pengo exactly what he was interested in. Referring to what appeared to be an order list, Sergei said he was anxious to obtain state-of-the-art engineering software—expensive and sophisticated programs that fell high on the embargo lists intended to stop American and Western European high-technology products from landing in Eastern Europe. He asked, for example, if it would be possible for Pengo to obtain computer-aided-design software for designing chips.

Pengo deliberated for a few moments, then decided he should tell Sergei what a hacker really did. He explained that he was capable of traveling the world through a computer. He could hop from West Berlin to Pasadena in a few seconds, dipping in and out of foreign computer systems in the blink of an eye. But at the moment, Pengo complained, his means of transmitting information were limited. He had a modem, but it sent and received information at a paltry 120 characters per second. The electronic theft of the kind of software that Sergei wanted might take days at the rate Pengo would be able to gather it, making such an exercise far too risky.

Pengo began to make his pitch. "Of course, I could do it if I had the right equipment"—perhaps a high-speed modem, one that would allow him to manuever more swiftly, and a more powerful computer with plenty of disk space. It was, after all, his dream to hack on powerful computer systems, but he didn't mention that to Sergei.

Pengo hesitated. Sergei was silent. The teenager continued: How would the Russians like to set him up to do safe hacking from phone lines that couldn't be tapped, perhaps even to do his hacking from East Berlin? It would be to everyone's benefit.

Pengo sensed no sympathy, nor could he see even a flicker of comprehension on Sergei's part. His only response was to suggest that Pengo try to fulfill his requests, after which they could talk again. If Pengo needed to talk in the interim, Sergei said, he should feel free to come back. With that, the Russian invited his two visitors to take a meal with him. If not an unqualified success for the two West Germans, the visit was in its own way instructive: Pengo had established contact with a KGB agent and had learned what it might take to engage in espionage for the Soviets.

▲ ▼ ▲

Most eighteen-year-olds, no matter how interested in computers, wouldn't have had the chance to talk business with a KGB agent. If Pengo had been asked to describe how he felt about what he had just done in political or even ethical terms, he would likely have shrugged his bony shoulders, rolled another cigarette and passed the question off not just as unimportant but as annoyingly beside the point. Politics and ethics, he would say, had nothing to do with it. Hacking was the means and the end, the information and its destination secondary. He wanted a powerful machine to hack with. He couldn't afford one, the Russians could, and it was the Russians who seemed interested in letting him indulge his passion. He might even say that the purity of his purpose struck him as somewhat heroic; his goal was to be the world's best hacker. That he was living what he might have read in a spy novel made it all the more exciting.

Pengo was the son of middle-class Berliners whose lives were torn apart in 1961 by the wall that divided their city. In 1945, five years after his father, Gottfried Heinrich Hübner, was born, the first Western troops marched in and Berlin became a city administered jointly by the four Allied powers. It had been badly damaged, with only half its residents remaining and a fifth of its buildings destroyed; there was no electricity or gas, and drinking water had to be hauled in from the country.

In the late 1950s, when travel between East and West required only the price of a subway ticket, Gottfried went to the technical university in West Berlin to study engineering. Renate, small, blond and his high school love, traveled frequently from East Berlin to visit him, against strict orders from her father, an impassioned Communist Party functionary. Renate's trips to the west so enraged her father that he informed the East German authorities of the nineteen-year-old girl's hostile act against the Party. He banished her from his home. Renate continued her trips to visit Gottfried, assuming that she could return at will despite her father's opposition. Fate decided her future, however, during one late-summer visit in 1961. On the day before she was to return, a line was cut through the city's heart. The young couple went to the Brandenburg Gate and watched East German soldiers unfurl a twenty-eight-mile-long barbed-wire barrier. Within days, the Berlin Wall was under construction.

Gottfried and Renate settled in Schöneberg, an older section of West Berlin that became famous for its town hall, from the steps of which John F. Kennedy delivered his famed "Ich bin ein Berliner" speech.

Renate was studying graphic design and Gottfried was just starting out as a construction engineer. Both were pulled into the anti-authoritarian movement, the European equivalent of the American Left's counterculture that was sweeping through Europe. They married when Renate got pregnant, and Renate dropped her studies to become a full-time mother.

Hans Heinrich was born in July 1968 into a tumultuous time. Three months earlier, Rudi Dutschke, a brilliant sociology student and pacifist who had founded an opposition party in West Berlin, had been shot and nearly killed outside his apartment on Kurfürstendamm, West Berlin's main boulevard. Dutschke was shot by a housepainter who couldn't stand what "Communist pigs" like Dutschke represented. The incident triggered mass riots in the city, and the unrest in Berlin inspired student uprisings throughout West Germany.

Hans tended more than most children to give free rein to his fantasies. As a young child, he used the rubble of a partially reconstructed Berlin as an endlessly fascinating playground. Hitler's massive concrete fortifications, scattered throughout the city, were indestructible, so architects learned to design their way around them, occasionally just building new apartments on top of them. The abandoned bunkers beneath were appropriated by the child explorers of Berlin, and Schöneberg had one of the best. It stretched for half a block, stood several stories high and reached four stories below-ground into a catacomb that was filled with endless stretches of ramps and passages. Oblivious to the historical significance of their romping place, neighborhood children scouted its expanse for old helmets, uniforms and other remnants. The building was a real-life fantasy world, a secret domain perfect for role-playing adventures; perhaps it was even a first taste of the kinds of interlocking paths and channels that computer networks would come to offer.

Partly out of boredom at home, Renate joined a group of mothers who were exploring alternative education for their children. Every day after kindergarten the five-year-old Hans went to a *Kinderladen,* an experimental preschool not unlike the Montessori schools in the United States. Parents of *Kinderladen* children generally belonged to the anti-authoritarian movement of the time and believed that their children should decide for themselves what they would eat, when they would learn to eat with utensils and even when their diapers needed changing. Parents like Gottfried and Renate favored uninhibited self-expression in their children, encouraging them to resolve conflicts by themselves and to assert themselves in ways that had been forbidden to their parents.

Yet the Hübner family was drifting apart. Hans was still in school when his father and mother divorced. Hans stayed with Renate, and his younger brother, Ferdinand, moved in with Gottfried. Renate went to work as a dental technician building ceramic teeth, and Hans joined the ranks of latchkey children.

He wasn't much of a student. At ten, he was an awkward, slightly plump, bespectacled kid who showed little aptitude for penmanship, English or art, and none at all for sports. His performance wasn't enhanced by the breakup of his family. His teachers regularly sent home reports stressing Hans's potential, while complaining that he was lazy and disruptive. The only subjects in which he excelled were math and physics. When report cards were issued and Hans called Renate with the news, she would tell him to forget the grades and just read her the written remarks.

But Hans had a charm that tended to obscure his failings. From an early age, he displayed a sophisticated, relaxed and dry sense of humor. Whenever he got into trouble, he was able to talk his way out of punishment, inspiring awe among his more academically diligent friends. His teachers decided that what the young charmer needed was more intellectual stimulation: they decided to advance him by having him skip the eighth grade. But this backfired. The youngster did not throw himself into schoolwork in a more mature surrounding. Instead, he withdrew into his own world, alienated from the classmates for whom he felt little affinity.

When he was twelve, Hans discovered squatting. Squatting in Berlin started in the 1970s when artists, punks and runaway teenagers laid claim to abandoned industrial buildings, apartment houses and stores. By 1980, there were more than one hundred buildings in West Berlin occupied by squatters. Hans joined a group that took over an abandoned store in Schöneberg, where they set up a punk band. What the adolescent lacked in musical talent he made up in aesthetic sensibility: his six-inch spikes of hair dyed jet black soared above his head with the help of copious applications of soap. He completed the costume with black military boots and a heavy chain around his hips.

He was now growing quickly, and sharp angles were replacing the soft folds of childhood. The wild adolescent distressed his parents. First there were incidents of shoplifting—a cola, for instance, or the electronic parts to build an amplifier. When Renate went to the police precinct to retrieve Hans after the cola episode, officers displayed pity mixed with disgust at her for having borne such a son. She was so enraged that she slapped Hans across the face, but she succeeded only in stinging her own palm. And then there was the time Gottfried was called to the hospital to fetch his barely conscious son, who had gone to a rock concert, drunk too much and smoked his first appreciable quantity of marijuana.

▲ ▼ ▲

Personal computers arrived in Germany in the early 1980s, mainly by way of England. The first wave were toys that went into the bedrooms of teenage experimenters, children of middle-class homes in trim and tidy neighborhoods in antiseptic towns across West Germany. Parents gladly paid the money for an Atari or a Commodore 64, toys that seemed capable of serving an educational function, rather than see their adolescent children experimenting with drugs or joining the punk scene.

Hans was destined to become an electronics freak. In 1982, Sven, a friend since first grade, borrowed a hand-held computer the size of a large paperback from someone at his school and took it home. Hans began to program it immediately, as if he had been programming all his life. Though a miserable English student, he had learned enough to write his first BASIC program, a loop that produces an infinite number of *hellos*:

```
10 print ''hello''
20 goto 10
```

It was a rudimentary program, to be sure. Later Hans was to develop an elegant self-taught programming style. He had a logical mind, stimulated by the discovery of an ability to create something from nothing. He was transfixed. Squatting in abandoned houses had its pull, and spending time with girls was nice enough, but here was something that could command Hans's attention as nothing ever had.

It wasn't long before the hand-held computer yielded to mail-order kits for bigger computers that had to be soldered together. As their first big project Hans and Sven put together an entire computer—a Sinclair. Until the advent of the Sinclair, personal computers, even those assembled from kits, were out of the reach of most young pockets. But now, for $250, Sven and Hans could buy a complete computer. Even measured against the standards of the day, the Sinclair was exceedingly primitive. At that time Apple Computer, Commodore and IBM were already on the market with machines that look similar to what people now use. But all that the two young Berliners could afford was a kit that, once assembled, resembled a small box of chocolates plugged into a television set. The software for the Sinclair came on tape cassettes, not on the floppy disks of today. The machine had very limited memory, but it was enough to write a program or two. For Hans, the Sinclair provided a taste of the hidden power of the machine, of binary worlds to be explored.

Even as the computer lay in pieces in his room, waiting to be assembled, Hans was getting ready to start programming it. He had taken the programming manual with him on a vacation to southern Italy that summer with his father and his father's girlfriend. Culture just annoyed him, and if the two adults coaxed him into accompanying them to museums and ancient churches, he sulked. Finally, they gave up and left him in the rented villa, inseparable from the book.

When Hans returned from Italy, the two budding computer experts spent day after day in Hans's room, a dark little chamber covered with graffiti, putting together the Sinclair. Oddly enough, although Hans had never done much with his hands, and wouldn't have known what to do with a hammer, his instincts for putting together the computer were perfect. But after he had put it together, his desire for a better computer was immediate. As soon as he could, Hans ordered a more powerful model, the Spectrum. It had the same design as the Sinclair, but twice the internal memory.

Circumventing rules became an early obsession for Hans. His main activity in programming was piracy—he attempted to bypass the copyright restrictions on the commercial software cassettes, mostly games. Hans wrote little programs that allowed him to load a game into the computer, save it and put it onto another cassette. It was a challenge to the young hacker, not an ethical question. If the publishers tried to build anticopying devices into their programs, they just had to expect hackers to try to unravel them.

Hans and Sven turned into computer-game freaks. Their favorites were those with good graphics, arcade-style games that involved some relatively simple task, such as shooting lasers at successive waves of bug-eyed aliens. These electronic games really came to life when played on the colorful, large screens at a video arcade. Hans was spending much of his free time at the video arcade around the corner from his school. The arcade was run by a shifty character who hired teenage boys willing to work in exchange for the privilege of communing with the electronic gadgetry. For doing minor repairs on the games, renting out videos (mostly pornographic) and generally watching over things, Hans had the run of the place. He also set up something of a pirate's copy center, making illegal copies of Sinclair Spectrum software for all his friends.

The arcade also provided the setting for his first true hack. Hans discovered that the small strikers used to light gas stoves created the same effect, when their caps were removed, as coins dropping into slots. One electric spark from the striker next to the coin slot of a video game and twenty free plays appeared out of nowhere.

The lighter enabled Hans to spend hours exploring the various games in the arcade. His favorite was *Pengo*. It was simple and addictive. The player assumed the role of Pengo, a penguin who pushed ice blocks up and down and back and forth on the screen, aiming them at malicious monsters called sno-bees. The trick lay in destroying the sno-bees before they destroyed the ice blocks and left the heroic little penguin defenseless. Hans battled the sno-bee enemies for hours on end, oblivious to the passage of time. Often he stayed at the arcade through the night and stole into his bed at 6:00 A.M., half an hour before his mother got up to go to work. At 7:00, she would tap lightly on his door and tell him to get ready for school. He got up, got dressed and headed straight back to the arcade. The ruse worked until the police visited Renate one day to tell her that not only was her son a chronic truant but he spent all his time at a video arcade, some of it lending out pornographic videos.

Hans was soon to stumble into an even more compelling world. His introduction to computer communications came in the person of Barnim Dzvillo. A friend from school, Barnim had a Commodore 64 with an acoustic coupler modem. Even though Hans disliked the Commodore 64 because it bespoke an affluence that Sinclair owners despised by reflex, Barnim's modem was an irresistible enticement. Hans had first seen such a device when he watched his physics teacher use one to call the school's central computer to retrieve class schedules. But he had never used one.

One evening Hans went to Barnim's house and Barnim showed him how to use the modem to dial into an electronic bulletin board in West Berlin. Before logging on, Hans had to choose a name for himself, a handle. Pengo was the obvious choice. By the end of the evening, Hans was Pengo, and Pengo was hooked.

Barnim also gave Pengo his first taste of electronic adventure by showing him how to sneak electronically into a mainframe computer by going through Tymnet, a commercial data network owned by the McDonnell Douglas Corporation with dial-up ports (computer access points) that touched down in more than seventy countries. A computer user with a modem can call the number of a local computer, then by typing the correct address, connect to another computer anywhere in the world. However, networks such as Tymnet have an unavoidable security flaw: they publish their telephone numbers widely, making it possible for unauthorized outsiders to get onto the network with one local telephone call, then hop to computers attached to the network. To log on to the network, all one needs is a user identification and a password. Soon Pengo discovered 3M, the American manufacturer of Scotch tape. To make communications easier for customers at 3M, the company had set up a help system accessible from Tymnet. Typing "3M" and "Welcome" established Pengo's connection to the company.

Attached to his Commodore 64, Barnim had a primitive modem that was excruciatingly slow, retrieving just 300 bits of information—or about 30 characters—every second. At this rate, it would take almost a minute to read a single page of text being transmitted by a remote computer. But it didn't matter to Pengo that it was slow. It seemed that the whole world was opening up to him. Pengo and Barnim began to wade through screen upon screen of information. Mostly what they saw was a detailed description of other 3M computers they could reach—computers in West Germany, France, Great Britain, Mexico and Chile. They stored the information on a floppy disk and printed it all out.

Being inside a global computer network gave Pengo the same adrenaline rush that playing a video game did, magnified many times over. Once he had entered a computer network, he was no longer playing a game; he was master of real machines performing real tasks. He could be nowhere and everywhere at the same time. From in front of a computer screen, he could open doors and solve problems. He was able to get things. The stuff on the screen wasn't anything more than electrons hitting phosphor, but it was nice to imagine that there really was something there, around the back somewhere. The networks—unconstrained by conventional geographic boundaries—had become a self-contained universe known to a growing number of computer researchers as cyberspace. It was inevitable that a young nonconformist like Pengo would take up residence there. A world that his parents didn't even understand well enough to forbid was an irresistible outlet.

Pengo had to have a modem of his own, so he went straight out and bought a kit. Unfortunately, his homemade device was primitive and required intensive manual labor to operate. Whenever he reached a remote computer and received a special connect tone, Pengo hooked his acoustic coupler to the telephone and a set of loudspeakers and tuned the coupler like a radio in order to match the connect tone. The modem worked well for only a couple of hours before its printed circuit board heated up, and Pengo had to unhook it and hang it out the window for a few minutes to cool it down.

▲ ▼ ▲

NUI, pronounced "NOO-ey," stands for "network user identification." It was other people's NUIs that sustained many a West German hacker's career. A NUI opened the door to Datex-P, West Germany's govern-ment-controlled computer data network. Once you were into Datex-P, you were on a bridge into the equivalent Tymnet network, and from there into thousands of computers throughout the United States. Unlike private network services such as Tymnet and Telenet in the United States, European data networks are usually run by the government. The West German government, in particular, regulated all communications very closely. The West German postal and telephone administration, the Bundespost, owned and controlled the Datex-P network and issued all NUIs. Modems and answering machines had to be registered with the Bundespost. Installing your own equipment without registering it with the Bundespost meant inviting a hefty fine. None of these restric-tions existed in the U.S. or even in most other European countries. The Bundespost argued that the heavy regulation was necessary to protect the integrity of the telephone network. But in a way, the conservative Bundespost was asking for trouble. It was little wonder that smart young rebels should want to skirt the system, to assert themselves by doing something as relatively harmless as spurning the slow and outmoded state-approved modems in favor of speedier brands, or even as brash as wandering into computers where they had no right to be. And it wasn't surprising that West German hackers made a sport out of stealing NUIs.

In the mid-1980s, most stolen NUIs were obtained at the Hannover Fair every April. Stealing them was simple. It was just a matter of looking over a preoccupied exhibitor's shoulder as he typed in his NUI and password while logging on to a remote computer. Stolen NUIs quickly made the rounds of the West German hacker community.

In the mid-1980s, when Pengo began to break into systems in earnest, many were invitingly easy to enter. Digital Equipment Corporation equipped each VAX computer it shipped with three built-in accounts, each with its own password. There was the "System" account, with *Manager* as the password; the "Field" account with *Service* for a password; and the "User" account, whose password was, conveniently enough, *User*. It was up to customers to disable those accounts and passwords once they had established their own. Often they didn't, leaving a back door open to anyone who happened along. Once a single computer was breached, if the computer was linked to others, network "cruises" were next. If the computer was on more than one network, it could be used as a springboard into many other computers on other networks. Network hopping gave Pengo a certain amount of protection from pursuers who would have to trace him back through many connections and different time zones.

Breaching computers combined the challenge of solving a giant puzzle with the thrill of breaking the law. Pengo would come home in the afternoon, turn on his computer, find some others to "chat" with on electronic bulletin boards and trade some information, perhaps a NUI in exchange for a way to break into protected files. Then, his mind racing around the world while his body remained in Berlin he would start to probe the network, sometimes with a specific target in mind, sometimes without. The farther away the computer, the better. And seeing that fifty users were logged on to the machine he reached gave rise to a special thrill of being inside something extremely large and powerful.

There was a lot for Pengo to familiarize himself with, most of which he taught himself. He figured out how to write scanning programs, automatic dialers that were programmed to run through the night, dialing number after number in search of a high-pitched modem tone on the other end. And he learned how to probe, ever so tentatively, inside a system once he had broken in. Sometimes he couldn't resist trying to engage a system manager in a conversation. Such was the case in early 1985 at SLAC, a high-energy-physics research center at Stanford, which Pengo reached one day at 4:00 A.M. through a Tymnet connection. He got into a chat with a system manager there, who seemed happy to exchange pleasantries with a German hacker. But not long after that first friendly chat, Pengo encountered a second manager on the same system at SLAC who told him to scram. Pengo said he had no intention of going away, and if they tried to get him off the system he would crash it. When the manager replied that that was nonsense, Pengo carried out his threat. He wrote a small recursive program and sent it to the computer to execute. The program worked like a chain letter. It created two copies of itself, each of which, when executed, created two copies, and so on. This was one sure way to exhaust the computer's resources quickly. In less than a minute, overwhelmed by the flood of work to be done, the computer was gridlocked. In the vague ethical code Pengo had fashioned for himself and his hacking, he knew he had just crossed the line. On the other hand, he had felt compelled to be true to his word.

Moreover, crashing a computer at a big and important place like SLAC was something to tell others about.

This was electronic machismo for Pengo and the others he met on line, hackers with handles like Frimp and Task and Treppex and Zombie. On the one hand, hacking could be a solitary sport. But at the same time, once a breakthrough was made, once a huge Digital VAX computer at a research laboratory or company somewhere in the electronic universe was penetrated for the first time by one of the West Germans, it was an occasion to strut. Telling others of the hack would earn points, but it would also oblige the successful hacker to share the password. And that carried with it the risk that dozens would descend on the computer in question. Soon after that, whoever was running the computer would usually lock the intruders out by changing passwords. But it was difficult to resist sharing a triumph with others who would appreciate the significance of the deed.

Pengo was one of the first to penetrate CERN, the high-energy-physics laboratory near Geneva. A consortium of researchers from fourteen nations, CERN was locked in a race with such U.S. laboratories as SLAC and Fermilab in Chicago to discover the top quark, one of the fundamental entities of which nuclear particles are made. To that end, CERN was planning the construction of a mammoth accelerator, seventeen miles in circumference, a ring passing under villages, farms and the nearby Jura Mountains. CERN was in a difficult position. Not only did the hundreds of scientists at the laboratory have to grapple with a confusing array of languages and cultural gaps, but their work had to be accessible to thousands of researchers outside the facility. Sharing information quickly and spontaneously lay at the heart of the consortium's charter. Closing down the network connections to protect against some electronic delinquents was out of the question.

Soon after the first electronic intruders discovered CERN, its computers were inundated. The laboratory became the place to post messages and to loiter for hours at a time, waiting for friends to log on. CERN was also an effective springboard to other centers, since it had computer connections all over the world.

Different system managers had different reactions to break-ins. Some were even friendly. One manager at a European research lab, who subscribed to the notion that hackers might even serve a useful purpose opened his system to a West German teenager who offered to identify security weaknesses. Others didn't mind getting into the occasional chat with an uninvited visitor. The following conversation took place in 1985 between a hacker from Hamburg and an operations manager at CERN:

```
VXOMEG::SYSTEM hallo HUHU. Do you have a
little time???
VXCRNA::OPS Yes, we're back from dinner
VXOMEG::SYSTEM dinner? You think about
eating all the time, too?
VXCRNA::OPS we're here from 1600 to 2300 and
we have to eat!!!
VXOMEG::SYSTEM Of course I'm always hungry
for more VAXEN!!!! HAHAHA
VXCRNA::OPS Why the VAXEN at CERN and not a
bank? HAHAHAHA
VXOMEG::SYSTEM No! I'm a hacker and no
criminal or industrial spy!!!!
VXCRNA::OPS Yes I understand . . . but isn't
it more fun with a bank?
VXOMEG::SYSTEM A bank is ok but it has to be
a databank!!!
```

But mostly, system managers just got annoyed; they had better things to do than chase juvenile trespassers out of their computers. In the middle of 1985, Alan Silverman, a CERN system manager, posted a message warning colleagues of the hackers inside the system. He urged his colleagues to avoid obvious passwords and not to keep password lists sitting in easily accessible files. Silverman pointed out that security bugs in the system had enabled a lot of hackers to get in, and he urged his people to eliminate the bugs.

To explore the computers at CERN, it was necessary to have some working knowledge of VMS, the operating system running on most of CERN's computers, especially those connected to the network. The very features that made VMS so attractive to the tens of thousands of Digital Equipment Corporation users made it attractive to hackers too. Digital promoted VMS vigorously as its operating system of choice for the VAX, the company's most powerful computer. The smallest VAX could take care of all the computing needs of a small business while the Sinclairs and Commodores that Pengo and his friends had access to were good only for games.

Although the VMS operating system couldn't be used, for instance, on an IBM computer, it was the most common operating system in the Digital environment, running on any VAX, from the smallest to the largest. VAX VMS could be very solicitous of unskilled commercial users. For example, VMS had a help facility, obviating much of the need for manuals. For intruders, the generous help feature was yet another way of exploring a system's possibilities.

By the end of 1984, sixteen-year-old Pengo was paying scant attention to his schoolwork and even less attention to his school friends. He left his job at the video arcade and spent more and more time by himself, with his computer and his modem. Pengo's mother was puzzled by what he was doing until dawn, and completely mystified by his constant use of the telephone. But unlike many parents, Renate was determined not to interfere with her child's interests. She didn't ask him for an explanation and he didn't offer one. He didn't appear to be making long-distance phone calls, so she couldn't complain about the bill. In West Berlin, one call of unlimited duration to the local Datex-P access point cost about fifteen cents. Pengo's computer friends were coming over more frequently and working on the computer well into the night. Although they tried to keep quiet, the close quarters made it all the more difficult for Renate to ignore the mysterious goings-on in the next room. She needed sleep. But like many parents, she was glad her child was becoming computer-literate. Parents all over the world believed educators who argued that every child should learn to use computers at the earliest possible age. Renate and Gottfried weren't the only parents who didn't stop to think about some of the less attractive consequences of computer literacy. For adventurous children with little respect for rules, computer literacy was likely to lead to electronic juvenile delinquency—computer trespassing. So that Pengo could continue his hobby, Renate and Gottfried exchanged sons and Hans moved into a private room in the back of his father's large apartment across town.

Gottfried's only ground rule was that the phone remain free until 8:00 P.M., so Pengo began his network excursions after 8:00 and kept them up until his father appeared at his door at midnight to tell him to go to bed. The rap on the door signaled his father's bedtime and another six hours of uninterrupted work, punctuated by occasional trips to his father's refrigerator. During intensive sessions, Pengo's preferred nourishment was yogurt, not so much for its healthful aspects as for the fact that it was a clean and efficient food with no crumbs that could fall into the crevices of the keyboard. His headphones affixed to his ears, Pengo tapped the keyboard to the strains of Kraftwerk, the West German synthesizer band that sang paeans to pocket calculators and home computers in the monotone of automatons. He would listen to one album for weeks, until he couldn't stand it any longer.

It was during one of Pengo's all-night sessions that he met up with Obelix.

Obelix was a round-faced nineteen-year-old from Hamburg with a skittish edge. He had discovered computers when he was about fifteen. He began programming on a Commodore VIC 20 in physics class and before long he was better than his teacher. Games didn't interest him much; it was the idea of exploring the forbidden territory of computer networks that was really captivating.

Pengo and Obelix met electronically in late 1985, when the two happened to be logged on to a computer somewhere in Munich at the same time. Obelix introduced himself as a hacker from the Chaos Computer Club in Hamburg. Chaos members had become notorious as the "VAXbusters." Pengo had already heard of Chaos, and the Chaos hacker and his new friend struck up an electronic conversation. Obelix invited the seventeen-year-old stranger from Berlin to the upcoming second annual Chaos meeting in Hamburg. After barely passing his final high school exams—to the astonishment of Gottfried who had predicted utter failure—Pengo headed for the Chaos meeting.

▲ ▼ ▲

West Germans came late to computer hacking, in the early 1980s. To the general public in Europe hackers were the perpetrators of computer break-ins and they had little in common with the original hackers from the Massachusetts Institute of Technology in the 1960s and 1970s, for whom hacking meant extremely clever computer programming that extended the limits of technology.

The Chaos Computer Club was formed in 1984 by Hewart Holland-Moritz, an elfin, thirty-two-year-old computer programmer and fervent radical with a boisterous sense of humor who shortened his name to Wau (pronounced "vow") Holland for its simplicity. He chose the name Chaos, on the other hand, for its pure shock value. In reality, his club was the very picture of meticulous organization, with a hierarchy of officers and subofficers, some of whom came to fall just short of celebrity status. Wau was soon joined by Steffen Wernéry, a jittery and intense twenty-two-year-old high school dropout full of bravado and a natural ability to organize. Like a seasoned PR man, Steffen could be articulate and outspoken without saying anything of substance. When it came to difficult questions, he deferred to Wau.

Wau's approach to computers and to hacking was, he claimed, driven by ideology. A crusader for privacy rights, he had written articles for the left-wing Berlin newspaper *Tageszeitung* criticizing West Germany's controversial 1983 census as an official conspiracy to pry into people's lives. He also attacked as threats to privacy the computer-readable identification cards that the West German government was developing, and computer data-gathering systems designed to sniff out terrorists. On its face, Wau's defense of privacy seems to have conflicted sharply with his adherence to the Hacker Ethic, the well-established, if informal, code of conduct observed by earlier generations of computer hackers: not only should all information be free, but access to computers and the information they contain should be unlimited.

As Wau saw it, however, there was no tension between his theory and practice. Like a lot of computer people, Wau was a libertarian when it came to these issues. That is, he believed government and other large institutions shouldn't be meddling in people's personal lives, but individuals should be able to get at anything. The systems Wau hacked into were owned and operated by authoritarian institutions, forces that hardly championed an individual citizen's right to privacy. Hackers like Wau thought of themselves as modern Robin Hoods. Through hacking, they could expose gaping holes in computer security and heighten public awareness of security loopholes. Paradoxically perhaps, the club was trying to appeal to both the West German love of law and order and a concern for civil liberties. If he broke into computers, Wau argued, it was to make the point that West German institutions were wrong if they thought their computers were safe from outside meddling. And it was with the 1984 hack of the Bildschirmtext system that Wau sought to prove his point.

Bildschirmtext was an electronic information service that was supposed to make the computer terminal an indispensable fixture in every West German home, third only to the telephone and television. Simple telephone lines would link home terminals with huge data banks offering information and services for sale. In 1984, the Bundespost made bold predictions that by 1985 one million West Germans would be using Bildschirmtext, or Btx for short, for everything from calling up train schedules to ordering opera tickets. But it was expensive for consumers, and for those not familiar with computers it was intimidating. Getting comfortable with the Btx electronic marketplace meant getting comfortable with technology as a routine part of life. West Germans carried on an uneasy relationship with technology. It could have been the memory of the horrifying consequences of Germany's superior military technology in World War II, or it could just have been the same ambivalence that everyone in the industrialized nations felt toward technology. So a national queasiness about the information age had a habit of emerging in ways both subtle and obvious. For some West Germans, computerization carried the specter of lost jobs; for others, it raised new and unsettling threats to *Datenschutz*, or data protection. West German citizens hardly welcomed new ways of gathering and interpreting ever more information about them. It came as little surprise to most West Germans when, by 1988, just 120,000 households in the Federal Republic had signed up for Btx. Eventually, the Federal Republic would concede that its $450 million investment in Bildschirmtext had been an extravagant failure.

Wau and Steffen were convinced that the Btx service was far from foolproof, and they used the Hamburger Sparkasse, Hamburg's largest bank, to prove it in 1984. The trick was simple: once the two had gotten their hands on the bank's Btx identification code and password, they set up an automatic dialer and programmed it to call the Chaos club's $6-per-call Btx information service continuously through the night—charging the calls to the bank. By night's end the bank's bill had come to $81,000, but Wau and Steffen didn't try to collect the money; instead, they held a press conference to announce their coup. The bank stunt brought the fledgling club instant notoriety. Germans became convinced that their bank accounts were helpless victims in the hands of electronic hoodlums. Their children signed up with Chaos.

▲ ▼ ▲

Pengo's first Chaos conference was an enlightening experience. Seeing so many kindred spirits in one place was like being released from solitary confinement. The most remarkable part of going to the Chaos congress in the Hamburg suburb of Eidelstedt was staying up through the night and hacking with others. Nests of cables connected computers to modems and modems to telephones. Mattresses were strewn in random rooms throughout the building to accommodate waning stamina. Wau took seriously the possibility that the authorities might disrupt the meeting, so he stationed guards and metal detectors at the door. Outsiders took an immediate interest in Chaos. They viewed the club as a symbol of harmless dissent in West Germany. Chaos seemed the very picture of clean fun when compared to the dread Red Army Faction, terrorists who advocated violence as the only effective instrument of change, or Berlin's *Autonomen*, outlaw punks specializing in a less directed brand of violence, or even the young Neo-Nazis, who signaled that Germany could be in danger of losing its collective memory. Chaos welcomed the attention, using any opportunity to hold a press conference, and the 1985 Hamburg congress was no exception. Newspaper reporters and television crews from throughout Germany, filmmakers from England and sociologists from Berlin's Free University nearly outnumbered the computer enthusiasts. German television crews stalked the conference, meters of cable snaking behind them. The nightly news carried reports of the latest gathering of technological wunderkinder.

Part of it might have been simple one-upmanship, part of it might have had something to do with the dynamics of the group, but it was during the Hamburg conferences that some of the most daring feats took place. At one point, for instance, Pengo and a small group of others slipped electronically into a development computer for the Ottawa Police Force; they stayed on the system for several hours, looking around what seemed to be a data-base program for criminal searches, and stopped only when a system manager discovered strange activity on one of the accounts and shut it down.

At age seventeen, Pengo cut a striking figure. The rangy young man was dressed entirely in standard-issue Berlin black to match his jet-black dyed hair, chic Girbaud pants billowing at his thighs. He quickly established his credibility. As the others looked on, he broke into a Digital Equipment computer somewhere in the U.S. and, by writing a small program in DCL, a programming language used for automating common command sequences, he built a bulletin board specifically for congress attendees. People sitting two feet apart in the cramped room in Hamburg could log on to the computer overseas and have a conversation. It was an impressive feat, and it was clear to the onlookers that they had a true talent in their midst.

On a more personal level, Pengo found that despite their electronic rapport, once they met face to face he and the butterball Obelix had little in common beyond their computer fixation. Though both teenagers came from middle-class families, Obelix cultivated many of the values that Pengo had come to scorn. Obelix was a budding conservative with strong anti-Communist sentiments and a desire for wealth that eclipsed his other traits. He squired friends and visiting hackers around town in his mother's Mercedes, claiming that it was his. His role models were

Malcolm Forbes and Heinz Nixdorf, two avowed capitalists from relatively humble beginnings. Obelix had every intention of becoming a
self-made millionaire and was forever coming up with electronic inven-
tions, each the basis for a new company that would flourish overnight.
Pengo didn't take many of Obelix's get-rich schemes seriously, but he
did take seriously Obelix's love of things technological. When he first
arrived at Obelix's house in Hamburg, Pengo saw that Obelix owned a
1200-baud modem, which was capable of retrieving more than five and
a half pages of typed text each minute. In Pengo's eyes such a device was
certainly state-of-the-art, and it won his immediate respect.

▲ ▼ ▲

Pengo had wanted to meet Wau, and though he decided Chaos's founder
was interesting, his age and politicized discourse made it hard for the
teenager to relate to him. But then he met Karl Koch, a tall, wiry
twenty-year-old from Hannover with a gaunt and troubled face. When
Karl first introduced himself, he opened his laptop computer and pulled
from its battery compartment a brick-sized chunk of hash. The consumption of copious drugs was just one aspect of Karl Koch's life. His preferred
name, he explained to Pengo, was Hagbard Celine.

Hagbard told Pengo he had come to hacking by way of *The Illuminatus!*
Trilogy by Robert Shea and Robert Anton Wilson, an eight-hundred-
page conspiracy-theory novel whose hero, one Captain Hagbard Celine,
is out to fight the powerful secret society, the Illuminati. The Illuminati,
dating back to eleventh-century Islam, ruled the world. In the book,
Hagbard Celine infiltrates the Illuminati in order to defeat them. Not
only did the Hagbard living in Hannover in 1985 believe the Illuminati
were still around and responsible for plotting to rub out everyone from
John F. Kennedy to Ian Fleming, he believed that he was in fact Hagbard
Celine and that his mission was to stop the Illuminati from mastermind-
ing a nuclear holocaust. Hagbard of Hannover was convinced that the
world's computer networks were a conspiratorial ruse, and that the people in charge of them were, accordingly, running the world. Hagbard's
idea was to insinuate his way into the computer matrix and carry out
electronic subversion.

Hagbard didn't feel a need to learn how to program. He could work
just as well without doing so, as long as he had the right telephone
numbers and passwords. He would leave it to the others to write the
fancy little programs for collecting passwords and escaping detection.

Pengo sensed a deep intuition in Hagbard. He admired an imagination that could follow such a steady course on such an outlandish trajectory. Pengo used computer equipment to patch into cyberspace but Hagbard truly lived there. Pengo identified with his own father—cynical with a logical, mechanistic approach to things—and he knew that explained his fascination with electronics. He understood that Hagbard's Illuminatus theories were nothing more than the construct of a troubled mind. Even so, their fanciful and poetic qualities were exactly what Pengo liked and envied in the young man. He knew that the drugs played no small role in warping Hagbard's perceptions, but at the same time they enhanced Hagbard's already fertile imagination.

Hagbard and Pengo took the train back to Hannover together and during the ride Hagbard told Pengo that if he were ever to commit suicide he would build an atomic bomb, climb to the top of the World Trade Center in New York City and set it off. His tone was so serious that Pengo felt he meant it. Entranced, he wanted to find out more about this charismatic, deranged new friend.

Karl Koch's family, it appeared, had come unglued early in his life. His father had left Karl and his sister alone with their mother when Karl was small. When his mother developed cancer, it was the young Karl who watched her die. Karl's father, a high-profile journalist for a Hannover newspaper and a heavy drinker, also died of cancer when Karl was sixteen. His inheritance of 100,000 marks, or about $50,000, was enough for him to buy a revamped Porsche, rent a nice apartment in Hannover and support an expensive drug habit. In high school, Karl had been something of an activist. Like a lot of other West German students, he had been an outspoken critic of nuclear power. But after school, and after his father's death, hash and LSD shunted politics aside. After reading Illuminatus!, he was only Hagbard Celine.

Hagbard took a quick liking to the hacker from Berlin who was willing to smoke joint after joint with him and show him new computer tricks. Hagbard was bisexual, and he asked Pengo if he was as well or, if not, if he was at least interested in experimenting. When Pengo declined, Hagbard accepted his decision graciously. It flattered and intrigued Pengo to have such an avid admirer. Pengo had believed himself to be on the fringes of life in West Berlin, a technological rebel in a city famous for its rebels. But in Hagbard, Pengo saw a character far closer to the edge than any he had ever met and far closer than Pengo had ever gone himself, even in the shoplifting days of his youth or the hours he had spent hanging around the sleazy arcade. Pengo tagged along with Hagbard, happy to meet the small but close circle of Hannover hackers that called itself Leitstelle 511, after the area code for Hannover. They met weekly at a bar called Casa Bistro for a hackers' *Stammtisch*—a group gathering that takes place regularly at the same table at the same bar or restaurant. These new acquaintances were a world away from what Pengo had come to know of the hackers he had met at the Chaos meeting in Hamburg, and light-years from the likes of his friends back in Berlin.

Among those young people who grew up surrounded by the reborn opulence of Western cities such as Munich and Hamburg, a sizable percentage demonstrated considerable ignorance about the "other" Germany, East Germany. Many West Germans regarded East Germany as a foreign country. Interest in the history that confronted West Berliners at the end of every road was barely perceptible among young people from the rest of West Germany. World War II had become grandfather's war, not father's war. A sanitized version of history had been taught to the generation born directly after the war. By the time a new generation of teachers came along from the sixties movement, insisting that the history of the Third Reich be taught in West German schools, indifference had taken hold among the students. They had heard too much, and they wanted to close the book on it. The passage of time had left young people, if not with a subconscious tendency to explain away Hitler's crimes, then with a curious blank spot. While Pengo, like most Berliners, had contempt for the self-satisfied myopia of West Germans, his obliviousness to the political or legal consequences of what he did was a curious contradiction in his character. Perhaps his being exempt from the draft, as were all other West Berliners, had helped create his political blind spot—politics had never intruded in his life in a real way.

In the Hannover crowd Pengo finally found a group whose anti-authoritarian yet apolitical frame of reference came surprisingly close to his own. It was a scene where drugs and computers were so interlaced that Pengo began to understand Hagbard's brand of reality. In Hagbard and his friends, Pengo had finally discovered a world that seemed to intersect with his vision of how he wanted to live and think; it was a world where the only reality was a reality of one's own choosing. In Hannover, he had happened upon a group of people who appeared to share in his obsession. Parties meant all-night sessions at someone's apartment in front of his computer, while hash provided inspiration, and cocaine and amphetamines chased away fatigue.

It was during his first trip to Hannover that Pengo met Hagbard's cronies, including Dirk-Otto Brzezinski, nicknamed Dob, and Peter

Carl. They and others would gather, break into systems and take drugs together on and off for the next few years. They were brought together and held together by a common passion, though they wouldn't necessarily have sought each other's friendship under other circumstances.

Dob was another example of the link between computers and drugs that was new to Pengo. The twenty-six-year-old goateed computer programmer had a taste for good hashish and expensive meals. By appearances, Dob was remarkably ordinary; his casually corporate garb and wire-rimmed glasses hardly indicated anything rebellious, or even free-thinking, though he had spent his youth as an outcast among schoolmates in Kenya, where his parents worked in international development.

Through geographical maneuvering, Dob had cleverly avoided his compulsory stint in the West German army, not so much out of political conviction as out of a simple desire to skirt the service. So he was forced to live the life of an itinerant, commuting back and forth between Hannover and West Berlin and taking on contract work. He was highly intelligent, with an astounding facility for absorbing vast amounts of technical information. When asked a technical question, no matter how obscure, he invariably knew the answer. He was an expert at programming large Siemens computers, the computers of choice for most West German government agencies and large corporations. Like IBM's mainframes in the United States, however, Siemens computers were not well liked by programmers, who found the machines' operating system cumbersome and old-fashioned. Someone with Dob's talent, therefore, was in great demand. In a good month, Dob could make $12,000.

When he was in Berlin, Dob took a room for long stretches at the Hotel Schweizerhof, a luxury hotel in the center of the city. While there, Dob didn't venture far from a three-block area, which had everything he needed—restaurants serving heavy German and Czech fare, and the Belmont bar, a refuge where he could drink himself into a stupor, then stumble back to his room. Berlin offered a counterpoint to everything that Dob found bland about Hannover: its soulless provincialism, its reconstructed orderliness, its implicit curfews. Berlin coincided more closely with Dob's mental geography. He was known to go through periods of dark depression, when he would do nothing but sit in the Belmont, drink Glenlivet whiskey and play backgammon. Later, he owned a handgun, which he kept loaded with one bullet.

Then there was Peter Carl, a fast-talking croupier at a Hannover casino. When Dob lost his driver's license after a drunk-driving incident in 1988, Carl became Dob's chauffeur. Slight and youthful for his thirty-one years, Carl fascinated Pengo. He was part of a world that Pengo, for all his escapades, had seldom encountered in Berlin. Carl tried to live the life of an urban sophisticate, but he was essentially small time. His job in the casino paid about $2,000 per month. To augment his income, he occasionally transported cars to Spain, a country for which he had a special affinity, or engaged in petty drug trafficking. In the summer of 1985, Carl was arrested for smuggling hashish from Amsterdam to West Germany. For the offense, he got nine months of probation.

Peter Carl grew up dirt poor. He had been raised in an orphanage near Frankfurt. Carl managed to get some training as an electrical installer, and later he made an unsuccessful attempt to get through a Siemens *Fachoberschule*—a vocational school—but dropped out after a year. From that point on, he would accord undue respect to technically proficient people like Dob, people who had managed to accomplish what he couldn't.

Before Hagbard introduced Pengo to Peter Carl, he told him of Carl's unpredictable behavior, that in a fit inside his Hannover flat Carl had shot up the rooms, and that on one occasion Carl had flown into a rage, ripped a kitchen cupboard from its mooring on the wall and thrown it out the window. Pengo was prepared to meet a lunatic; instead, he found Carl interesting and friendly, if slightly off-balance.

On a later visit to Hannover, Pengo met the fourth and final member of Hagbard's Hannover band, twenty-four-year-old Markus Hess, a beefy Hannover University physics student. He seemed friendly enough, if slightly aloof, and more of a solid citizen than the other three. He didn't have a parent's death in his past, and he hadn't flouted his military obligation. Markus Hess was a product of suburban life, middle-class, a conventional achiever. While Markus was at university, a friend got him a job programming part-time at Focus, a small software company in Hannover specializing in UNIX-based software.

Each of the five new acquaintances had come to computers via a slightly different route. Peter Carl knew next to nothing about computers, but he liked to keep company with those who were technically savvy. For Dob, more programmer than true hacker, the computer was a refuge not unlike the Cafe Belmont. Hagbard's Illuminati-inspired paranoid fixation both engendered and sustained his obsession with hacking. Pengo's was an obsession with the idea of living in cyberspace and with a new ambition to be recognized as the world's greatest hacker. Markus Hess, whose *bürgerliche* childhood provided few outlets, had an element of rebel in him. When he discovered computers, he found a fully absorbing diversion from the straight and narrow path blazed for him by his parents. And he could enjoy the tension that existed between his proper job as a programmer at Focus and the secret life he was living with his computers.

Focus was in some ways a displaced American-style Silicon Valley start-up. Its six founders tried to run the company as loosely as they could in a largely bureaucracy-bound country. Udo Flohr, a young linguist and programmer, was Focus's president. The company started specializing in UNIX as early as 1984, when the AT&T operating system was still relatively obscure in settings outside of universities and research labs. On the strength of its expertise, the company quickly grew to fifteen people, winning many of its customers in the U.S. One of Flohr's most talented programmers brought Markus Hess into the company. The eager young Hess learned all that he could about UNIX.

The first time Flohr saw Markus Hess he was dressed in a tuxedo, on his way to a formal event. Unlike Flohr's other technically minded peers, Markus appeared to pay attention to certain of life's enjoyments not ordinarily associated with someone who spends the better part of a day in front of a computer. Markus had a busy social life, and such highly developed political opinions that at age nineteen he went so far as to join the youth organization of the Christian Democrats, West Germany's conservative party, which had been in power under Helmut Kohl since 1982.

Markus was a hard worker. He spent so much of his time at Focus, in fact, that his studies suffered. Twice he took the exam that would have enabled him to continue in physics and both times he failed. Flohr was just two years older than Markus, and the two were poles apart in their political views, yet the Focus president adopted an avuncular attitude toward the young programmer. He counseled Markus not to leave the university, but Markus dropped out anyway and enrolled at a correspondence school, majoring in computer science.

Markus came from a close family. The eldest of four children, he was a model son. His father was a career bureaucrat at a car parts manufacturer in the state of Hessen, his mother a medical secretary. Markus had always done well in school, finishing *Gymnasium,* the most advanced level of secondary school in Germany, with excellent marks on his final exams. Markus's parents had high hopes for their son but understood little of his passion for computers. The interruption in Markus's studies apparently unnerved his father, who paid a visit to Flohr to tell him to hire his son full-time or let him go. Offended by the elder Hess's demands, Flohr refused to be forced into a commitment, asked him to leave and told his young employee he would rather not see the elder Hess around Focus again.

Hess was a fine programmer, but he was also hardheaded and occasionally intractable. If Flohr came to the software team with news that he had sold a concept to a new customer, if not necessarily a finished product, Markus would throw up his hands and charge Flohr with selling air. Flohr, Hess would complain, was demanding the impossible. Inevitably, Markus's more flexible and imaginative partner would intervene and a compromise would be reached. Finally, despite such fundamental disagreements, in 1987 Flohr did hire Hess full-time on the small research and development team.

Hess discovered the Hannover hackers' group and Hagbard in 1985 by way of a friend who told him about a hacker who could do amazing things. This young man, Hess was told, had logins and passwords to dozens of military computers in the United States. If Hess wanted an introduction to system cracking, his friend told him, Hagbard was the one to meet. Hess was strictly a programmer; he had heard about the hacker clubs in Germany, but hadn't been particularly interested in finding out more about them. But when he met Hagbard, that changed. The exceedingly thin young man with dark blond hair and a soft voice seemed to cast a spell over others. He spoke as if he and only he were keyed in to life's deepest secrets. Ordinarily skeptical, Hess was taken with Hagbard, and soon he too became convinced of a certain *Durchblick* —unusual insight—emanating from the young man. Hagbard shared with Hess much of what he knew, and for the first time Hess realized that computers were not just isolated tools for programming—together with a modem, they were a flexible set of lock picks.

When Markus met Hagbard, it was as if a seeker had suddenly happened upon a seer. From Hagbard, Markus learned about stolen NUIs, chatting on line and coaxing computers open. Most important, he learned the value of patience and persistence. Hess saw that Hagbard was not idly boasting when he spoke of getting into the Arpanet.

Hagbard certainly cultivated some bizarre ideas. He believed that the power of what could be done with computers was limitless. Hackers, he proclaimed, embodied the future; their ability to release "soft bombs" into the world invested them with unspeakable power. If he, Hagbard Celine, were ever caught, he told Markus, he would become a "martyr for peace, disarmament and information freedom." Like Hagbard's other friends, Hess dismissed the Illuminatus theories as so much nonsense, but like everyone else he was drawn in by Hagbard's soft-spoken zeal.

Unlike the other three hackers, Hess had no use for the drugs that he saw at once sustaining and destroying Hagbard. Hash was the mainstay, but LSD appeared to be the drug of choice for Hagbard and the major inspiration for his paranoid fantasies. Hess's only real vice was his chain-smoking, accompanied by the occasional beer. And whereas he maintained a semblance of order in his two-room Hannover apartment, Hagbard inhabited chaos. He lived in a surround of overflowing ashtrays, heaps of dirty laundry and other signs of the necessities of daily life not just neglected but transcended.

As one of the few on the West German scene specializing in UNIX, Hess was something of an oddball, and he wanted nothing to do with organized hacking, so he had no use for the "VAXbusters" from Hamburg. Further, their area of expertise, VMS, was of no interest to him. Besides, Hess had a real job at Focus to attend to; he didn't have Hagbard's freedom to devote every waking moment to hacking. Even if he had, he wasn't sure he would be able to sit as Hagbard did, with a meditator's stillness, oblivious to outside distractions for hours on end.

However, it was such an infectious passion that before long Hess was using any spare moment he had exploring networks. In a way, it was Hess's own break from the predictable way of life he had always known, with its built-in constraints and expectations. Hess may not have been interested in group hacking or VMS, but he was interested in hacking into UNIX systems. Most of the time, he stayed late at Focus and worked from there. Sometimes he used Focus's NUI, sometimes he used his own, legitimately acquired NUI. Only occasionally did he resort to the standard practice of using stolen NUIs. That, as far as Hess was concerned, was straying too far toward questionable ethics.

One of the most useful lessons Hess learned from Hagbard was that this pursuit required endless patience. It meant sitting for hours on end, dialing and redialing. Once you had reached a system, and the friendly log-in prompt at a computer's portal asked for an account and password, it meant more time expended in getting in. Once you were on the machine, exploring its crevices and testing the limits of your powers took still more perseverance. Although Hagbard was no programmer, and was fully dependent on others to write programs for him, his infinite patience and utter single-mindedness made him a more effective cracker than many of the others.

Part of the tremendous attraction of computer networks for Hess, Hagbard and Pengo came from knowing that beyond their underpowered little personal computers, or even beyond Focus's comparatively more powerful machines, there was always a larger, faster machine somewhere to be explored. If a log-in prompt appeared, and a password worked, the next task would be to try to guess the type of the machine and how powerful it was. Privileges, sometimes referred to as superuser status, were frequently a goal. Unlike Hagbard, Hess and Pengo could try some creative programming once they were inside a computer. They could use operating system commands and write programs to find and exploit weaknesses in operating system security.

Once an intruder obtained superuser status, it meant that the machine was entirely under his control. It was possible to read and change people's files, peruse their electronic messages, destroy their work or even do their work for them. The ultimate benign hack would be to send someone a note announcing that the intruder had just visited and had found—and fixed—a bug in his program. Nothing could be protected from a superuser. And hacking was far more efficient than rifling through a roomful of filing cabinets, scrutinizing every scrap of paper for something interesting. The power of the computer could easily be turned back on itself. The machine could be instructed to show a list of all its documents that mentioned a key word or phrase, such as *classified* or *nuclear*, or whatever an intruder was seeking.

Like the others, Hess was transfixed by the sheer number of possibilities. The Arpanet alone linked thousands of computers. It in turn was part of the Internet, a network that linked so many computers that no one was really sure how many there were.

Pengo's first visit to Hannover ended when, at a party at Hagbard's, Dob said he was on his way back to Berlin and asked Pengo if he wanted a lift. It was five in the morning. Pengo had been away from home for two weeks. He had run out of money and clean clothes and he welcomed the ride. When they got into Dob's sports car, Dob produced a nugget of hash the size of a walnut. The two smoked it as they sped through East Germany at over one hundred miles an hour. Dob told Pengo he made the trip all the time, always at high speed, and always in an altered state of mind. Dob liked to carry out logistical chores such as a trip from Hannover to Berlin in the most efficient way possible, even if it meant placing his life in danger.

Pengo enrolled at the technical university in Berlin to study computer science, but he couldn't stay away from Hannover too long. He started to travel there frequently to spend a few days at a time with Hagbard. A visit usually started off with a round or two of hash and progressed to all-night hacking sessions. Once or twice, they went to Focus to visit Hess and watch him hack from there.

▲ ▼ ▲

It would have been natural to select Cliff Stoll out of a lineup of suspected computer fanatics. Skinny Cliff had a maniacal edge to his personality, cultivating what might even be a parody of the eccentric scientist. Wild corkscrews of brown hair projected six inches from his head; he had the gait of a pogo stick, a yo-yo always in hand and in perpetual motion. Stoll's speech was punctuated by a parade of exclamations—"Hot ziggity!" and "Holy smokes!," "Jeez" and "Really, really neat!" If his computer kept him waiting for more than half a second before executing a command, he would yell "Commie!" at it. In fact, "Commie!" was the mocking charge leveled against any recalcitrant inanimate object.

Stoll's political views leaned appropriately leftward, in part because it never occurred to him to think otherwise. He went to college during the 1960s, and he opposed the Vietnam War, but he wasn't much of an active crusader for the Left. In fact, he considered himself a nonideologue who resisted left-wing dogma. He operated more on his principles. It was on principle, for instance, that he made certain that all of his employers were involved in nothing but the purest of research. He couldn't bring himself to work for the Lawrence Livermore Laboratory, a government research center whose bread-and-butter projects include designing nuclear warheads and Star Wars weapons for the U.S. military, to say nothing of the National Security Agency, whose computer scientists were, by definition, spies.

So in 1986, Stoll took a postdoctoral job at Lawrence Berkeley Laboratory designing mirrors for the Keck Observatory in Hawaii. High atop the Berkeley campus, LBL is Livermore's sister laboratory thirty miles to the west. LBL is one of the United States' national research labs and the site of broad-based unclassified scientific research. When Stoll's grant money ran dry in late 1986, he was forced to look for something else to do.

For an astronomer in the midst of the Reagan era it wasn't an unusual fate. Federal money to support the basic sciences was shrinking. The money was over the hill, literally, at the weapons laboratories. As it happened, the computer knowledge Stoll had picked up in high school and college gave him a certain advantage over other astronomers who lost their funding. In August 1986 he became a system administrator, in charge of maintaining the laboratory's dozen mainframe computers. This meant being responsible for everything from backing up important data and taking care of computer security to helping the scientists who were his customers use the powerful machines more efficiently. Although it wasn't exactly the work he was looking for, it allowed him to stay in Berkeley.

One of his first assignments seemed simple enough: to reconcile a small accounting error that had shown up. LBL used some home-brewed accounting software, and the patchwork of programs, written by summer students over the years, had come up with a seventy-five-cent discrepancy between the normal system accounting and the lab's own charging scheme. Cliff stayed at work until midnight puzzling over the mysterious seventy-five-cent error, which he suspected might be a computational rounding error.

After careful examination, he discovered it wasn't a rounding error, but the work of an unauthorized person from outside the lab using the account of an LBL researcher who had left several months earlier. With characteristic gusto, Cliff became a self-appointed one-man SWAT team. He set up traps that captured the hacker's every keystroke on a printer and alerted him every time the intruder was in the computer. He kept a detailed logbook, and he wrote a software program that tripped his pocket pager whenever the trespasser logged on. Before long, he was doing little else but tracking the uninvited guest. Occasionally he even slept in his sleeping bag on his office floor to keep a constant vigil over the hacker.

Finding an intruder on a computer system is often as serendipitous as guessing the correct password to break into a system. Detecting a break-in can be a matter of timing, perseverance and ultimately luck, especially if the trespasser takes steps to cover his tracks. It requires the same skills that the hacker himself employs, and it can often mean getting inside the hacker's head to anticipate his next move. A system manager like Cliff Stoll hunts down an intruder not only because computer security has been broken. It's also a matter of wounded pride, and a threat to the institution's ability to keep its doors to the outside world open and keep hackers out at the same time.

Stoll gradually began to understand the hacker's strategy. By exploiting a mistake in the way LBL's computer managers had installed certain software, the intruder had managed to give himself privileges on the system that are usually reserved for system managers. As a result, the intruder was able to create accounts with names such as Hunter and Jaeger, assigning them such passwords as *Benson* and *Hedges*. And the hacker was careful: every few minutes, he typed the command "who," which listed everyone using the computer. Evidently, he knew how to log out in a hurry. If the hacker thought he detected a legitimate system operator on the computer, or someone else with full privileges, in a single keystroke he could instantly disappear back into the electronic beyond.

Stoll didn't know, of course, whether it was indeed just one hacker plaguing his computers or a gang. Some empirical evidence seemed to support the case for just one hacker, but he wanted to be certain. So he set up a scientific probe, to see if he could discern typing patterns from among thirty colleagues at the lab. Once he decided he could, he applied the same test to the hacker's typing rhythms and discovered that most of the time the typing came across the telephone lines in methodical, evenly spaced strokes. Only occasionally was the typing more random, as if someone might be hunting for the keys. He didn't pause to consider that by the time the keystrokes had been transmitted through intermediate computers and data networks, all of the information that could identify the typist had long since disappeared.

The most obvious solution to the problem would have been to lock the hacker out entirely. And that would have been simple. Stoll needed only to change all the lab's passwords and tweak a piece of software called *GNU Emacs* that ran on LBL computers. A powerful programmer's text-editing program, *GNU Emacs* was used by nearly everyone at LBL.

Because LBL programmers had installed *GNU Emacs* on the lab's computers in a way that gave a user special privileges, any user could access any file on the system using a command called "movemail." In effect, LBL created a hole in computer security—an effective window into normally inaccessible areas within the computer, and the hacker had discovered this. Stoll was beside himself. He decided that instead of slamming the door and keeping the hacker out, he would let him roam through the system with relative freedom and catalogue his every move —and then trap him. He reasoned that by keeping the system open he could get the hacker to stay on the line long enough to let the phone company trace him back to his lair.

What really was the danger of an interloper browsing around anyway? The information in the system was often personal but, as far as the LBL system was concerned, not vital for national security. The hacker could peruse grant proposals and information about the computer system, as well as electronic mail carrying gossip, news and love letters. Stoll's salary and résumé were open for browsing, too.

To Stoll the problem was not so much the hacker's rummaging around the LBL computer as his use of that computer to ricochet his way to other computers over the Arpanet—to virtually anywhere in the Internet. With the simple command "telnet," the hacker could instruct the LBL computer to connect to Internet computers anywhere in the world, at military bases and Pentagon contractors and research laboratories. He had only to come up with an account and password for those computers. In fact, the hacker was losing interest in the Berkeley machines, and was just using them as a launchpad to others, especially those on Milnet, the unclassified military network connecting Defense Department installations.

Given Stoll's studious avoidance of military matters, he had no real idea what went on in those computers, but their locations alone sounded like serious business: Redstone Missile Command in Alabama, the Jet Propulsion Laboratory in Pasadena, Anniston Army Depot, Navy installations in Virginia and Florida and the Air Force Systems Command Space Division in El Segundo, California. Not only was the hacker logging on, but he was directing his searches in an ominous way. He told the computer to seek out files containing such words as *stealth*, *nuclear* and NORAD. He latched on to files about the space shuttle's secret missions.

Stoll had heard a lot of stories about hackers from colleagues at Stanford, fifty miles south of the San Francisco peninsula, and Fermilab, near Chicago, but they had struck him as harmless pranksters. This interloper seemed different. Stoll was stunned by the hacker's ability to traverse the nation, nudging computers open just by guessing obvious passwords. The nation's computers appeared to be surprisingly unprotected. A guest account, for instance, might have *guest* as its login and *guest* as its password. Or the login might be a user's last name, the password his first. So the hacker could ask the computer to show him a list of users on the system, then try breaking in simply by typing in people's names. On the handful of occasions that the hacker managed to give himself superuser status, he would create his own accounts or change passwords of existing accounts for his future use.

When the hacker hopped over the network to a computer at Lawrence

Livermore, Stoll panicked. He called Lawrence Livermore and told the system manager to shut down that machine.

▲ ▼ ▲

In 1986, thoughts of breaking into computers for money were beginning to waft through the air. One Hamburg district attorney had developed a theory of how young hackers might unwittingly allow their computer expertise to be exploited for purposes of industrial espionage, or even by the East Bloc: Communist agents, he reasoned, already expert at manipulation, would have no trouble engaging the minds of wayward teenagers. Not only did nervous West German authorities and imaginative computer crime experts have their own finely crafted theories about the potential for exploiting impressionable young hackers, but the hackers themselves were beginning to see what was possible.

Sometime in early 1986 the West German group decided to try in earnest to market its talents. Back in Dob's Hannover apartment, under the influence of many pipes of hash passed back and forth, Carl, Dob and Hagbard had a night of intense discussion. The first question, of course, was how to execute the initial contact. One idea was to go up to the door of the Soviet embassy in Bonn. Another idea, lifted perhaps from the pages of an espionage potboiler, was for one of them to slip a note written in code into his passport, a missive that would alert a border guard to his mission. It was a fine idea, all agreed, but no one was quite sure what the note should say. Peter Carl didn't have the technical talent of his colleagues but he had made up for that with sheer bravado and a sense of the possible, so the intrepid Carl readily agreed to travel to East Berlin and do a little cold calling. The idea was simple enough: they were hackers who could get into some of the world's most sensitive computers. From those computers they could extract sensitive information, information they knew would interest the Soviets. What was more, they could provide the Soviets with some of the software they needed to catch up with the technologically more advanced West. Why shouldn't the Soviets want to do business with them? Of course it was illegal. They all knew that. But in selling the Russians military and scientific information, they argued, they would be doing their part for world peace. A name for the project? Equalizer.

The idea was not to teach the Soviets how to hack themselves but to keep them dependent on the group somehow, to tell them just as much as they needed to know to stay interested, but not so much that they would be able to hack on their own. If the hackers were going to sell the access codes and hacking knowledge itself, then it would have to be a onetime deal and their price would have to be steep. A million marks, they decided. For that they would tell the Soviets about networks, and throw in the lists of logins and passwords for computer systems world-wide. The night wore on, and so did the hash. More hacking business ventures came to mind. Why restrict themselves to doing business with the Soviets? Why not include the Chinese? That idea was discarded when the group agreed that China wasn't enough of a player in the struggle between world powers. Project Equalizer had to stay focused.

Hagbard and Dob took on the job of putting together a "demonstra-tion package" to show to their future business partners. Hagbard assem-bled a list of logins to computers, including those at SLAC in California, a computer at the Department of Energy and the U.S. Defense Depart-ment's Optimis computer. Once Hagbard had opened a computer, he left it to Dob to do research inside the computer, looking for material that might interest the Soviets. One document with the title "Radioac-tive Fallouts in Areas 9a and 9b" sounded good. So did one called "Propellants of ICBMs." Once he had downloaded the data, Dob trans-ferred it onto diskettes and made printouts of each file, being careful to delete any clues to passwords and break-in methods. All told, they gath-ered material from thirty different computers.

In early September 1986, Peter Carl drove across East Germany from Hannover, left his car in West Berlin and took the subway to the Soviet trade mission in East Berlin at Unter den Linden. He approached a guard who was seated behind a glass partition, introduced himself as Peter Carl from Hannover and asked to speak to someone from the mission, as he had a business proposition to discuss. He assumed it was obvious that he was there to speak to someone from the KGB. The guard told him to take a seat. After a thirty-minute wait, a man appeared in the waiting area and asked Carl what he wanted to propose. Carl explained that he was a hacker from the West with access to interesting information, and he wanted to propose a business deal. The man nodded and disappeared. Ten minutes later, a tall, dark-haired man emerged from the building's recesses. He introduced himself to Carl as Serge, the French pronuncia-tion of the Russian name Sergei, and showed the visitor into a sparsely furnished conference room.

Sergei asked Carl what he had in mind. Carl explained again that he was a member of a group that could get its hands on interesting infor-mation. Sergei was only vaguely familiar with the term *hacker*, and asked

Carl to explain. Carl described most of what he knew. He said hackers were people who could break into computers and retrieve information and programs quickly and surreptitiously. Sergei still appeared to be confused, but his interest was piqued and he asked for more specifics. This particular band of hackers, Carl said, had the means to break into dozens of Western computers and get everything from information on high-energy-physics research to proprietary banking data. Carl said he would like to offer the Soviets a package of West German hacking know-how, including logins and passwords to dozens of military computers in the United States. His price: one million marks— more than half a million dollars. Sergei raised his eyebrows but remained silent.

Carl kept talking. He wanted very badly to appear to be someone worth one million marks. He had a salesman's delivery and a salesman's confidence even if the terms he tossed around were not quite accurate. He said he didn't have the demonstration package with him, that it was in West Berlin, ready for him to pick up and deliver to Sergei. He hadn't wanted to bring it with him on this first trip, as there was no telling who might search his bags.

Sergei must have been amused and curious. If nothing else, the appearance of a computer hacker at the trade mission in East Berlin had no precedent. As a rule, when it came to gathering technology the Soviets had a long tradition of doing their own fieldwork. Since most of what they were interested in, especially technology for advanced computing, was on a list of highly restricted technologies maintained by a consortium of Western nations known as COCOM, the Soviets had long since resorted to extralegal means of procuring hardware and software. The FBI liked to maintain that Northern California's Silicon Valley, where much of American computer innovation resided, was crawling with KGB agents. The FBI claimed that one of the primary missions of the Soviet consulate in San Francisco was to funnel U.S. technology into the Soviet Union. The consulate building was suspected of having a hidden forest of antennae and other surveillance equipment on its roof—all targeted at capturing sensitive telephone calls in Silicon Valley.

Through the years, a smattering of Soviet espionage cases had become public, but only because the spies were caught. The Soviets acquired advanced computers by hiring agents to set up dummy companies, order whatever was needed and then fold quietly. For the most part, the Soviets had a long-standing practice of reverse-engineering technology based on what they were able to gather. Of course, there was no way the

Soviet Union was going to build a technological infrastructure by that method. Nevertheless, it was a system the Soviets continued to use partly because of trade restrictions, and partly from a simple predilection for doing things that way. All told, their software was a motley collection of retooled operating systems and compilers, roughly equivalent to the American originals but transposed into Cyrillic. Their hardware was based principally on VAX and older IBM 360 designs and they were always on the lookout for good VAX software, especially VMS source code.

Given the ease with which American computer networks can be entered from a safe redoubt outside of the United States, it isn't hard to consider the possibility of a sophisticated Soviet intelligence-gathering operation targeting the vulnerable computers of commercial American high-technology ventures and nonclassified U.S. military systems. In the early 1980s, officials in the Reagan administration expressed alarm at the existence of a circuitous computer link that would have allowed Soviets in Moscow to log directly in to American computers: an out-of-way international research center located outside of Vienna and known as the International Institute for Advanced Statistical Analysis was connected by a commercial computer network to the United States and had a direct computer tie to a research center in Moscow. The institute lost its U.S. funding as a result of the computer link.

Some American officials argued that even though nothing secret was accessible from the center, it was conceivable that the Soviets could use the power inherent in computers to scan quickly through vast amounts of information and piece together a clearer picture of classified data. But there was no evidence to support this scenario, and several years later U.S. funding for the Vienna institute was quietly restored.

Whether the Soviets truly had designs on young computer outlaws is unclear. That the Soviets would have called upon a select group of young experts to rummage around inside U.S. computers for them, or would have dispatched someone like Sergei on a recruiting spree to the Chaos meetings in Hamburg, was probably a notion that existed only in the minds of nervous Western officials. But with this self-described hacker on his doorstep promising tapes filled with digital delicacies from the West, it wasn't surprising that Sergei considered the idea worth pursuing.

The Soviet official made it clear to his visitor that while he was by no means uninterested in what Carl was suggesting, he could hardly agree to hand over a million West German marks for something he not only hadn't seen but didn't quite understand. And Carl himself, still struggling with some of the technical concepts with which his cohorts seemed so comfortable, was in no position to deliver an extemporaneous lecture on the navigation of data communications networks, the computers that resided on those networks or the specific information they contained. Like that of any marketing man, Carl's job wasn't so much to understand what he was trying to sell but, by virtue of his unerring enthusiasm, to convey the value of his product. But even Sergei's most basic questions were too tough for him. Sergei asked Carl to return in the next few days with his demonstration package. It would be sent to Moscow for a thorough analysis and, if it were deemed to be worth a million marks, then a million marks would be forthcoming. Sergei then asked to see Carl's passport. He took some notes and left the room briefly. When he returned, he told Carl that the next time he came, as long as he used the border crossings at Friedrichstrasse and Bornholmer Strasse, the guards would let him pass freely.

Two days later, Carl took Dob's car and drove to the border at Bornholmer Strasse. After a brief glance at his passport, the guard waved him through. At the trade mission, he asked for Sergei. This time, Carl had with him the demonstration package: an index to computers around the U.S., with Pentagon computers at the top of the list. Under each heading was an index of what was contained in the individual computers. Account names and passwords had been carefully deleted. Sergei remained polite but skeptical. This time, Sergei gave Carl 300 marks for his expenses and made out a receipt. He also gave Carl a telephone number in East Berlin where Carl could reach him. He told Carl to learn the number by heart and to call only in an emergency. Carl used part of the money to fly back to Hannover. With the rest, he went straight out and bought a small Casio electronic notebook, where he entered Sergei's telephone number. When he spoke on the telephone to Hagbard or Dob about his trip, he imparted his information in an appropriately cryptic manner. *Paris* meant East Berlin. *Teddy Bear* stood for Sergei, Russia and the East Bloc. *Equalizer*, of course, was understood by all as the code name of the operation.

A week later, on instructions from Sergei, Carl appeared at a building on Leipzigerstrasse, a main thoroughfare for traffic in and out of East Berlin. He took a rickety elevator to the fifth floor. From what Carl could observe, the office served as a business dealing in heavy machinery and railroads. Sergei greeted him, and this time they had a general conversation. Sergei wanted to know more about Carl's background.

Carl could only assume that Sergei hadn't yet received word back from Moscow on the demonstration material. From then on, Sergei said, their meetings would take place at the Leipzigerstrasse location.

At the next meeting the following week, Sergei told Carl that he had gotten a response from Moscow. While the package contained some interesting information and bore out Carl's claim that the group could get into certain interesting computers, it wasn't exactly what they were looking for. More to the point, a million marks for the demonstration package was out of the question. However, Sergei was interested in certain things that could bring the hackers some money. While he wasn't interested in general hacker know-how per se, he wanted to know if Carl's group could produce information about radar techniques, nuclear weapons and the Strategic Defense Initiative. Moreover, he said, the source code for VMS and UNIX, compiler programs and programs for computer-aided design and computer-aided manufacturing could bring the West Germans a tidy sum. Sergei said his customers back in Moscow also wanted software from the American firms Ashton-Tate and Borland, two highly successful personal computer software companies.

Such an exchange wasn't precisely what Carl had in mind. He had imagined that he would present Sergei with a menu of sorts, courtesy of Hagbard and Dob, and that Sergei would tick off the various computers he was interested in. Then Carl would dispatch Hagbard to root out whatever he could. But Sergei appeared to have a different notion of what a hacker could provide. Not only was Carl not altogether certain he could get what Sergei needed, but he wasn't altogether certain what Sergei was talking about. Source code and compilers? These were things to consult with Dob about. It was beginning to look as if the things Carl had to offer Sergei didn't want, and the things he wanted Carl couldn't offer. Nonetheless, Sergei remained sanguine about the prospects of what Carl and his friends might be able to supply. This time, he gave Carl 600 marks, or $300, for his expenses and took him out to eat. Over lunch, they made small talk. Carl learned that Sergei was married with children, and that he liked to go fishing. But when Carl asked him to explain precisely what his job was, Sergei declined to answer.

Sergei gave Carl photographs of a young woman and a small child, along with the woman's name, address and telephone number. In the event that his frequent trips to East Berlin should prompt Western authorities to question him, Sergei said, Carl should tell them that he went to visit his girlfriend, with whom he had a daughter.

Sergei seemed to keep at least two lists containing names of data bases and software. It appeared to be a priority system of sorts, as the lists had a numbering scheme and items that were crossed out. By the fifth meeting or so, Sergei asked Carl to tell him something about the others in the group. Carl told him about Hagbard and Dob while Sergei made notes in a black binder.

And so the meetings went, once every week or so, through late 1986, always starting at noon, always consisting of a meal in a restaurant at which Sergei smoked Marlboros, drank glass after glass of orange juice and at the end handed Carl 600 marks for his expenses. Even when Carl had nothing to deliver, he went anyway, just for the 600 marks. Occasionally Sergei gave Carl small presents: a nice cigarette lighter, a bottle of spirits, some Russian caviar.

In spite of Sergei's measures for protecting Carl and his generally solicitous manner, the Soviet didn't appear to be particularly satisfied with what he was getting. Most of the material, he told Carl, consisted of indexes to information rather than the information itself. And that was often available only on microfiche. So Sergei brought his own computer expert to one of the meetings, but the expert could speak only English and Russian. He and Carl could barely communicate. Carl was frustrated. He didn't want the entire business going sour just because of a communications problem. So he asked Dob to accompany him on one of the trips to East Berlin. Carl told Dob that he would be able to clear things up because he would be able to understand exactly what Sergei was looking for. It took some persuading, but Dob finally agreed to go. During the meeting Sergei explained once again that he was not going to buy the hacker know-how for a million marks, and that he wasn't satisfied with the material that had been delivered so far. His interest, he emphasized, was in information from U.S. military computers, source code and compilers. Dob knew exactly what Sergei was saying, but the meeting was a disappointment. He didn't see that he would be able to get source code, nor could he get very excited about the idea of making a lot of small, low-paying deliveries over an extended period.

Then Carl told Sergei about Pengo. He told him that Pengo was a particularly capable hacker specializing in VAX computers, and that he could get good material. The Soviet expressed a great deal of interest in what Pengo might have to offer. He said he wanted to meet him in person and make his own assessment. Carl said he would bring him by.

▲ ▼ ▲

Meanwhile, Cliff Stoll's vigil wore on. He was obsessed with his hacker; all but the most basic housekeeping for the lab's computers was shunted aside. And the hacker seemed to get even more single-minded. It became clear to Stoll that this was no mere computer science student romping in an electronic playground—he had developed a keen interest in things military. Now he seemed to want to see files pertaining to intercontinental ballistic missiles and the Strategic Defense Initiative. Stoll watched him try with a persistent series of educated guesses to get into a computer at the White Sands Missile Range:

```
login: guest
Password: guest
Invalid password, try again
login: visitor
Password: visitor
Invalid password, try again
login: root
Password: root
Invalid password, try again
login: system
Password: manager
Invalid password, disconnecting after 4
tries
```

If it seemed that the hacker could cause some harm in the system he was poking around in, or if Stoll thought that the people in charge of the computer should know there was a hacker poking around in their data, he would call them. Perturbed and incredulous at first, they would close the hole that the hacker had used to climb in. So far the hacker hadn't found any sensitive national security data, or at least Stoll didn't think he had. But it was almost certainly on his agenda.

The hacker plaguing Cliff Stoll didn't seem to be much in the way of a computer genius. He was seldom inventive. In fact, his most remarkable trait was his plodding persistence: he created connection after connection, then, like a dog trained to sniff out drugs, systematically searched each system for military information. Long after Stoll had become overcome with fatigue, the hacker would continue twisting doorknobs. Stoll began to think the intruder might not be human at all. Could it be a robot, a computer programmed to look for military information? Stoll decided that it wasn't, simply because whatever it was made spelling errors.

At first, Stoll believed the intruder to be somewhere on the Berkeley campus. But there was evidence against that theory. The hacker was very familiar with UNIX, but his behavior showed that he didn't know anything about Berkeley UNIX, a variation on UNIX that was de rigueur in Berkeley. Instead, he was using the traditional UNIX commands first developed at AT&T's Bell Laboratories. He was speaking UNIX with a strong AT&T accent.

From the high keep of various universities, Stoll had never had much cause to interact with the outside world. His view of things left little room for shades of gray. As Stoll saw it, scientists who engaged in pure research and divorced themselves from the military were on the right side. The CIA, NSA, FBI and military establishment were, by contrast, sinister and shifty, not to be trusted. And so was this hacker. Not only was he inside computer systems that he had no right to be using, but he was robbing Stoll of his time, time to spend on the work he had been hired to do: to help the lab's astronomers use computers to design their telescope. If his more forgiving colleagues were entertained rather than upset by teenagers who broke into computers, Stoll saw nothing to excuse. In something of a contradiction to his self-consciously liberal way, Cliff Stoll was, in the end, a bit of a crank. He saw the hacker as a dark foe and he wanted to see him behind bars.

If he was going to catch the hacker in the act, Stoll knew he would have to trace the telephone calls. And in order to trace them, he would need a search warrant. So Stoll did something that was at odds with his political sensibilities: he called the local FBI office. He explained that there was a hacker in his computer who seemed to have a taste for military information. He was surprised and upset by the FBI's response: the agency had far bigger things to worry about than a loss of seventy-five cents.

The second call across the great academic divide to the nasty world of bureaucratic authority proved more successful. The Oakland district attorney's office was immediately interested. Stoll explained that the hacker was coming in through one of LBL's Tymnet links. Tymnet's network ran all over the United States; the hacker could be calling LBL from almost anywhere. In order to trace the call farther than the Oakland Tymnet connection, Stoll needed help from the phone company, and the phone company required a search warrant. The Oakland DA took care of that.

Pacific Bell, the local telephone company, traced the call from the Tymnet node to McLean, Virginia, and from there to Mitre Corporation. An MIT spin-off, Mitre is a Pentagon-funded research center. When Stoll confronted computer security managers at Mitre with the news that a mysterious hacker was using it on his way to supposedly secure military and university computers all over the United States, the Mitre officials swore it couldn't be true; their computers were absolutely impenetrable, secured from the outside world, they claimed. But it turned out there was an enormous hole. The hacker was using a local area network (these networks tie computers within a building, permitting them to communicate locally at high speed) at Mitre to slip around the side of the company's computer security protections. He would dial into a pool of modems at Mitre that were shared by the local area network and then use the same modems to dial back out again.

The hacker used the loophole as his personal bridge to reach other computers. It was costing Mitre thousands of dollars in long-distance telephone calls because when the hacker called from Mitre he stopped network hopping and started dialing directly into other computers. He had also planted a Trojan horse program in the Mitre system that captured passwords as others typed them in and copied them to a hidden file he could retrieve later. Mitre's computer people were shocked by this discovery. They pleaded with Stoll to keep the whole matter secret. For the public to discover that a defense contractor engaged not just in classified work but with contracts to build secure computer systems had fallen prey to a hacker would have been a disastrous turn. In exchange for that promise, Stoll coaxed Mitre out of its past months' telephone bills. From scrutinizing the telephone bills, he discovered that the hacker had been active for several months before Stoll had even detected him —for far longer than Stoll had thought. Stoll had already counted close to thirty computer systems the hacker had broken into. The number of attempts was at least ten times that.

Soon after the phone company had completed the Mitre trace, Stoll watched the hacker jump from LBL into the Milnet Network Information Center, a computer network directory information service, where he discovered four network addresses and telephone numbers for CIA personnel. The hacker wasn't actually inside a CIA computer. That would have been far more difficult, because those machines are not directly connected to public computer networks. But he seemed to be getting closer. Stoll wrestled for a moment with his conscience. Why consort with the Establishment and alert the CIA to the electronic spy in its midst? Why not just drop the matter right there and let the hacker romp unchecked? But before he could do much more ruminating, Stoll was reaching for the telephone and calling the phone numbers the hacker had discovered. In contrast to the tepid response Stoll had gotten from the FBI, the CIA immediately dispatched four people to Berkeley to discuss the matter.

▲ ▼ ▲

It hadn't taken much to convince Pengo to join Project Equalizer. At first, Dob flattered him by letting him know how much they needed him and by telling him that Sergei had expressed a particular interest in meeting him. Hagbard had too many limitations. Given enough phone numbers and passwords, Hagbard could browse for hours inside VMS systems. He knew how to log in and find a directory. He knew how to search for keywords and files. He knew what a disk was and where electronic mail was stored. But beyond that, he was lost. Ask him, for instance, to determine which machines were connected to the system he was on and he was utterly confused. Often, he was too stoned to work very efficiently. He needed too much time to think of the next step. A gifted hacker worked as much on spontaneous leaps of intuition as on anything else. And for all his grandiose theories, Hagbard couldn't program. The group needed a VMS expert who could program. Pengo was the natural choice. Carl made it clear that it could be a lucrative venture.

Pengo had theories of his own on how to get money for hacking. He developed a three-point strategy. One idea was to sell a certain amount of know-how and charge a more reasonable sum, perhaps 150,000 marks, or $75,000. Another suggestion was that he organize some seminars for the East Bloc, to teach the Soviets about technology and hacking. The third idea, which Pengo seemed to latch onto with the most enthusiasm, was to convince the Soviets to set him up to do "safe" network prowling from East Berlin—that is, to supply him with a top-of-the-line VAX computer with plenty of storage capacity, a high-speed modem for fast data transfer and secure telephone lines that couldn't be traced. Of the three ideas, the last one seemed to have the most potential: it would give the Soviets the software they needed while providing Pengo with some easy money. That, Pengo decided, was the idea he would emphasize with Sergei.

But he could hardly meet Sergei empty-handed. At Carl's urging, Pengo looked for something to take over as a first enticement. Any software he could get his hands on, as long as it seemed impressive, would probably do. First he managed to procure a magnetic tape containing some software from previous forays—a chip-design program called a PAL assembler from a hack into Thomson-Brandt, the state-owned French electronics manufacturer, and smaller programs for the VAX. But to get something worthwhile from the networks, he would have to stay logged on to a computer system for a long time. That meant choosing a computer on which he knew the security was lax. He had been inside Digital Equipment Corporation's Singapore computer center before and he knew it fairly well. As far as he could tell, there was little security at the Singapore facility. The system manager seemed to be asleep at the switch, seldom checking to see who was logged on to the computer. It was easy to get on the Singapore VAX with full privileges, and Pengo knew it would be easy to stay on for a long time. He logged on late one night and found exactly what he needed. It was a security program for VMS called *Securepack*, developed at Digital in 1983 for internal use on Digital computers. The program, which allowed system managers to alter levels of privilege on a computer, would make an interesting nugget for the Soviets. Pengo downloaded the program and put it onto several diskettes. He also made about thirty pages of printouts, taking care to delete any information as to how the computers were actually penetrated.

On the night before the trip to East Berlin, Pengo, Dob and Peter Carl spent long hours hacking from Dob's room at the Hotel Schweizerhof. Hash was burning in abundance, and although it helped to keep spirits high, it also produced one unfortunate side effect: when he was high, Pengo had the habit of dwelling for hours at a time on one problem, which undermined his overall productivity considerably and augmented the phone bill. Every few hours, Peter Carl would poke his head into the room to check on Pengo's progress, and each time he threw up his hands in disgust. Pengo was costing them hundreds of marks in connect time, and from what Carl could tell, he wasn't producing anything.

Pengo decided to put a digital record of the session from the Schweizerhof onto several diskettes and take those over too. He and Carl emerged bleary-eyed from the hotel the following morning and started out for Sergei's office while Dob slept off the excesses of the night. Their loot stored in Carl's briefcase, they boarded the subway at Wittenbergplatz, a short walk from the hotel. They changed to a different line and boarded the train that traversed the eastern portion of the city, speeding past abandoned stations. It was a familiar trip to Pengo. He had been to

East Berlin many times when he was younger. On Christmas and Easter, when his parents were still together, the Hübners from West Berlin went to visit the Hübners in East Berlin. And even if Pengo hadn't stopped to consider the quirk of fate that brought him into the world in the West, it's likely that his indifference to borders was a sign that he understood, if only subconsciously, Berlin's provisional nature, a place where, in the end, allegiance was arbitrary.

Just fifteen minutes from the time they set out, they had reached the Friedrichstrasse station in East Berlin. With forty-five minutes to spare, they took a slight detour to Alexanderplatz. Carl produced a fat joint and lit it, muttering that he needed it. Pengo could only laugh and wave his hand at the offering. It was too early in the day to get stoned. Besides, he was nervous.

When Sergei greeted them in his office, Pengo already had an inkling that this was not going to be an easy sell. He had to convince the Russian that he was a valuable asset, someone worth investing in. Dob had portrayed Sergei as someone who knew nothing about computers and could only recite from a list of things the Soviets might be interested in: compilers, source code and information from military computers. Pengo could see that he might not get much further that day. But where Dob tended to be passive, letting things happen to him, Pengo was more aggressive and outspoken. He told Sergei that he was well versed in VMS, and that he could bore his way into many different computers. By way of example, he named Mostek, a U.S. semiconductor maker; Teradyne, a Boston high-technology company; Thomson-Brandt; Philips in France; and Genrad in Dallas. He went on to list more conquests: SLAC, Fermilab, MIT, Union Carbide. If Sergei were interested in individual accounts and logins, Pengo said, he would sell them. For an account at the Jet Propulsion Laboratory, for instance, he suggested a price of 150,000 marks.

Sergei didn't seem to be responding, so Pengo came out with the rest of his plan. First he suggested that he conduct hacking seminars for the Soviets. An alternative would be that the Soviets set him up to conduct safe hacking from East Berlin. Sergei said that didn't interest him. He took out his order list and read aloud from it. His "customers" back in Moscow, he said, needed UNIX and VMS source code and compilers. VMS version 4.5 alone could bring the West Germans 250,000 marks ($125,000), compilers another 30,000 marks each. As for the thirty or so pages of computer printouts the two young visitors had brought along

Sergei said he had no idea what to do with them. Nonetheless, in addition to the customary 600 marks, Sergei handed Carl an envelope packed with 100-mark notes and invited his two visitors out to a meal.

As he and Carl left East Berlin and headed back to Dob's room at the Schweizerhof, they counted the money. It was 3,000 marks, of which Carl gave Pengo 1,000. Pengo felt less discouraged than self-important. He had made progress. A Russian KGB agent had listened to him talk for an hour. The Russian hadn't said yes, but then again he hadn't said no, and when he examined more closely the information Pengo had already provided, Sergei would be ready to deliver the VAX. Project Equalizer was becoming Pengo's own first step toward becoming a paid hacker—not just any hacker, but the best hacker in the world. He knew that he could get what the Russians wanted, if only he had the right equipment.

▲ ▼ ▲

It was Markus Hess, the one who liked to hack alone, who turned out to be Cliff Stoll's intruder. When Markus and Hagbard had first met, Hagbard had told of being in Fermilab and CERN. With a little guid-ance, Markus learned to explore around West Germany and Switzerland and the United States. Soon, he found a gateway into the Internet through University College, London. He called Hagbard immediately to have him come over and watch. The London computer was just the kind of springboard into the Internet they were looking for. From there they found a Tymnet node, and from there Hagbard discovered a way into a bank of modems at Mitre Corporation. Neither knew precisely what Mitre was, or even where it was, but it was a rich find. The modems at Mitre, it seemed, saved the last number called, and Hagbard and Hess could easily redial those numbers. That was how they first happened upon a computer at the Anniston Army Depot and Optimis, a U.S. Defense Department computer data base with information about military studies. Optimis gave access to anyone who typed *anonymous* as a login and *guest* as the password. *Guest* and *guest* did the trick at Anniston.

In the middle of 1986, Hess and Hagbard discovered SLAC in Cali-fornia. Hagbard was content to stay there and poke around a bit, engag-ing the system managers in occasional on-line conversation and chatting with others who had also found a way into the SLAC computer system. But Markus wanted to see where he could go from there. From SLAC he soon found a path that led to the University of California at Berkeley, and Lawrence Berkeley Laboratory. The Berkeley university computers foiled Markus's attempts to get in, but LBL was wide open. The laboratory liked to encourage outside researchers to use the LBL computers, and passwords at LBL were often the same as user names. The situation in Berkeley, in fact, seemed too good to be true. Security seemed to be a joke to these laid-back Californians.

Hagbard, with his pedestrian approach and his lack of programming skill, didn't fully appreciate what the LBL computers had to offer. Markus, however, could spot bugs in programs and exploit holes in the system. He had been playing around with the *GNU Emacs "movemail"* program on the LBL computers and it dawned on him that the program had been installed to run with superuser privileges. It was a major discovery. It freed him to wander through the LBL system as much as he liked. He began poking around in people's directories and looked for an account to appropriate that had lain unused for some time. It was always a better idea to use an existing account than raise eyebrows by creating a new one. He found that someone named Joe Sventek hadn't logged on for months and decided to become Sventek for a while. With superuser privileges, he could change his password to anything he pleased. *Benson* was one choice. *Hedges* was another. When he noticed that one of the people whose directories he was peeking into was getting suspicious, he logged out at once. From then on, he made certain to check to see who might be on the system whenever he logged on, just in case a real system manager was surveying the system's activities.

Sometimes Hess connected straight to America; other times he went via the University of Bremen, another security sieve, where he added a new account for himself called Langmann and used the university's modems to dial out to the Datex-P international data network. From there he got to LBL. Others discovered the fuzzy security in Berkeley. In the early summer of 1986, a lot of different hackers broke into LBL, partly because it was easy to get into, and partly because it was so easy to go from LBL into other computers. And from LBL Hess could explore a rich array of other computers. Military sites offered a special thrill. He never logged on without checking a computer for its military information. Before long, Markus had developed an expertise of sorts in U.S. military acronyms.

▲ ▼ ▲

By late 1986, Stoll had settled into a routine based on the hacker's movements. Every time the hacker appeared, Stoll's pocket pager sounded. Whatever the time of day and wherever he was, in the shower, on his bicycle or sitting down to breakfast, Stoll would drop everything, call Tymnet to start a trace, then run to a computer, log on to the LBL system and watch the intrusion. But the hacker wouldn't stay on the line long enough for a trace to be completed. One Saturday in early December, Stoll came a step closer. A Tymnet trace showed one of the calls coming from a transatlantic satellite, and to the satellite from the Datex-P network in West Germany. From somewhere in Germany, it seemed, the hacker called into Datex-P, asked for Tymnet and then connected to U.S. computers.

By now, Stoll was locked in a Faustian embrace with the CIA, the National Security Agency, the FBI, the Air Force Office of Special Investigations and the Defense Intelligence Agency. But even as he probed his conscience, he was caught up in the excitement. As far as Stoll was concerned, any international links to the hacker smacked of intrigue. He envisioned spies muttering to one another in darkened alleys. Yet back in Berkeley, there wasn't much he could do. For all his vigilance, he was powerless. His success in tracing the hacking depended entirely on the cooperation he got from those authorized to do the tracing. He kept watching the hacker's every move. The wheels of the Establishment forces turned slowly; all Stoll could do was conjure up his own visions of "the other side"—and try to minimize the damage.

If the hacker began to delete files or tamper with a system, Stoll could use the UNIX "kill" command to disconnect him immediately. And when the hacker seemed to be getting into computers containing sensitive information, or tried to download sensitive files, or what Stoll could only surmise were sensitive files, Stoll employed a low-tech but effective solution: he took his keys from his pocket and dangled them next to the wires connected to the hacker's line, shorting out the circuit for just an instant. To the hacker, it looked as if his connection was being interrupted by simple line noise and he would try again. Again came the keys. Eventually the hacker would give up.

The next step was to trace the call within Germany. Stoll stood by as the network experts at Tymnet negotiated with the Bundespost authorities to put on a trace. Finally, he got word that the call had been made from the University of Bremen. The German authorities informed the university that its computer systems were being infiltrated by an outsider. Flustered by the news, the university shut down all outside connections for three weeks. But that didn't seem to stop the hacker. The next trace showed he was coming from Hannover. But even that wasn't conclusive proof that he was a German. Who was to say that Hannover wasn't just a computer stopover from a flight that started in Botswana, Islamabad or anywhere at all? The only way to continue the trace was to get a search warrant, and that meant negotiating an agreement between American and West German authorities. For Stoll, it meant more waiting.

The New Year, 1987, came and went. Stoll's frustration was mounting. As a research scientist, he knew all too well the virtues of patience, but this research project had yet to yield any results. His main constraint was his utter dependence on outside authorities and their cooperation. And the fact that foreign authorities were now involved could only complicate matters.

▲ ▼ ▲

Hess began to realize that something was going on when he met Hagbard and Dob for a beer one night in October 1986. "What have you been doing with the LBL account?" Hess asked his friends.

"Not much," Hagbard insisted.

But they did appear to have a new scheme. They spoke of a Sergei and a "Teddy Bear," and allowed that they had ways of making money with computer accounts now. Hess was afraid that those two might abuse the LBL account and cause it to go the way of so many other computers locked tight when intruders were detected. He decided that even if Carl and Hagbard hadn't done much with LBL so far, this was one hacking sandbox he wanted for himself. A week or so later, he changed his LBL passwords to just one: *LBLHACK*; he banished *Benson* and *Hedges* from the system and didn't tell Dob or Hagbard. He wasn't concerned about Pengo. If Pengo wanted to get into LBL he could probably do it on his own.

Two weeks later, Dob proposed a business deal to Hess. He needed him to make a copy of the Berkeley UNIX source code that Focus had. There would be money in it for Hess. Hess agreed to do it. It seemed like an insignificant favor. Berkeley UNIX, the variation of UNIX used at most universities and research labs, was widely distributed and easy to license. Making one copy didn't seem like a major breach. It took Hess about a week to pull what he could of the UNIX source code together. Carl picked up the software and Hess thought nothing more of it until nearly a month later, in November, when Hagbard and Dob asked him to take a walk with them one night outside the Casa Bistro. In a matter-of-fact tone, Dob came out with what was going on. "The UNIX source code was sold to the East," he said. "And that means you're in it with us now." Hagbard stood a few feet back and said nothing.

It confirmed what Hess had suspected: Dob wasn't jesting. In fact, if anything, there was more than Dob was letting on. It seemed that this was part of an established operation. Apparently Peter Carl had been making regular trips to East Berlin to deliver the fruits of his friends' hacking and general software piracy. Fancying himself the group's *Windmacher*, the one who makes things happen, Carl, it seemed, had spent several weeks conducting a circumspect survey of local hackers on their willingness to supply him with material for the East. Hess realized that he had joined a core group composed of Carl, Dob, Hagbard and Pengo. Hess gathered that Dob, while not doing much of the hacking himself, had been over once or twice to meet Sergei as a technical expert. Moreover, Dob was the group's hub, the only one with a link to each member.

Hess went straight home and the scene outside the bar played back in his head. Great, he thought to himself, what a fine business you've gotten yourself into now. Carl had already paid him 500 marks for the software. Hess had consumed his share of spy thrillers, including generous helpings of Robert Ludlum novels, with their requisite abundance of derring-do, double agents and double-crossing. He suspected at once that he was suddenly subject to blackmail. And he didn't need to consult a book of West German law to know that he was now part of something not only illegal but extremely serious. Nonetheless, Hess had no intention of interrupting his hacking routine. He had just seen the movie *War Games* for the first time, catching the film on German television. Thoroughly inspired, he had made it his goal to do what the movie's young protagonist had managed to do: to get into NORAD, the North American Air Defense Command in Colorado. He still had the LBL computers to himself, and they stood so wide open and so widely connected to other computers that he was certain he would be able to find a way into NORAD from LBL, given enough persistence.

Hess would maintain to the end that hacking for him was merely a thrill-seeking game. His interest in military sites had only to do with their being the most forbidden fruit on a network, and nothing to do with his being an errand boy for the Soviets. Hess continued to hack.

▲ ▼ ▲

It was Cliff Stoll's girlfriend, Martha Matthews, who came up with a brilliant ruse to catch the intruders. Martha was a twenty-four-year-old Berkeley law student headed for a Supreme Court clerkship, her calm bearing an ideal counterweight to Stoll's manic edge. If this rogue was so persistent in his pursuit of military data, she argued, then they should use his insatiable appetite to trap him. The idea was to round up volumes of government data, disguise it as secret military information, plant it in the LBL computer as bait, then entice the hacker by naming the false files something irresistible like "SDInet."

The two of them set about collecting hundred of pages of excruciatingly dull and technical government documents they found lying around on the computer system and put them into a single account on the LBL system. They created new titles for each file, spicing up the language so that it looked as if the documents described a new computer network that would coordinate research on the Strategic Defense Initiative. In one of the files they even included a letter inviting people to send for further information on SDInet. For the price of a first-class stamp, interested parties could get documents on "SDInet connectivity requirements," "SDInet management plan for 1986 to 1988," or even a "classified SDInet membership list." Inquiries were to be sent to Mrs. Barbara Sherwin at LBL, a name of Stoll and Martha's conjuring. Because the information was so voluminous, they explained, it would have to be sent via the U.S. mail rather than over computers. To speed things along, they added that requests for documents must be received by the lab no later than January 30, 1987. If the hacker jumped at that bait, they might even get a return address.

Stoll set up the SDInet file so that only he and anyone posing as a system manager would have access to it. The next step was to sit back and wait for the intruder to log on.

A few days later, the hacker was back for a routine cruise of the LBL system. Within minutes, he noticed the SDInet file. And sure enough, he stayed interested for more than an hour. Soon thereafter, Stoll got word that the trace had been completed to a certain residence in Hannover. But he wasn't given more details, and certainly not the hacker's name.

Then, as if to provide positive proof that espionage was involved in this hacker's activities, a few months later, well after the January 30 cutoff date, the lab received a letter addressed to Barbara Sherwin. The stationery letterhead said Triam International in Pittsburgh, Pennsylvania. The author of the letter was one Laszlo Balogh, and he asked for specific classified information that had been listed in the bogus SDInet file. Stoll decided that Laszlo Balogh must have had some connection with the hacker, since Stoll and the hacker were the only two people in the world who could get at the SDInet file. Stoll's first call was to the FBI. He was told to find a glassine envelope, presumably to preserve fingerprints, and mail the letter at once to FBI headquarters.

▲ ▼ ▲

At 6:00 in the evening on June 27, 1987, it was business as usual at Focus Computer in Hannover. In the company's second-floor offices, most of the employees were still engrossed in their work. Udo Flohr, Focus's president, was preparing to leave when the office bell rang. When Flohr opened the door, he was greeted by seven people—two police investigators from the Bundeskriminalamt, West Germany's FBI equivalent, four local police investigators from Bremen and a Bremen district attorney, who presented Flohr with a search warrant. Computer fraud was the charge, and the officials wanted to be shown to Markus Hess's office. Flohr was too taken aback to be cordial; his first reaction was to be defensive, even rude. He knew Markus to be an adventurous spirit, and he wouldn't have been in the least surprised to hear that Markus had done some hacking. But a lot of programmers hacked around sometimes. It was to be expected. It was even part of Hess's job to test security systems. Flohr couldn't believe that Hess's offense had been so odious that it required the presence of seven people. He showed the group up to the fifth floor, where Hess was working in his office. As he left flanked by three of the officers, Hess looked more surprised than panicked. Flohr immediately called the firm's attorney and asked him to go over to Hess's apartment.

The remaining officers spent the next several hours at Focus, searching through everything in sight. When they began poring through stacks of old newspapers, Flohr lost his composure. "What are you looking for?" he demanded.

"You tell us," one of the officers replied, as if Flohr were somehow involved in a well-orchestrated cover-up scheme, and kept rummaging through the newspapers.

"Well, whatever it is, it can't be in there," Flohr snapped back.

After nearly four hours, finally convinced that they had searched everything possible, the police left Focus.

Hess had never been one to panic, and the sudden appearance of a clutch of police officers at his office door hadn't given him any reason to break his calm. He said nothing during the fifteen-minute ride to his apartment, and once they arrived, he insisted on waiting for his lawyer before a search could commence. Within a few minutes, the Focus attorney was on the scene, and the police investigators grew discernibly more cordial. As the police searched through every nook in sight, Hess made a pot of coffee and he and his lawyer waited for them to leave. After two hours the search ended, and, since they weren't quite sure what they ought to be looking for, the officers left with Hess's two home computers and dozens of papers tucked under their arms. Markus returned to Focus, where the search was still in progress, picked up a few of his things and went straight to the Casa Bistro, where the weekly hacker meeting was just getting started. He said nothing about what had just happened. He was back at work the next day.

Hess's attorney promptly filed a complaint on behalf of both Focus and Hess, charging that a search had been conducted based on insufficient evidence.

The surprise visit had been enough to scare Hess away from hacking, but not enough to stop him from supplying Peter Carl with software for Sergei. The money was so easy to get that it was difficult to turn his back on it. Since late 1986, he had received reasonable pocket money— always in cash and always in 100-mark bills—in exchange for the various programs he had handed over to Carl, many of which he had simply copied from Focus computers. That, Hess reasoned, was a completely separate matter from the hacking. Hess was less taken with the conspiratorial, adventurous side that the others seemed to enjoy so much. He knew that Pengo and Dob had accompanied Carl to East Berlin, but he had no desire to go. Aside from a family vacation to Yugoslavia he had made as a child, Hess had never been to an East Bloc country, and he had no intention to go now. He was content in his role as the software provider.

▲ ▼ ▲

It wasn't until Peter Carl produced the UNIX source codes from Hess that Sergei made a first substantial payment of 25,000 marks—about $12,500—in 100-mark notes. Carl noticed that on the lists he kept, Sergei had crossed out "UNIX" and written "25,000" next to it. Carl also noticed that a notation for IBM's VM operating system had been crossed out and "50,000" had been scribbled next to it. Curiously, "VMS" was also scratched out, even though Carl hadn't yet been able to deliver it. Acknowledging Carl's confusion, Sergei remarked, "You have competition."

After that, the payments amounted to 3,000 marks here and 5,000 marks there for bits of software that Sergei thought were worth something. Carl's practice was to keep at least half of the money for himself and dispense the balance to the provider of the goods. After all, he figured, why not keep at least half the money when he was doing the dirty work? Carl had lost his job at the Hannover casino in the midst of a highly publicized table-fixing scandal, so the income from the Soviet business was becoming important to him. Still, no one was getting rich from the scheme, and by the time the whole business was over, Sergei would dispense roughly 90,000 marks to the group.

After his meeting with Sergei, Pengo was hacking as often as he could. He broke into the Japanese branch of Lotus Development Corporation, but found nothing. Every time Carl dropped by on his way to see Sergei, Pengo had to tell him he had nothing to offer. His ability to hack uninterrupted was a function of his access to passwords and NUIs, and by the end of 1986, stolen NUIs had a much shorter useful life. The Bundespost had developed far more sophisticated means of tracking down abused NUIs. In fact, once a NUI was discovered, if it was shared, it disappeared with a few days. So hackers had stopped sharing NUIs. Pengo, too, guarded the ones he used closely. With a NUI in hand, he could stay on line for at least two weeks, doing nothing but hacking. Without one, he stayed off line, stoned most of the time, programming here and there, occasionally hitting up Dob for a good meal and some hash.

Pengo was beginning to be extremely careful with his work. He tried to maintain as low a profile as possible. When he was breaking into computers on the Easynet, Digital Equipment's internal electronic mail network and the place where he thought he might find VMS source code, he probed only as much as he knew was safe without being detected. But it was a frustrating exercise. Aside from the source code and compilers, Pengo suspected Sergei wanted not just any military material, but information on sensitive military networks, along with detailed information about the logical setup of the U.S. military machine, in order to infiltrate it. But to find that kind of information, he wasn't sure where to begin. He became less and less willing to share information with others, such as those from the Hamburg/Chaos scene. He had already made the mistake of telling Obelix and the others about the Digital computer in Singapore, and within a few days the machine was locked down. Now that he was to make a livelihood from hacking, the stakes seemed higher.

It wasn't ideology that made Pengo want to be a paid hacker for the Soviets. Despite what he had said to Sergei, Pengo didn't feel particularly strongly about politics. He hadn't exactly lied; he did come from Berlin's leftist scene. And he was sympathetic, at least in vague terms, with what Gorbachev was trying to achieve in the Soviet Union. He had developed a strong sympathy for West Germany's eight-year-old Green Party, which stood for the preservation of a threatened environment and opposition to nuclear weapons and nuclear energy, and was, for a time at least, gaining a stronghold in Berlin. And Pengo was an avid reader of the left-wing newspaper *Tageszeitung*. Occasionally, he even fancied himself an anarchist, a fitting image for a young man from Berlin. But he was skeptical of the others' talk of world peace through hacking, their justification for Project Equalizer.

Pengo suspected that Carl and the others were motivated mainly by money. But Pengo had other reasons for what he was doing. Being a paid hacker, traveling over to East Berlin and dealing with the Russians were simply extensions of his life in front of the computer, the same life that alienated and mystified his family. Pengo believed that he was upholding a commitment to hacking as a thing in itself. Part of this idea had come from reading *Neuromancer*, an intense and chilling science fiction novel populated with high-tech lowlifes. Written in 1984 by William Gibson, the book came to define what later was called cyberpunk. The novel's protagonist, a drug-addicted computer cowboy named Case, is presented with the opportunity to save his rapidly decaying life by breaking into computer networks and stealing data. *Neuromancer* became Pengo's personal cyberpunk primer. After reading it, Pengo decided that if he hadn't already established himself with a nickname, he would have chosen Case. In Pengo's reality, working for the Russians held its own justification. It was something Case would have done. Pengo knew vaguely what was written in the German law books—that selling so much as a page torn from a West German phone book to the KGB was espionage—but he didn't really connect that to what he was doing. He was doing what he had always done: hacking. And now someone was acknowledging his talent.

Late in 1986, a few weeks after Pengo's trip to East Berlin, Dob had given Pengo another sizable chunk of money, thousands of marks to pay his phone bill. The money was meant to defray the costs of hacking with a legitimate NUI. But Pengo was still using stolen NUIs. He used the money to buy a Sony Walkman, a telephone answering machine and some records. In fact, he set up an impressive computer center in his bedroom at his father's apartment. He was teaching himself UNIX and had put together a computer, a decent modem and a printer.

It was a purloined NUI that got Pengo into trouble a few months after his meeting with Sergei. When the Society of German Engineers noticed that its Bundespost bill for Datex-P use was a hundred times the monthly average, the association complained to the Bundespost, which traced the unauthorized use of the NUI to one Gottfried Hübner in West Berlin. One night shortly after the trace had been completed, Pengo's phone suddenly went dead. He called the Bundespost to complain. The line was connected again. On December 1 at 9:00 A.M., Pengo was roused from his sleep by three officers from the West Berlin police. He remained perfectly calm and asked to see their search warrants. After several hours of searching every possible corner of Gottfried's apartment, the police confiscated all the computer-related evidence they could find. But they forgot one vital piece of equipment: the hard disk where all of the computer's information was stored. A few weeks later they returned for it, but Pengo just shrugged his shoulders and said he didn't know what they were talking about. He was eventually accused of concealing evidence and was fined for his unregistered modem. It had been a close call, but Pengo's father was more upset than Pengo—not at his son's transgression, but at the invasion of police into the apartment.

As 1987 went by, Pengo kept working for Sergei, but he wasn't making much progress in finding things that the Soviets might want. Once a month he got a call from Peter Carl, who was on his way to his regular meeting with Sergei whether he had something to take with him or not. Showing up empty-handed was embarrassing. "Have you got anything?" Carl asked Pengo each time. "No" was the response more often than not. Pengo knew he wasn't living up to his part of whatever haphazard arrangement he was involved in. But breaking into systems was getting more difficult, mostly because NUIs were so much harder to get. So he asked Peter Carl to give him Carl's own NUI, explaining that without a NUI he could always rely on he was at a distinct disadvantage. Carl agreed. What Carl didn't expect, however, was that Pengo would use the NUI for weeks on end and run up a bill of 4,000 marks—or $2,000 —in one month.

Carl's patience with both Pengo and Hagbard was wearing thin. Carl hadn't fully forgiven Pengo for spending an entire night at the Hotel Schweizerhof the previous September, talking a big line, smoking lots of Dob's hash and coming up with nothing particularly worthwhile. Sergei, moreover, was pressing Carl to cut Hagbard off from the group because of his heavy drug addiction and his habit of talking too much. Dob, too, had proved a disappointment to Sergei. Dob's motivation was flagging and he hadn't produced any source code for the Siemens computers he was so expert in. Carl had even taken Dob to East Berlin a second time so that Sergei could try to get him motivated, but he was increasingly indifferent.

Markus Hess was the only one who seemed to have Sergei's respect. Sergei told Carl he was glad that Hess didn't take drugs and pleased that Hess had come up with the source code to Berkeley UNIX. Most impressive to Sergei was Hess's expertise in U.S. military computers.

Indeed, Hess was the only one who came through with any regularity. Gradually, it seemed, the scheme had deteriorated from the great hope of the million-mark coup to just seeing what they could get away with. Some of Sergei's more gullible purchases, in fact, originated with Hess. As soon as Minix, an operating system with the look and feel of UNIX, came into the Focus office in late 1987, Hess copied it and handed the copy to Carl. Sergei handed Carl 4,000 marks ($2,000) for the software. When Sergei later discovered, apparently from reading an American computer magazine, that Minix source code normally sold for 120 marks, he was furious. But that didn't stop Hess. A few months later, Hess filled a disk with UNIX software, freely distributed at the last European UNIX User Group meeting, and Carl sold it to Sergei for 2,000 marks. When Sergei found out that it was all public-domain software, he warned Carl never again to sell him something like that. Carl's reaction was indifferent. At least Hess was producing. Pengo couldn't even come up with a list of passwords.

Dob, too, was losing his patience with Pengo. More than anything, Pengo just seemed irresponsible and self-absorbed. For months, Pengo had been using Dob's Rainbow computer, a Digital computer similar to an IBM Personal Computer. When Dob left for a vacation in Nairobi in the summer of 1987, he asked Pengo to retrieve from the computer a bill for work Dob had done and send it on to his employer. Despite profuse promises, the matter slipped Pengo's mind and the bill remained untouched in the computer. Dob flew to Kenya via Amsterdam, but on his way back from Kenya the flight made a stop in Hannover. He was arrested almost immediately at the airport on charges of evading his military obligation. Dob sat for weeks in prison in Hannover and it wasn't until he saw his funds dwindle to nothing that he discovered Pengo hadn't done the small favor he had asked of him. It translated to a loss of nearly 10,000 marks for Dob. To further erode their friendship, when the Rainbow broke, Pengo neglected to have it fixed.

Then, in the late summer of 1987, the hack that was to focus world attention on the Chaos Club happened. The NASA hack, as it came to be called, had been started innocently enough by a group of Chaos hackers as a simple, if all-consuming game of seeing how many computers could be reached on the space agency's vast SPAN computer network. The NASA hackers did it by inserting an ingenious piece of software of their own invention into VMS version 4.5. VMS 4.5 was approximately the twenty-eighth iteration of Digital's operating system. Each time an operating system is updated and new features are added, new security flaws almost inevitably arise. Often, they go unnoticed. When a small group of Chaos hackers discovered a flaw in VMS 4.5, they seized on it and wrote what they called "the *Loginout* patch." It was a clever little Trojan horse program embedded in the VMS source code designed to collect passwords on any computer they entered. And it was the very piece of code that Kevin Mitnick and Lenny DiCicco would chance upon a year later and use to infiltrate scores of Digital's own computers.

In two computers named Castor and Pollux at NASA headquarters in Washington, the group had found shuttle proposals and reports on booster rockets. This was not particularly sensitive information, but NASA didn't necessarily want its files opened up to a pack of curious West German computer hackers.

Obelix, who was part of the group, had been keeping a list of all the computers they penetrated. But there came a time when the hackers lost count of the machines they had infected with their *loginout* patch—it could have been 150, or it could have been 500. They began to fear that the Trojan horse could run out of control. Someone more malicious than curious could find it, copy it, install it in another computer and cause some real damage. It was time to notify someone official. Their first idea was to inform Digital itself and leave the problem in the company's hands. But after thinking it through, they decided that it might not be the best idea: Digital was likely to respond with a call to the police. Their next thought was to alert the American ambassador, but they rejected that idea because, they decided, he probably wouldn't understand what they were talking about.

The NASA hackers decided to tell Wau Holland and Steffen Wernéry. The two Chaos leaders weren't sure how to react, so they called two television journalists, Thomas Ammann and Matthias Lehnhardt.

At thirty-one, Thomas Ammann was well established as a free-lance television journalist who did most of his work for "Panorama," a West German television magazine. Ammann specialized in technology and, of course, the German hacking scene. The thirty-seven-year-old Lehn-hardt was Ammann's terse and cynical partner. The journalists sat down with the Chaos leaders, who told them the story and said they were planning to inform the authorities. Eager to save the story for themselves, the journalists tried to discourage them. They argued that it would just bring an unwelcome investigation and put Chaos in a bad light. But Wau and Steffen thought the material they were sitting on was too sensitive to withhold. Eventually, they reasoned, it would come out anyway. A few days later, they called a contact who, they believed, might know whom to call at the Bundesamt für Verfassungsschutz, a West German agency roughly equivalent to a domestic CIA, and they asked their contact to inform the agency. As it turned out, the agency itself wasn't sure what to do with this odd, indirect confession. Without informing DEC or NASA of the problem, the Verfassungsschutz simply sat on the matter. Ammann and Lehnhardt decided to force the matter into the open and rushed to put together a program for "Panorama" about the hack.

NASA refused to comment. Digital public relations, holding firmly to a policy of refusing comment on sensitive matters, didn't return the journalists' telephone calls. After a dozen attempts to rouse someone at the Digital office in Munich, Lehnhardt finally spoke with a public relations woman and challenged her directly. "This doesn't look very good for DEC," he said, hoping to chisel loose a response.

"I know, but we can't say anything," came the answer.

On September 15 the show aired, declaring that NASA's top-secret network had been violated. File footage of the shuttle hurtling into space filled television screens throughout the Federal Republic, as if to imply that as a result of their wanderings inside the NASA computers, the hackers could manipulate a shuttle launch. The journalists even interviewed Obelix, citing him as an outside computer security expert. Obelix commented on the hackers as if he had no idea who they were. The program threw Digital's Munich office into some disarray. West German hackers discovering a gaping hole in Digital's operating system was humiliating enough. The fact that they were romping undetected for months was even worse. Finally, NASA confirmed that hackers had been in the SPAN network but insisted that they had only browsed through insignificant information.

For his part, Pengo had kept his distance from the NASA incident. A year had passed since his trip to East Berlin and he had stopped trying to hack for Sergei. He had gotten no response whatever from his first delivery, none of the feedback he was hoping to see. It was a big disappointment, especially for someone whose ambition had been to become the world's greatest hacker. Moreover, Carl's visits and phone calls had tapered off, a signal that he wasn't placing much stock in Pengo's contributions any longer. There had been no formal ouster, but Pengo figured he was no longer part of the group.

It appeared that the whole espionage affair was dying down. It even looked as if the hackers weren't going to get caught. Pengo's attention had turned to other things. He was looking for a way to go straight. He had just started his own small company with a friend in Berlin named Clemens. Called NetMBX, the small firm would write networking software, do some consulting and run an electronic bulletin board. Starting out with Clemens's equipment and the merest whiff of start-up capital, the two budding entrepreneurs set up shop in a cramped little office on the outskirts of West Berlin. Pengo was still enrolled at the technical university, but he neglected his studies to flirt with the possibilities of capitalism. He joked that he was becoming a "yoopie."

▲ ▼ ▲

Stoll's frustration was mounting. Five months had gone by since the first successful trace and the authorities were as reluctant as ever to let him in on their side of the investigation.

Finally, in late June, Stoll got a call from the FBI telling him that the hacker had been apprehended, and his home and office searched. But Stoll still couldn't pry loose a name. With the hacker caught, LBL no longer had to lure him in with its flimsy security. That day, Stoll's lab changed its passwords and tightened security. The attempts at the LBL system stopped. Stoll was able to pat himself on the back and at the same time expound to his colleagues on what he considered to be the moral questions raised by this act of illegal computer joyriding. These computer networks, Stoll argued, which scientific researchers, computer scientists and students depend on for sharing information and for cooperative work and even for sending electronic love letters, assume a certain level of trust. And as the reach of the networks grows, so too must the trust. How, then, can computer scientists expect to build and maintain open networks such as the Internet if someone like the Hannover hacker is going to abuse the networks? For Stoll, the incident highlighted an insoluble dilemma: security versus information exchange.

Stoll began to think about publishing the story of how he chased the West German hacker. In early 1988 he sent out a book proposal and prepared a technical paper for a computer journal on the various entrapment methods he had used. But in April, a few weeks before his technical article was to appear, he was preempted when a story about the Hannover hacker came out in *Quick* magazine, a West German hybrid of *People* and *Vanity Fair*. The magazine identified the hacker as Matthias Speer, a play on Markus Hess's name based on the fact that both names —Speer and Hess—were also those of prominent Nazis. Mysteriously enough, the story relied heavily on Stoll's logbook, the very logbook that he had turned over to the FBI and the CIA months before. Someone, Cliff decided, most likely the authorities in Bremen, must have leaked the contents of his logbook to *Quick*. Shortly after the *Quick* story appeared, *The New York Times* put the story of the West German hacker on its front page, mentioning espionage and the Pittsburgh letter for the first time. The laboratory then held a hurried press conference.

Within a few days, the story of the hacker who had roamed freely inside sensitive U.S. computers was everywhere. One diligent reporter even managed to root out Hess himself and journalists began to call his house.

Hess took cover from the press. After his bust the previous June, he had stopped hacking entirely. But he hadn't stopped supplying Peter Carl with software to take to Sergei. In early 1988, Hess sent over a magnetic tape containing a copy of X Windows, an advanced software program, with the GNU Emacs program—the very software he had used to gain superuser status at LBL—as an extra bonus. He was rewarded with 2,000 marks. During his June interrogations, he figured out that his LBL excursions had been traced. But he was taken completely by surprise to learn that he had been monitored so closely. And he was shocked to see his photograph in *Quick*, a picture that had been taken through the window of his apartment while he sat at the computer. The Balogh letter, in particular, created quite a stir in Germany.

▲ ▼ ▲

As it turned out, Laszlo Balogh was another strange character.

Howard Hartmann's wife had long since decided that Laszlo Balogh was a "charming snake." But Hartmann was more understanding. After all, he had known Balogh for nearly twenty years, and had even hired the Hungarian émigré at his small geological surveying firm in Pittsburgh. The thirty-seven-year-old Balogh had immigrated to the United States in 1959, settled with his family in Pittsburgh, attended local technical schools and, through the sixties and seventies, worked at local firms as an engineering technician. The attributes Mrs. Hartmann thought unsavory Hartmann decided had more to do with Laszlo's "other" life.

Laszlo Janos Balogh had always cultivated an aura of mystery. When Hartmann hired Balogh as an $8-an-hour technician in 1984, the tall and dark-haired Hungarian, who bore a slight resemblance to Omar Sharif, told Hartmann that there would be periods when, because of his work for the U.S. government, he would have to leave town for two or three weeks at a time. Laszlo's résumé stated that from 1966 to 1985, he worked for the FBI and the CIA "as a consultant on counter intelligence."

Hartmann already knew a little bit about Laszlo's other life. A year earlier, Laszlo had been in the Pittsburgh newspapers when he blew the whistle on former associates who had been planning to sell millions of dollars' worth of stolen computer parts to Soviet agents in Mexico. The suspects' defense attorneys had disputed Balogh's credibility and accused him in turn of stealing diamonds. The three defendants in the illegal exports case said Balogh had boasted to them of being a CIA hit man, a black belt in karate and a bodyguard for two Kuwaiti princesses.

Laszlo also had his own small firm, called Triam International, which provided security and surveillance services. Triam had European connections, Laszlo told Hartmann, so he would be making occasional calls overseas from Hartmann's business telephone, but he would promptly pay the firm's bookkeeper for the calls. Laszlo also hired Hartmann's daughter, Linda, to type occasional letters for Triam. Laszlo promised that he would conduct all of his own work outside of normal office hours.

The arrangement seemed perfectly acceptable to Hartmann. Balogh was a hard worker and a competent technician. As long as Laszlo was working in the interests of the U.S. government, Hartmann saw no reason to object to the unconventional arrangement. Hartmann had only one stipulation for Balogh.

"I don't want anything that would bring any discredit to my firm," Hartmann warned.

"Oh, that's no problem at all," Balogh responded.

Hartmann had trouble figuring this complex character out. Why would he work for so little pay when his other life offered so much excitement and material reward?

"There's only so much excitement I can take," Laszlo responded. He told Hartmann that he didn't care much about the money, and he just wanted to keep a low profile.

Vagueness was the essence of Balogh's life. He never invited colleagues to his house in a middle-class neighborhood of Pittsburgh, where he lived with his wife and three daughters.

As far as anyone could tell, Laszlo always carried a gun, either in a shoulder holster under his suit jacket or strapped to his ankle. But when Hartmann asked him about it, his response was terse.

"Packin' a rod, Laz?" Hartmann once teased.

"Yup. Always," was the only response.

And whatever Balogh was doing for the government had its material rewards: he owned two Mercedes and a Jaguar, and he dressed in tasteful, expensive clothing. When asked a question about anything personal, he responded in the most circumspect of terms. His refusal to answer directly the most straightforward questions even started carrying over into his work at the firm. If Hartmann asked Balogh whether he had performed a certain calculation, hoping for a simple yes or no, he got a convoluted and confusing response instead.

There was also something slightly cartoonish about Balogh, an awkwardness that, together with his secret life, put Hartmann in mind of actor Peter Sellers's Inspector Clouseau: smooth and charming in some ways, clumsy and foolish in others. Once, for example, when Hartmann was suing a former client who refused to pay his bill, he sent Laszlo out to have a talk with the delinquent customer. A tape recorder was hidden beneath Laszlo's coat. In the middle of the conversation, the microcassette came to the end and the recorder shut off with a piercing electronic tone.

There were some people who mistrusted Laszlo, and they told Hartmann to watch himself. They were sure he was involved in shady activities, but no one ever had a scrap of evidence. When Hartmann got curious enough to ask a friend who was a police sergeant to run a search on Laszlo, it came up completely clean—perhaps too clean, the sergeant told Hartmann. Someone could be protecting Laszlo.

When Hartmann got a subpoena for his telephone bills in February of 1987, Laszlo was the first reason to come to Hartmann's mind. And the federal prosecutor in charge of the investigation confirmed his suspicions.

"It's not you we're looking at. It's an employee of yours, Laszlo Balogh," the prosecutor said. He added that he would probably call him again for Laszlo's personnel file. But he never did.

In early 1988, Laszlo went to work for another engineering firm in the area. In April, when news about West Germans and a mysterious letter to a Berkeley research laboratory surfaced, Hartmann maintained his loyalty to Laszlo. Balogh told one newspaper reporter who tracked him down in Pittsburgh that he had been answering an advertisement in a trade magazine when he wrote the letter to LBL. He told the newspaper that he had been a consultant to the FBI since 1966 and that he worked as a weapons broker to such countries as South Africa and Saudi Arabia.

Despite overwhelming evidence to the contrary, Hartmann wanted to continue to believe that the matter had something to do with Laszlo's work for the U.S. government, and that as long as he was working for the right side he deserved some support. When reporters found Hartmann and started asking him questions, he would say only that Laszlo had been an excellent employee. Laszlo called him to thank him for his loyalty.

"You're a true friend," Laszlo said.

"Well, Laszlo," Hartmann replied, "I couldn't discuss things I had no absolute knowledge about. The only thing I know for certain is that you're definitely unique."

Nonetheless, the incident with the letter had piqued Hartmann's curiosity sufficiently that he asked his daughter about it. She had typed about ten of Laszlo's business letters on Triam letterhead, and she remembered having typed the letter to Barbara Sherwin at LBL, but she couldn't recall much beyond that.

"Do you remember what material you were referring to when you were typing out the request for specific information?" Hartmann asked his daughter.

She thought about it. "I think it was a magazine," she responded. But she couldn't be sure. "Laszlo took lots of material from magazines."

It wasn't until Hartmann stumbled upon what he considered to be a strong link between Laszlo and a theft of equipment from Hartmann's firm two years earlier that Hartmann realized just how elastic Laszlo's loyalties may have been. Now he wanted to find out all that he could about Laszlo's side work. But when he had his lawyer call the assistant U.S. attorney who had once asked for the phone bills, the attorney would only say that the case was still under investigation.

The finer points of Laszlo's life didn't matter much to Cliff Stoll. The very appearance of a letter from someone with what appeared to be shifting allegiances was enough to convince Stoll that there was something strange going on. If what Laszlo had told reporters was true—that he was an FBI operative—then what was Laszlo doing responding to a notice that only the hacker could have seen and presumably delivered to the Soviets? After that one brief interview Laszlo disappeared, and the FBI maintained a curious silence on the subject of Laszlo Balogh. Everyone had a theory, but no one would ever know for certain how much of what was said about Laszlo was true and how he found out about SDInet. His links to the hacker who had been in the LBL computers were never fully explained. And the only clue to what he was doing was on Hartmann's telephone bills. On April 21, 1987, the same day Laszlo wrote the SDInet letter, he made several calls to Bonn, West Germany, to a residence near the American embassy.

▲ ▼ ▲

If it hadn't been for Nixdorf, the entire telex incident might not have happened. And to Pengo it seemed that if the telex incident hadn't happened, the entire spying business might have faded away.

In early 1988, the local police in Munich needed to set up a high-speed data link between Munich and federal police headquarters in Wiesbaden for all their telex traffic. It was an unrewarding, time-consuming programming job because of the ad hoc nature of telex communications. The police asked Nixdorf, a large West German computer company, to take it on, but Nixdorf was too busy, so Nixdorf farmed it out to a small software company in Berlin. That company had qualms about working for the police, so it subcontracted the work to NetMBX, Pengo's start-up company, for 40 marks an hour. Suffering from lack of business, little NetMBX was in danger of folding before it had a chance to get off the ground. Clemens and Pengo were only too happy to take on the work. Pengo enjoyed the irony of it all. Not only was he a prominent West German hacker whose name was known to the authorities, but he was one with leftist leanings who had been doing business with the Soviets. And now *his* firm was going to be responsible for setting up communications between two state police headquarters.

Mindful of the problems Pengo might encounter if he were so brash as to go to police headquarters in Munich himself, in late March he dispatched Clemens. Clemens spent two days in Munich working on the link. When he had finished, he asked the police in Munich to send a series of sample telex transmissions, which he then copied onto tape and took with him back to Berlin for use in testing. When he got home, Clemens printed out the telexes and showed them to Pengo. Pengo was highly amused. Among the missives were a death threat allegedly sent by the Red Army Faction terrorists to West Germany's research minister and the travel schedules of two top police officials, complete with the security measures planned for the trips. When Hagbard came to Berlin to visit a few weeks later, Pengo couldn't resist showing him the telexes.

Hagbard wasn't doing quite as well. Since early 1987, he had been in and out of psychiatric hospitals and detoxification centers. He was recovering from a difficult love affair with an American diplomat who had returned home. And he was more convinced than ever of the Illuminati's grip. But now he was turning the world's problems into his own scourge. He believed that the AIDS virus had originated with him as a means for annihilating the Illuminati, that the letters were in fact short for "Anti-Illuminati Destruction System." To make things worse, Hagbard's inheritance had long since run dry, and his drug dependency was acute. So when a new set of journalists from Hamburg approached him while working on a story about computer break-ins, he told them that he was one of the most powerful and talented hackers they would ever encounter, but that his story came at a price.

He told them he could steal his way into most anything. They wanted to see proof, so they paid him 500 marks to come to Hamburg and stay for a week in a good hotel. They even paid for Pengo to come, too. Germany's computer hackers were accustomed to seeing some reward for revealing their secrets. Usually, these were informal arrangements—a filling meal in a nice restaurant, round-trip train fare in exchange for a window into hacks at CERN or Thomson-Brandt. The two journalists told Hagbard and Pengo that if they could publish the story in a German magazine such as Quick, substantially more money would be forthcoming.

While they had Hagbard there, the two journalists sat him down with his back to the camera and interviewed him about what it was like to be a hacker. They also had him compose a hacking manifesto of sorts, complete with his theories on Hagbard Celine and the Illuminati. In his seven pages of rambling prose, Hagbard disclosed that the National Security Agency was operating a secret department called OSAD, the Offensive Software Applications Department. There, Hagbard said, NSA was preparing for a war of the future, a computer war fought with "soft bombs," computer viruses. Hackers, therefore, represented the pivot of the world's fate. "Yes," Hagbard concluded, "the computer war, our soft war, has begun."

But when the journalists put Hagbard in front of a computer, he had trouble backing up his extravagant claims. He tried and failed to get into Bolt, Beranek & Newman, a research center and think tank near Boston, and again into the Jet Propulsion Laboratory. Hagbard was simply going to have to produce something more convincing.

So the telexes came in handy. When Pengo showed him the telexes that Clemens had collected, Hagbard seized the opportunity. He persuaded Pengo to let him have some of them and presented them to the two journalists as a product of hacking, claiming that he had gotten them from Pengo, who had broken into the police computers in Munich.

By now it was July 1988, three months since the "Matthias Speer" story. But rather than diminishing with time, the story of Markus Hess and the system manager in Berkeley was still making headlines in West Germany. The publicity was beginning to frighten Pengo. It wasn't so much the *Quick* story itself that worried him as the mention of the Laszlo Balogh letter. Pengo knew nothing of SDInet or Laszlo Balogh, but it made Pengo queasy. Everyone in the country was talking about espionage. Pengo had a brief conversation with Hagbard about it, and he asked Hagbard how much he thought the authorities knew about their role in the entire affair. But it was useless. Hagbard was crazier than ever. He was beginning to talk to too many journalists about too many things. Pengo had a brief conversation with Dob to discuss Hagbard's loosening lips. Dob assured him that no one believed anything Hagbard said anymore.

But Ammann and Lehnhardt, the two television reporters who had broken the Chaos NASA-hack story, had heard about Hagbard's telex story and were on their way to Berlin to question Pengo about it. Ammann was also curious about Balogh. He was convinced that something was going on involving the East. And when the head of "Panorama" suggested to Ammann that he look into this story about police telexes, Ammann agreed immediately. The perplexing telex story was reason enough to go to Berlin for the day and talk to Pengo.

Pengo didn't show up for the interview. When the journalists called him, he apologized and said he had forgotten about it, so they made a date for later that night. Ammann, Lehnhardt and a third reporter, Gerd Meissner, a gifted young free-lancer, took Pengo to a crowded and smoky café, a student hangout near West Berlin's technical university. Ammann was struck by the streetwise ways of Pengo, who did not conform at all to the picture of hackers Ammann had developed back in Hamburg. Instead of a social maladroit, Ammann was greeted by an amiable and appealing nineteen-year-old. After ordering one of the most expensive dishes on the menu and falling on his food like a starving man, Pengo set the record straight about the telex incident. Fine, the journalists said, but now they wanted to know about Markus Hess, a mystery in that his name had never before surfaced in connection with West Germany's hacking scene.

"What about the Hess story?" Ammann asked. "How well do you know Hess?"

"I don't know him at all," Pengo responded, fairly shouting to raise his voice above the surrounding din.

"Well, what about hacking and espionage?" Ammann asked.

"I don't know anything about hacking and espionage," Pengo insisted.

Ammann wouldn't give up. "What about those keywords, NORAD and SDI?"

Pengo shrugged.

Ammann changed his tack. "What about the letter from Balogh?"

By this time, Pengo had consumed three tall beers and he was beginning to loosen up. The question about the Balogh letter sent him over restraint's edge.

"Okay," he said calmly, as if he had hit upon an afterthought, "it's true that some software has been passed to the Soviet Union. But it didn't amount to much." He told them that a contact had been made in East Berlin with a Soviet agent named Sergei. Lehnhardt said nothing but Meissner nearly slipped out of his chair. Ammann's first reaction was not to believe it, and he asked Pengo to back up his claim. So, maintaining his cool, speaking as if he might be noting a change in the weather, Pengo told them the story of Helmstedt.

Helmstedt, a quaint border town in West Germany, was the police checkpoint to the East German highway leading to Berlin. It was where Westerners traveling to Berlin by car were stopped and their papers scrutinized by a lineup of stiff young border guards. As Pengo had heard the story from Peter Carl, when Carl wanted to make his initial contact to agents in the summer of 1986 he slipped a small piece of paper inside his passport at Helmstedt on which he had written something in code expressing his desire to talk to someone about technology transfer. When he got his passport back, more than an hour later, he found a second note inside containing a telephone number, the number of a contact at the KGB.

The Helmstedt story, of course, had been another of Carl's grand illusions, but Pengo believed it. So did the journalists. It was classic, the journalists decided, and it must be true. What, Ammann asked the surprisingly calm young hacker, did he plan to do now? Pengo looked as if that were the first time he had considered the question. Well, he said, he had been out of the action for more than a year now, and he was hoping the whole matter would simply die on its own.

This was one story Ammann didn't want to lose. But it also presented a tricky situation: he could hardly base a potentially explosive story on the word of one nineteen-year-old, but if he started sniffing around, he could endanger Pengo. Moreover, it hadn't taken any grand effort to pry the story loose from Pengo. Given the fact that he had forgotten about the interview at first, Pengo probably hadn't intended to bare his soul. There was no telling, given a few beers, whom else he might entrust with this information. And as a free-lancer, Ammann didn't have the resources to post Pengo to Elba or another suitably isolated island until he could confirm the story.

On the one hand, Ammann was surprised that something like this hadn't happened earlier. He and other journalists who followed the hacker scene had often discussed the likelihood that hackers would become spies. But Ammann saw that the confluence of friends, circumstances and drugs had created a perfect setting for what Pengo was describing. Haghard seemed to be dependent enough on drugs to look for any route to easy money, and Pengo struck Ammann as a quirky mix of naïveté and shrewdness. Aside from seeing Pengo each year at the Chaos congress, and observing his steady rise to the top of the hacking scene, Ammann didn't know much about the smooth and accomplished hacker from West Berlin. Mostly he knew him as a bright, opinionated young man, something of a would-be anarchist, somewhat distanced from the rest of the Chaos group. Now he seemed at once oblivious to the consequences of what he had done and hardened in a way that Ammann hardly expected from a teenager. It didn't surprise Ammann to learn that this youth had been traveling to East Berlin to deliver the fruits of his electronic marauding in America and Western Europe. If he was telling the journalists this much, it was highly likely that there was a lot more to the story.

The three journalists told Pengo he should think carefully about what to do, but that just letting the situation sit was probably not the best idea. By the time they left the café, Pengo was trembling visibly.

A few days later, Ammann and Lehnhardt returned to Berlin, and by then Pengo was truly anxious. Pengo told the journalists that he wanted their help in getting out of the entire mess. One way Ammann knew to assure an exclusive story was to gain Pengo's trust, even to go so far as to put him in touch with the right people. Ammann called Ulrich Sieber, a law professor, attorney and computer crime authority in Bayreuth, a small university town in southern Germany best known for its annual Wagner festival. The thirty-seven-year-old Sieber hardly seemed the most natural choice to defend a hacker spy from Berlin. He was well respected in the Federal Republic as an attorney who represented corporations deprived of their rightful profits at the hands of software pirates. But Ammann also knew him to be fair and, in his own way, open-minded. When Ammann called Sieber, the journalist was circumspect in his description of the case. He explained simply that there was a hacker who had a problem and needed advice. With no specifics in hand, Sieber agreed to devote a Saturday to discussing the case in person with Ammann and the unnamed hacker.

Gerd Meissner was put in charge of chauffeuring Pengo from Berlin to Bayreuth, a six-hour drive. He had to pick Pengo up at his father's house at 5:00 A.M. in order to get to Bayreuth in time. When they arrived at Sieber's university office, Ammann was already there. A relentlessly cheerful and proper man, Sieber was friendly and gracious to the threesome, and he made clear his desire to help with whatever the problem could be.

Pengo appeared calm, perhaps even a bit uninterested in the whole affair. "Yeah, there are some problems," he told Sieber as he sat down, and he gestured to his two companions. "These guys say I should talk to you about it." Sieber asked Pengo to tell him exactly what had happened. Pengo was reluctant at first, even slightly distrustful of Sieber, whom he knew to be no proponent of computer hacking. But Sieber assured him that his story would be safe with him. So, rolling cigarette after cigarette, Pengo came forth with the story. He told of Carl's contacts with Sergei, of his own visit there and of the various pieces of software that Carl had delivered to East Berlin. He described his own deliveries: the security program, the assembler from Thomson-Brandt and the session logs. And he described what he thought the others had sent over. He told of American military sites, of source code and of logins to sensitive computers throughout the world.

When Pengo was finished, Sieber sat back in his chair. The two journalists spoke first. They wanted to know what the options were. They asked what Pengo's rights were and what loopholes, if any, Sieber could find for him in Germany's computer crime statute. Pengo inquired about the possibility of trying to forget about the whole thing, and disposing of evidence.

Expert though he was, Sieber wasn't quite sure where to start. He was certainly riveted by the tale. As a youth, he had taken his English studies partly by way of Ian Fleming's James Bond stories, and in at least one small aspect of his work he had managed to find a connection between the world of international intrigue and the relatively obscure realm of computer security. In his otherwise dry writings on the topic of computer crime, Sieber had warned a decade earlier that hackers could easily fall into the hands of the KGB. Few people believed Sieber at the time, and some of his most ardent detractors had dismissed him as a publicity seeker. So for Sieber it was a clear vindication of his theories to learn of such a case.

First, he issued a polite but firm warning that if Pengo had any intention of destroying evidence, he had come to the wrong lawyer. Then, thinking out loud, he expressed his first reaction to Pengo's predicament. It was clearly not so much a hacking case as one of espionage. It was patently criminal, and a serious offense at that. As for Pengo's idea to forget about it and let the whole thing blow over, Sieber said that was one option to consider. But as a pragmatic lawyer he advised against it, explaining that it could hang over Pengo's head for years and could someday be used by the Soviets to blackmail him, even ten or twenty years hence, when Pengo had a family and a job to protect. Another problem was that some of Pengo's friends might already have informed others of what they had done, thus placing Pengo in a great deal of danger. Pengo said he thought it was entirely possible that Hagbard had already said more than he should, and that worried him.

Sieber's second suggestion to Pengo was that he inform the authorities himself. As it turned out, West German law contained an amnesty provision for espionage. If a West German citizen-turned-spy gave himself up to the state before the crime was discovered, and if turning himself in prevented further damage to the Federal Republic, he could receive full immunity from prosecution. Sieber was familiar with the provision because he had tried and failed to convince the German parliament to adopt a similar provision for hackers when it adopted its hacking law in 1986. No sooner had Sieber mentioned it than Pengo latched onto the idea. It seemed like a perfect escape hatch. It had been a long time since he had called Dob and Carl his friends. Hess he hardly knew, and Hagbard seemed too confused and unpredictable to protect in any case. So he had no qualms about betraying them. Sieber warned that if Pengo did turn himself in, amnesty wasn't necessarily guaranteed.

There would always be the risk that the authorities would choose to prosecute him anyway.

Sieber told the group he knew of someone from the Bundesamt für Verfassungsschutz, the German domestic secret service. As it happened, Sieber had been at a conference some time earlier where he had met a Verfassungsschutz officer and sometime security consultant who had given the lawyer his card. Sieber asked Pengo if he would like him to try to reach the man at home. Pengo nodded, and Sieber disappeared into the next room to make the call. He reemerged ten minutes later to report that he had reached the officer and, just as Ammann had, explained the problem without mentioning a name. The secret service contact needed some time to think things over and would call back within the hour. So the group settled down to wait for the return call. Ammann found the incongruity of the scene entertaining. Here was one of West Germany's leading experts on computer crime, in his conservative blue suit, seated across a desk from one of West Germany's most prominent hackers, clad in his Berlin black sweatshirt and black jeans, nervously rolling cigarettes. And Sieber seemed more than willing to take on this unpredictable young man's case. It was in some ways, he decided, a perfect hack.

Only once in Sieber's questioning of Pengo did incredulity register on the lawyer's face. "Did you have any scruples about what you were doing and did you consider whether it was unethical?" Sieber asked Pengo.

"I don't care about ethics," came Pengo's flat response. "If it's Russian interests or Western interests, I don't care about that stuff."

After a pause, Sieber suggested that such sentiments would hardly constitute a solid defense, and he recommended that Pengo not stress that line of thinking in his talks with the authorities.

An hour later, Sieber's telephone rang and the secret service contact said he was willing, just as Sieber had been, to devote a day of his weekend to hearing about the case without knowing any names. He said he and two colleagues would drive to Bayreuth from Cologne the following day, Sunday. Ammann and Meissner drove home, and Sieber suggested he get Pengo a room in a nearby hotel.

Hartmut Pohl, one of the Verfassungsschutz's computer security specialists, showed up in Bayreuth early the next day with two colleagues. One of Pohl's companions was a senior officer and more of a generalist. The other, an old hand at intelligence work, was to become Pengo's regular contact. Sieber met with them privately first and, still withholding his client's name, attempted to strike a deal with them. Finally, it was agreed that if Sieber's client agreed not to dissemble in the slightest, then his chances for amnesty would be excellent.

And so the interrogations began. The first session that Sunday lasted four hours. Sieber wasn't one to underestimate the reach of the KGB and the East German secret service when they wanted someone out of their way. Concerned for his client's safety, when the interrogation was over Sieber insisted that Pengo not travel by train back to West Berlin, a trip that would take him across East Germany, and that he fly instead.

▲ ▼ ▲

Pengo sensed he was in for a difficult time. Now that he had exposed the story, he knew things were somehow beyond his control. At the same time, in finally coming out with it, he had unburdened himself and, in a way, he felt better. But this meant that a new burden had landed on his shoulders: the weight of having set things in motion, of having betrayed the others in order to save himself. Yet he wasn't spending much time worrying about that. He was somehow sure that the others had their own plans for getting out of the predicament.

Pengo entered into a form of receivership. The matter was an official case, with a specific, no-nonsense agent from the secret service assigned as Pengo's official contact. And Pengo was expected to comply with any of the contact's demands. If the contact called and told Pengo to get on a plane the next morning for two days of interrogations in Cologne, Pengo was expected to put aside all other plans and do so. And the three television journalists who had pried his confession loose in the first place had taken a proprietary attitude toward the situation. Pengo got the impression that the journalists wanted to have at least one of the trio at his side during his every waking moment, lest another journalist should start sniffing around the story.

But as a practical matter, nothing really changed. Pengo's little company continued to take on sporadic work, and, much to his amusement, even did some consulting for the German commands of the armies of the United States and France in Berlin. He moved out of his father's place and into a small studio apartment of his own in Kreuzberg, a neighborhood known not only for its dense population of Turkish immigrants but also as the focal point of Berlin's counterculture. It was a scene with strong appeal for Pengo; he had always felt that Berlin was the only place to live, and Kreuzberg had become the only place to live within Berlin.

In fact, it seemed that those around Pengo—the German officials, the three journalists and Sieber the lawyer—were exhibiting far more concern over the matter than Pengo himself. As if to thumb his nose at the entire Western Establishment, just two weeks after his confession, to the great consternation of those in the government working on the case, Pengo accepted an official invitation to consult for the Bulgarian Cybernetic Institute in Sofia. Before the trip, Pengo had to visit the institute's branch in East Berlin for a meeting. Remembering Sergei's invitation to come by whenever he liked, Pengo decided to try Carl's passport trick with his own. Hoping to bypass the process of getting the one-day advance permission that was required of every West Berliner, Pengo went straight to passport control at Friedrichstrasse, handed the guard his passport and said he had an appointment. Unimpressed, the guard pushed the passport back at him and told the young West German to get the prior clearance. Either the trick worked only for Peter Carl, Pengo thought, or the Soviets had become alarmed by the attention Hess had attracted and now considered the espionage operation closed.

The trip to Sofia was uneventual enough, at least as far as the work itself was concerned. Pengo donned a "yoopie" outfit of jacket and tie, and spent the week making formal presentations about computer networks to the Bulgarians, who seemed hungry for knowledge of any Western technology.

Several months after Pengo's unexpected trip to the prosecutor's office and his confession, he discovered just how fortunate his timing had been. It turned out that Hagbard had done exactly the same thing at the advice of his own lawyer a few weeks before Pengo did. The handling of the two cases was strikingly parallel. Hagbard's lawyer, too, had invoked a section of the amnesty provision in the espionage laws and had his client turn himself in. It was a close call, to be sure. If Pengo had waited even another week, the authorities might have been ready to open a formal case against him, and might not have offered him the promise of lenient treatment.

A few weeks after Pengo's initial confession, the interrogations in Cologne became more intensive. His trips there usually consisted of an evening flight from Berlin to Cologne, where his contact would pick him up and deposit him at a hotel outside of the city. He was under strict instructions to make no telephone calls from his room. Promptly at 9:00 the next morning, the questioning would begin.

It wasn't so much the idea of being beholden to the authorities that began to fray Pengo's nerves, but rather the psychological pressure his interrogators exerted. They played out a typical good-cop, bad-cop routine with him: one would cajole while another threatened. They refused to believe that he hadn't done it for the money. Pengo could see they didn't understand that he had simply wanted to become the world's greatest hacker. He tried to explain what hacking was all about. He suggested they read *Neuromancer*. Bearing in mind that his chances of getting off were vastly improved if he told the complete truth, Pengo tried to follow the Bayreuth lawyer's suggestion and be as forthcoming as possible, but there were times when his recall was less than perfect. Not only had two years elapsed since Pengo's one and only delivery to Sergei, but apparently the years of hash smoking had taken something of a toll on his memory.

One pivotal subject on which Pengo was less than direct was that of the magnetic tape he had given to Sergei. Since he hadn't had the proper equipment for making such a tape, he had had to ask someone else to make it for him. That would have meant calling a friend and asking him to copy certain files from a computer onto the tape. Pengo had faced the basic technical problem of transferring information out of a computer in a portable form. And for that, he had needed the cooperation of someone with access to the right equipment.

At first, Pengo told his interrogators that he thought he had obtained the tape from Obelix in Hamburg. But as soon as he had suggested Obelix as a possible source of the tape, Pengo reversed himself and denied it. Perhaps he had second thoughts about implicating his friend, who could face harsh penalties if he were found to have collaborated, even unwittingly, with Pengo's spying adventure. Not only could Pengo not recall who had made the tape for him, but, he claimed, he wasn't sure what was on it. Pengo tried to explain that just as he had no tape mounting equipment at his disposal, he also had no tape reading equipment, so once he got the tape, he had no way of seeing what was on it. But the men questioning him were skeptical. Perhaps to see if Pengo's sudden bouts with amnesia might be cured with a little gruff treatment, an agent from the army's Secret Service branch attended one of the interrogations and scowled through the session. When Pengo's memory faltered, the military officer threatened that if Pengo continued with his evasive answers, he would flush the young hacker down the nearest toilet. Pengo was unperturbed, whereupon the Verfassungsschutz officers, more embarrassed than anything else, apologized to Pengo for their crusty military colleague.

Still, the young Berliner was hardly relaxed about his fate, which now lay in the hands of the very authorities he had been trained to mistrust. But he had to laugh at the conspiratorial manner in which his interrogations were occasionally conducted. During one trip to Cologne, he was driven from the airport to a secluded patch of woods where a second vehicle awaited him. Upon arrival, he was shown several photos of strangers and asked to identify them.

Often, after a day of exhaustive questioning the interrogators and their subject repaired to the hotel bar for a more relaxed probe into Pengo's political leanings. He was as frank with them as with anyone else. As Pengo saw it, these men were "black-and-white people." They viewed Communism simply as bad, while Pengo liked to believe he was more open-minded. Perhaps because of his upbringing, or perhaps because of his firsthand experience with the border life in Berlin, he recognized shades of gray in politics. If anything, he told them, he was a leftist. The authorities seemed always ready to listen to what Pengo had to say, if not quite ready to be converted to what they considered to be the young man's wayward political ideals.

The interrogations continued through the summer of 1988, after which the Bundeskriminalamt—the investigative arm of the federal police and the rough equivalent of the American FBI—was on the case. For each new set of questioners, Pengo repeated his story. Time and again, he was asked about how the whole affair was conceived and who made the initial contact with the Soviets. He was asked repeatedly to describe the core group of five that eventually emerged. Pengo said that he had come into the group later than the others, and that after his own delivery to Sergei, the Soviets' lack of feedback had disappointed and discouraged him. His involvement had diminished gradually, and by the time he engaged Sieber, he had been out of the business for more than a year. But he couldn't say what the others were doing. And he refused to comply with what he saw as unreasonable and irrelevant demands. When his Secret Service contact asked him to get them some information about Dob's trips to a Berlin brothel, Pengo refused in one blunt phrase: "I'm not your spy."

In response to repeated demands for more details about the precise workings of the group, Pengo insisted that he knew little about the others, with the exception of Hagbard and Dob. And since their falling out over the Rainbow and Dob's forgotten invoice, he had had very little contact with Dob. He said he knew that Dob owned a gun and could be dangerous. He believed that Carl had been over to East Berlin at least two dozen times, and he knew that Dob had gone at least once as well. He said that he knew Markus Hess only vaguely, and that he had scant knowledge of Hess's breaking into Lawrence Berkeley Laboratory. He said he might have been inside LBL computers himself once or twice, but he insisted that he knew nothing whatever of SDInet or the mysterious Laszlo Balogh from Pittsburgh.

To an outsider Pengo might have maintained that he was perfectly at ease during the interrogation process, and that he was comfortable with being in league with the authorities. In fact, if he were asked to come up with a moral principle that guided his actions, he would have responded that he had been working for one side and now he was working for the other. It was as simple as that; morality was not really the issue.

He continued to work at his little consulting company, which was in the process of merging with another small Berlin start-up. But the effect of the interrogations was beginning to show. Even without Dob as supplier, Pengo's dependence on hash increased. Throughout the autumn he was high more often than not, and he drank liters of coffee every day. He was a classic "stress smoker," as the Germans say, lighting a cigarette, stubbing it out after two or three puffs, then lighting another one. He made dates with people and forgot about them, or simply sloughed friends off. When he bumped into his ex-girlfriend at a concert one night he was skittish and jumpy and muttered something about being in trouble with the secret service.

If Pengo was beginning to have a hard time with life as a snitch, the interrogations with the authorities were taking a deeper toll on Hagbard. Each of his attempts to overcome his dependence on drugs had failed. He was living in a halfway house at Hannover and had no money. His ever-narrowing circle of loyal friends was at a loss for how to help him. When they were with him, he spoke of nothing but the great conspiracy that reigned over their lives. And there was little family support for Hagbard. His stepmother, his sole remaining parent, wanted to have little to do with him. Not only had he squandered his generous inheritance, but she was certain that he was selling some of the valuables he had inherited as well. More than once he had visited her and intimated that he would like to have certain paintings that had belonged to his father. She suspected that he planned to sell those as well, and told him she would no longer allow him inside the apartment if she was alone when he came.

Hagbard had explained his predicament to Johann Schwenn, the Hamburg lawyer to whom he had turned for help. To Schwenn, the calm man with the soft voice didn't seem any more disturbed than other young people in trouble with the law. Schwenn was a prominent civil libertarian with a busy caseload; he had little time to make deep inquiries outside of what his client told him. Though very thin, with skin so pale it was nearly translucent, Hagbard was polite, self-consciously neat and well spoken. And in a sign to Schwenn that this young client had some sense of right and wrong, Hagbard hadn't been eager to betray his friends. Schwenn didn't inquire much into Hagbard's personal situation, and the young man offered no details.

▲ ▼ ▲

In late 1988, the unusual case of the hacker spies landed squarely, if also fairly arbitrarily, in the hands of Ekkehard Kohlhaas, one of a handful of federal prosecutors at the attorney general's office in Karlsruhe who worked espionage cases. The diffident forty-three-year-old prosecutor had recently arrived at his post after several years in the public prosecutor's office in Karlsruhe, followed by a stint at the Ministry of Justice in Bonn, where he worked on hostage cases. He didn't know whether or not to consider himself lucky to have this high-profile case. It was a first for his small department; not only was there no precedent in the Federal Republic for such activity, but there was no precedent anywhere. Moreover, West Germany's attorney general was placing an unusual amount of significance on the case. It smacked of an entirely new brand of espionage. If indeed military secrets had been stolen from U.S. computers and delivered to the Soviets, as well as hundreds of logins and passwords, the case could prove to be explosive.

Kohlhaas, whose spare office betrayed his avoidance of anything so technical as even a typewriter, was rather intimidated by the technology and technical matters surrounding the case. He joked with others that spelling *computer* was challenging enough for him. Suddenly he was faced with the task of prosecuting a ring of young men whose esoteric endeavors—breaking into computers thousands of miles away and stealing information—he had trouble comprehending. But he decided to ignore, for the most part, his technical limitations, and focus on the espionage involved. After all, whether they had downloaded information destined for the KGB from computers in the Far East or photocopied it was immaterial. Espionage was espionage.

The Bundeskriminalamt and the secret service had distinct interests in the case. The last time the secret service and the police had been one and the same was during the Third Reich. To prevent even the possibility of another Gestapo, the two agencies had been completely separated. Not only is communication between the two entities minimal, but they tend to compete with one another. To the secret service, this was an interesting and novel case that it would have liked to keep for itself. The attorney general, who worked with the Bundeskriminalamt, simply wanted to see justice prevail. Unlike the secret service, the prosecutors were not about to promise anything like amnesty to an informant.

It was prosecutor Kohlhaas's assignment to build a case against the three suspects based primarily upon the claims of two somewhat unreliable informants. In fact, the credibility of the two young men was a matter of some concern. The one with the nickname of Hagbard was a drug addict from Hannover who had been in and out of psychiatric hospitals and drug rehabilitation centers. During his full-day interrogations, Hagbard could remain alert for half an hour at most before his head began to fall toward his chest. His hands trembled and occasionally he excused himself to take pills of some kind. Only the medication, coupled with copious quantities of sweets and coffee, helped keep him fully conscious. Moreover, he seemed to have trouble remembering dates and details of any kind. Eventually, if he was given long enough to concentrate, he could reconstruct an important event, or remember the name of a computer he had broken into. At the bottom of the transcript of one interrogation, the young man had signed a disclaimer stating that the medication he was on made it difficult for him to articulate his thoughts clearly.

Hagbard's friend Pengo, who had followed closely on Hagbard's heels to the West German authorities, had his own idiosyncracies. He was clearly very bright, more direct and lucid in his police interrogations, and certainly more in command of his faculties, but his manner was unspeakably arrogant. Here was a young computer hacker, under investigation for a crime against the state, who displayed no regret whatever for his actions. When Pengo was summoned to Karlsruhe to restate his police testimony before a court magistrate, the judge asked the young hacker who would pay for the thousands of marks' worth of connection time that had accrued from all the stolen NUIs. "I don't know," came Pengo's response. "It's not my problem." Kohlhaas suspected that it wasn't so much a moral imperative, or even a vague sense that he had done something wrong, that prompted Pengo to unburden himself to the West German authorities, but a fear that his friend Hagbard had already done so. More irksome still than Pengo's hubris was his curious lapse of memory when it came to the magnetic tape. The prosecutor simply couldn't believe that someone who fancied himself a superior hacker and who wanted to make a favorable first impression on a KGB agent would hand over a magnetic tape filled with data without knowing what the data were. Kohlhaas had little doubt that Pengo was protecting someone, most likely the Hamburg hacker who called himself Obelix.

And there were places where Hagbard's version of events differed markedly from Pengo's. Hagbard insisted that it was not just lists of computers that went to Sergei, but hundreds of computer logins and passwords to highly sensitive military computers as well. Pengo admitted that he had often given logins to Hagbard in the past, but he doubted that the information he gave to Hagbard eventually made its way to Sergei. And while Pengo insisted that his share of the money from Sergei had come to no more than 5,000 marks—or $2,500—at the very most, Hagbard claimed that Pengo had got at least three times that.

In order to bring some outside proof to bear on the claims of the two young informants, Kohlhaas decided to initiate some surveillance on Markus Hess and Peter Carl in Hannover, and Dob Brzezinski in Berlin. He also ordered telephone taps, but very little came from them.

From the start, perhaps to keep Pengo guessing, the West German officials wouldn't let him know what lay in store for him. Pengo's lawyer kept up a steady correspondence with the authorities, urging them to view his client's confession as a demonstration that cooperation between hackers and the state could yield positive results, in this case the containment of a new form of espionage.

The three journalists, in the meantime, were preparing their television broadcast on the hacker-spy case. They crisscrossed West Germany, interviewing everyone from the head of the secret service to members of the Chaos Computer Club. They dispatched a correspondent in the United States to interview Stoll at his Berkeley laboratory. They took their cameras to East Berlin and produced footage of the building that housed Sergei's office. And they even asked Pengo to sit for an interview. The authorities had warned Pengo time and again not to talk with television journalists. For once, common sense prevailed. Delighted though he usually was to talk, this time he turned them down.

By the end of 1988, Peter Carl sensed that he was being watched. During a visit to Sergei's just after the first of the year, Carl passed on detailed reports of the latest security incident that had swept through the U.S. It was the case of a computer virus released by Robert Tappan Morris, a graduate student at Cornell University. Morris had unleashed his program one evening in November of 1988. Within hours it had crashed computers at universities and research institutions throughout the U.S., alerting the nation to the vulnerability of its computer networks. The West German hackers assumed that anything that had been so quick and so devastating would be of considerable interest to the Soviets. In addition to their virus report, Carl gave Sergei a virus report written by Cliff Stoll.

At the end of their usual repast, the KGB officer told Carl that a pause might be in order. For one thing, Sergei said, Mikhail Gorbachev was coming soon to visit East Berlin. With so many players involved, the security of the operation was already low. The last thing the Russians wanted was to have an espionage operation exposed while the Soviet leader was in town. Sergei's second reason for halting their meetings was that he was fairly certain that Western intelligence had its eyes on Carl's movements. But he assured Carl that the operation was just being placed in temporary abeyance.

When Carl reported back to Hess that Sergei had called for a hold on the action, Hess was a little bit relieved. The whole affair had long since lost its attraction for him and the best he could hope for now was to see the entire thing disappear. There had already been a couple of close calls. First there was the initial scare in the summer of 1987, when Bremen police had traced calls to his telephone. Then there was the rush of publicity in April 1988, when the *Quick* story appeared, describing Cliff Stoll's hunt for the hacker inside the Berkeley laboratory. But since then, everything had quieted down. Hess had a new job as a programmer at a publishing firm in Hannover and it was going well. It just seemed that the money he had made from this entire KGB business —about $9,000—was too little to risk continuing.

▲ ▼ ▲

December came, and still no arrests had been made. For nearly six months, Pengo had been living in a state of official limbo. He had few people to talk with about his predicament, so for the most part he learned to live alone with the secret. When it came time to go to the annual Chaos meeting in Hamburg just after Christmas, Pengo welcomed the chance to get out of West Berlin.

The turnout at the 1988 conference was smaller than it had been in previous years. Nervous parents, by now well aware of the hacker culture, kept their teenagers away because of the club's increasingly sullied reputation. Breaking into computers was now against the law, and that was deterrent enough for some would-be hackers. What was more, the Bundespost's new security measures on the Datex-P network had made casual hacking much more difficult. Many of those who still hacked, working on the principle that solitary work was safer, kept their distance from the Chaos crowd. And for those who did attend the meeting, much of the time was taken up with ruminating over politics: a network for exchanging information on environmental protection was planned and the groundwork was laid for a push for a law that would mirror the U.S. Freedom of Information Act. To the hard-core hackers in the group, such talk just seemed to prove that Chaos was losing its way.

But for Pengo, going to the conference was a way to pretend that nothing had happened. In some ways, it seemed like old times. Obelix picked him up at the airport in his mother's Mercedes, and once he got to the conference, Pengo immediately felt better. In spite of the laws against hacking, nearly two years old now, those who counted themselves among the hard core couldn't help but stay up through the night, trying to get into what computers they could.

Pengo was now well known to the crowd. Those who didn't know him certainly knew of him. At twenty, he was already a veteran. Some self-styled hackers had little skill beyond collecting passwords from others. Pengo was famous as an elite hacker, one who could also program. He was famous for being from Berlin. He was famous for having stayed out of trouble. And the fact that when he did get into trouble he outsmarted the police by hiding his hard disk was the stuff of Chaos Club lore. The younger hackers were in awe of him—after all, he had the reputation as one of the first to crack Philips and Thomson-Brandt computers in France.

Throughout the three-day meeting, Pengo was followed around by reporters and television cameras. In fluent English, a language whose American vernacular he picked up on the computer networks and electronic bulletin boards he frequented, Pengo chatted amiably and self-confidently into the viewfinder of a BBC TV journalist doing a documentary on European hackers. When he talked about the Chaos VAXbusters, he said, "We thought we were the best. I don't know if we were." Pengo was a particular favorite among the West German journalists; he was always willing to sit down for a long interview with a local newspaper reporter, chain-smoking his hand-rolled cigarettes as he described some of his best hacks.

Pengo dipped in and out of the individual conference sessions, leaning casually against doorframes, his stance that of someone accustomed to situating himself next to the closest exit. At a session on the future of Chaos, with a note of irritation in his voice, Pengo spoke up. He told those in the room that he was unhappy with where he saw Chaos going. To a strictly technical hacker like Pengo, Chaos's political direction was unacceptable. Concentrating on things like environmental protection, he told them, was diverting the group from its technical origins. It was little wonder, he said, that the truly talented hackers were beginning to abandon the club.

It had nothing to do with the espionage investigations that were under way, but rumors were beginning to circulate among the Hamburg crowd that someone was betraying them, working for the West German government. Wau Holland suspected that there was a plant somewhere inside the Chaos Club's ranks. But no one had a specific reason to suspect anyone of espionage. And Pengo was relieved to see that Wau's suspicion was directed at others. Hagbard wasn't there this year and no one appeared to miss him.

▲ ▼ ▲

Peter Carl had been thinking for some time of moving to Spain and starting a computer company. There was nothing tying him to Hannover; he had been divorced for more than a year and his wife had custody of their ten-year-old daughter. Since leaving his croupier's job in 1986, he had been collecting unemployment. To augment the 880 marks he got each month from the state, Carl occasionally ferried cars to Spain for a local used-car dealer. And then there was the money from Sergei. But that business relationship had gradually diminished, and there was no telling if, now that Sergei was calling for a break, it would ever start again. Certainly it was nothing to count on for a steady income.

So in early 1989 Peter Carl began to give serious thought to moving to Madrid. He already spoke reasonably fluent Spanish, and Spain had always had a certain pull for him. So he traveled to Madrid, opened a bank account and looked into renting an apartment. In February 1989, Carl went to Berlin to discuss his prospective new business with Dob and Hess. He told Hess that he had money to invest in the new company and he wanted to discuss technical details with his two friends. Hess had come along to Berlin and was happy to offer advice, but he had no intention of uprooting himself from Hannover. He was happily ensconced in his new job, making a comfortable salary of 2,800 marks—or $1,400—a month. He was glad to serve as adviser to the other two and nothing more. Hess's days of risk-taking were behind him. For Peter Carl, one of the major attractions of a fresh start in another country was the idea of putting the spying affair behind him once and for all. It was certainly one way to ensure that he would never be caught. On the other hand, as time passed it looked more and more as if they were going to get away with the whole thing anyway.

▲ ▼ ▲

Dob was asleep on the floor mattress in his tiny one-room apartment in Berlin when he was awakened by a loud snap, the sound of someone breaking down his door. He sprang up in his bed to see four pistols directed at his head. Behind each gun was an officer in civilian clothing. "What is going on?" Dob demanded.

"Surrender your weapon!" they ordered back, although there was no gun in sight. He was arrested at once.

Markus Hess had just settled into a comfortable exercise regimen. Five mornings a week, he arose at 6:45 to go for a swim before work. On March 2, 1989, he was just returning, refreshed and cheerful, his hand in his pocket reaching for his keys to open the front door to his apartment, when he heard a quiet, polite voice at his back, the tone of someone who might be asking for directions. "Herr Hess?" He turned around to see eight well-dressed men staring at him. He knew immediately what was happening, and although he wasn't surprised when they informed him at once that he was under suspicion of espionage, Hess began to pray.

Upstairs in Hess's apartment were more officers. Hess insisted on calling his lawyer's office. When police had escorted Hess to his apartment in June 1987, his lawyer had been on the scene within a few minutes, and the lawyer's acumen had emboldened Hess to treat the situation with a cavalier attitude. But this time, the police were frighteningly well organized. When Hess called his lawyer's office, there was no answer. On his second attempt he reached a secretary, who told him his lawyer would be out for most of the day. Hess had no choice but to let the search begin. After a thorough scouring of Hess's apartment, the officers took him for an interrogation at police headquarters in Hannover, where the authorities had rented out an entire floor of the building for the handling of the hacker-spy case. It was Hess's first introduction to Ekkehard Kohlhaas, the prosecutor in charge of the case.

Markus Hess's first impulse was to admit to hacking, and in particular to breaking into the computers at the Berkeley laboratory. He admitted that he knew Hagbard and the others and that he had given Hagbard three or four logins for the purpose of hacking. He even admitted to providing software to Dob and receiving money for it from Carl, and to giving software later to Carl directly, then getting money for it. But he denied any espionage activity and any knowledge of it.

"I've never consciously worked for a hostile secret service," he declared in a carefully crafted phrase. "Carl never said for what or for whom he needed the software and I never asked."

"Are you familiar with someone by the name of Sergei?"

"That name is unknown to me," Hess lied.

"Sergei is believed to be from the trade mission in East Berlin. Sergei is believed to be the one who ordered this software."

"I have absolutely no knowledge of this," Hess insisted.

But as the questioning wore on, it became clear to Hess that someone in the group had been talking. The next question unhinged Hess completely. "According to information we have from the federal police, Sergei was mentioned openly in conversations among Brzezinski [Dob], Koch [Hagbard], Carl, Hübner [Pengo] and you, wherein the logins you procured were meant to go to Sergei. What do you say to that?"

There was no point in continuing to deny things. Everyone's name was out. "That's right," Hess replied. "Those people did speak of Sergei. And it was clear to me that Sergei was in the East and he must belong to the Russian secret service."

Hess went on to explain that after he had heard the others talk of logins going to Sergei, he immediately went back to LBL and altered the passwords into the Berkeley computer.

When his interrogators read him several lists—supplied by Hagbard —of dozens of systems for which Hess had allegedly supplied logins, including a computer at the Anniston Army Depot, the Optimis data base at the Pentagon and several computers of the Environmental Protection Agency, Hess denied it. In fact, from what Hess could tell, Hagbard had been feeding the authorities a damning array of confused stories. He was telling of logins to computers that dated back to late 1985, when Hess and Hagbard had hacked together. But these logins, Hess tried to explain, had nothing to do with Sergei.

The story that Hess told in those first few hours was the one he would stick with for the next year, even as the prosecutors built their own compelling case against him. By the time the ten-hour interrogation session had ended, Hess had confessed to delivering at least half a dozen programs to Carl with the full knowledge that the software was headed for the East. Mostly, he said, it was software he more or less chanced upon at Focus. But he insisted that he had never given computer logins

and passwords to Hagbard to give to Peter Carl, and that not once had
he given information from computers to Carl to sell to Sergei. Hagbard,
he claimed, had psychological problems, and that fact should be taken
into account when weighing the credibility of anything Hagbard said.

Unlike the search of the Focus offices in 1987, this time Hess sensed
he would remain in custody. He spent the night in jail. The following
morning, he was flown by helicopter to Karlsruhe for a deposition before
a magistrate. It was an uncomfortable and awkward ride, as Hess and
Kohlhaas were forced to sit side by side on a narrow seat. During the
thirty-minute trip, Kohlhaas sat silently and read through his papers.
The two men didn't exchange a word.

Because Markus Hess, a young man with a clean record and stable
job, presented no flight risk, he was freed later that afternoon. His first
call was to his father, who drove to Karlsruhe to pick him up. Hess spent
the weekend with his parents, an hour's drive from Karlsruhe in Fulda.
He had a lot of explaining to do, mostly as to why an otherwise perfectly
well-adjusted son, the very picture of decency and conservative values,
had engaged in such anomalous and unseemly behavior, placing his
bright future in jeopardy, and all for the sake of 15,000 marks. Hess had
always had a close and warm relationship with his parents, but he had
never discussed his computer break-ins with them, not only because they
were illegal, but also because his parents simply would not be able to
understand his general obsession with computers. In the end, his parents
made it clear that they would stand by him.

Peter Carl had been arrested the previous morning as he was pulling
out of a parking space near his Hannover apartment. He was just starting
out on one of his car deliveries to Spain and there was no telling if this
was to be his permanent move there, so the police acted swiftly. Two
unmarked cars blocked Carl as he began to drive away. One officer leapt
out of his car and into Carl's. The arrest took approximately thirty
seconds.

For both Dob and Carl it became apparent after an hour or so of being
questioned that Pengo and Hagbard had gone to the authorities. Eventually, both of them confessed to espionage. But they weren't to be
accorded the same leniency that Markus Hess got. Dob and Peter Carl
had previous arrests on their records, and both were considered flight
risks—Carl for his plans to go to Spain and Dob for his avoidance of
military duty. Both were taken into custody. Carl's ex-wife came forward
to say she would take out a loan for 1,000 marks for bail, but prosecutor
Kohlhaas felt uneasy about setting Carl free and urged the judge to deny

bail. Kohlhaas saw his case strengthen when, during the search of Carl's apartment, a Casio pocket calculator was found. It contained the telephone number for one Sergei Markov.

▲ ▼ ▲

It wasn't just three isolated arrests that took place that day. The Bundeskriminalamt was a model of preparedness. That morning, units of the state security branch of the Bundeskriminalamt were dispatched throughout cities in the Federal Republic. A total of fourteen house searches were conducted. Hackers and friends of hackers, some from the old Hamburg crowd and some from the Hannover hacking group, were rousted from their beds and taken to police headquarters in their respective cities and questioned about their knowledge of the espionage ring. It didn't take long for the news to circulate that everything had been set in motion. That morning, Obelix, who had also been taken in for questioning in Hamburg, had called the journalists to say that arrests were under way. The journalists in turn started the process of substituting the special half-hour "Panorama" segment they had prepared weeks in advance for the evening's regular programming. After the regular evening news, the North German television station broadcast a teaser for the evening program, hailing the upcoming show as a scoop not to be missed.

When the Panorama show aired at 11:00 P.M., three million West German television viewers were tuned in. The show opened with a dramatic shot of the Glienicker Bridge in Berlin, a classic point for exchanging spies from both East and West. It framed the hackers' images in silhouette and, switching to a world map thick with arrows traveling from Silicon Valley to Moscow, described this as the first major case in a new era of high-tech espionage. Cliff Stoll recounted the troubling keywords the LBL hacker had sought. He stared into the computer, concern etched on his face. "Somebody was inside my computer looking for Star Wars information."

▲ ▼ ▲

Pengo had suspected something might be afoot. Earlier in the week, his secret service contact had planned to fly from Cologne to Berlin to meet with him, but then had canceled the appointment at the last minute. Pengo had moved out of his father's apartment a few months earlier without filing the necessary registration papers with the local police to account for his whereabouts, a bureaucratic procedure required of every citizen. Sensing that arrests might be imminent, and worried the police might show up at Gottfried's apartment first, he warned his father that there might soon be a reprise of the December 1986 bust. Pengo's parents were only vaguely aware of the trouble their son had fallen into—and was attempting to escape from. They knew that he had an investigation pending against him for something rather serious, and related to his hacking, but they both shied away from asking specific questions. Renate had long since given up trying to understand what her son was doing with computers and had no way of gauging what the problem really was. Gottfried was distressed to hear that the police might be back.

"What are they going to do and when are they going to show up?" he quizzed his son.

"I don't know," Pengo replied with the fresh impatience that only an exasperated son can express to his father. "Excuse me, but I don't know. I don't know how many will come or when they'll come or whether they'll come and order a strip search or just say, 'Have a nice day.' " If nothing else, the conversation prompted Pengo to file his change of address with the police the following morning.

Three days later, at about 9:00 in the morning, he was awakened by the sound of his apartment buzzer. It was the West Berlin police, who were there to conduct a search and question him. They confiscated miscellaneous pieces of paper, some containing computer printouts, others with Hagbard's handwriting, but nothing terribly incriminating. It was a pro forma exercise, of course, and it was clear from the moment they began that Pengo would be spared a real arrest. The police also conducted a search at the offices of Pengo's company, which had grown to five people. Pengo was surprised by the extent of the police action. Still, he was tremendously relieved to have the whole episode burst open at last. As far as he could tell, his name hadn't surfaced. And the "Panorama" program had shielded him too, referring to him as Frieder Sell, a Berlin student.

During the broadcast, Joachim Wagner, the head of the "Panorama" program and the show's anchor, had described the discovery of the hacker spies as the biggest case since 1974, when Günter Guillaume, then Chancellor Willy Brandt's friend and aide, was found to be a captain of the East German secret service. The Guillaume scandal had toppled a chancellory. Gerhard Boeden, the head of the Verfassungs-schutz itself, declared: "What we have here is a new form of hostile infiltration of our data networks." And for the work his agency did in uncovering the operation and controlling the damage, a spokesman for the secretary of the interior called the arrests a "severe blow to the KGB."

The "Panorama" program set off a chain reaction of media reports. Espionage was something the German people didn't take lightly. And West Germans were particularly appalled by the reports of a new and insidious form of espionage, of teenage malcontents who had stumbled upon computer vulnerabilities and not just exploited the weaknesses they found but used them to endanger the military security of the West. Various officials and computer security experts stressed repeatedly that some of NATO's most sensitive military data resided on the computers allegedly penetrated by Hess and the others. Even if the data that had been taken from the computers were not classified, or even sensitive, security experts in West Germany and the U.S. argued that classified information could be inferred from unclassified information.

But it didn't take long for the ministry of the interior to retreat from its original pronouncement that the KGB had been delivered a severe blow. Once the initial sensation had died down, skeptical journalists began to question just how much damage to national security the group of hacker spies had done. Had highly classified material changed hands? Or was it merely some harmless public-domain software? Moreover, why the public outrage when the real menace of computer espionage is most likely not from the Soviet Union but in sophisticated schemes perpetrated by both foreign and American corporate spies secretly tapping into their competitors' systems in search of trade secrets and market information?

▲ ▼ ▲

Pengo didn't enjoy his anonymity for very long. Within a few days, rumors about who was involved in the espionage ring began to circulate furiously through the press and in various hacking circles.

Wau Holland, Chaos's founder and keeper of the ethical flame, was beside himself with anger. If it was true that Pengo was involved, it was an unforgivable betrayal of hacking in general and Chaos in particular. Wau's first impulse was not to believe such a thing could have happened, and certainly not that Pengo could have done it. He had always considered Pengo somewhat naive, but certainly not capable of doing something so unconscionable. It wasn't so much the selling of software to the Soviets that bothered him, but that the hackers might have been selling software that came from Chaos people. Worse, they had no doubt been informing on Chaos and its members to West German intelligence agencies. In Wau's estimation, both of those transgressions were far worse than what they were being charged with. To see for himself, he called Pengo in Berlin.

"Just say yes or no," Wau said when Pengo answered his telephone. "Is it safe to talk?"

"Yes."

"Is it true what is being said about you?" was Wau's next question.

"Yes."

"That's all I wanted to know," and with that Wau ended the conversation.

It took just a few days for true names to start surfacing to the general public. A few days after the "Panorama" show, *Der Spiegel*, the Federal Republic's largest serious news magazine, identified Hess and Hagbard by their real names. In a second article in the same magazine the following week, everyone's name came out. Then, more gradually, the fact of Pengo's and Hagbard's cooperation with the authorities inched its way out as well.

Pengo was immediately on the defensive. In response to the uproar over his spying, his betrayal of the others and his work with the authorities, he submitted a lengthy posting in English to the *Risks* forum, an international computer network discussion group about potential hazards of computerized technology. *Risks* is read by computer scientists and other technical experts throughout the world. Peter Neumann, an American computer scientist and the editor of *Risks*, was so surprised to see the blunt note in English coming over the forum's electronic transom that he decided to publish it:

```
Date: Fri, 10 Mar 89 18:09:25 MET DST
From: Hans Huebner <pengo@netcs.SMTP>
Subject: Re: News from the KGB/Wily Hackers

I have been an active member of the net
community for about two years now, and I
want to explicitly express that my network
activities have in no way been connected to
any contacts to secret services, be it
western or eastern ones. On the other hand,
it is a fact that when I was younger (I'm 20
years now), there has been a circle of
```

persons which tried to make deals with an
eastern secret service. I have been involved
in this, but I hope that I did the right
thing by giving the german authorities
detailed information about my involvement in
the case in summer '88.

For my person: I define myself as a hacker.
I acquired most of my knowledge by playing
around with computers and operating systems,
and yes, many of these systems were private
property of organisations that didn't even
have the slightest idea that I was using
their machines. I think, hackers--persons
who creatively handle technology and not
just see computing as their job--do a
service for the computing community in
general. It has been pointed out by other
people that most of the 'interesting' modern
computer concepts have been developed or
outlined by people which define themselves
as 'hackers'.

When I started hacking foreign systems, I
was 16 years old. I was just interested in
computers, not in the data which has been
kept on their disks. As I was going to
school at that time, I didn't even have the
money to buy an own computer. I enjoyed the
lax security of the systems I had access to
by using X.25 networks.

You might point out that I should have been
patient and wait until I could go to the
university and use their machines. Some of
you might understand that waiting was just
not the thing I was keen on in those days.
Computing had become an addiction for me,
and thus I kept hacking. I hope this clears
the question 'why'. It was definitely NOT to
get the russians any advantage over the USA,

```
nor to become rich and get a flight to the
Bahamas as soon as possible.

For punishment: I already lost my current
job, since through the publications of my
name in the SPIEGEL magazine and in RISKS,
our business partners are getting anxious
about me being involved in this case.
Several projects I was about to realise in
the near future have been cancelled, which
forces me to start again at the beginning in
some way.
                                  --Hans Huebner
```

It was a fairly bold statement to send out to nearly thirty thousand computer scientists, students and researchers—most of them American —who perused the *Risks* forum every day. Perhaps Pengo underestimated the number of people who would see his posting. They could hardly be expected to muster much sympathy for the self-confessed spy. Those *Risks* regulars who were willing to give Pengo the benefit of the doubt were impressed by his honesty. But the general reaction was one of horror. At best, this was someone who didn't understand what he was getting into, who was naive to the point of stupidity. If someone as unscrupulous and self-serving as this was frequenting the networks, what other dangers might lurk there?

Wau, too, wasn't to be swayed very easily by Pengo's defense of his actions. He saw it as a self-justification that was curiously lacking in any kind of moral reflection or thought as to how Pengo's cloak-and-dagger adventure might affect others. Wau wanted to divorce himself and his organization entirely from this young troublemaker. He instructed people from the West German hacking scene to hang up the phone if Pengo should call. A joke began to circulate that when Hans chose the sobriquet Pengo for himself, it was to prove more fitting than he knew: *Pengo*, that is, suggested not just the heroic penguin of the video game, but one who sprang from ice block to ice block, saving himself just as each was about to sink.

Pengo's parents, at least, were understanding. Renate, in particular, was quick to excuse her elder son's escapades as an understandable search for adventure. Moreover, both parents suspected that just as their son had always managed as a small child to pull his head out of the noose in the nick of time, somehow he would get out of this tight spot as well. But Renate's septuagenarian mother, whose early life had consisted of a series of unpleasant encounters with East Germany's Communist regime, was horrified that her grandson could be doing business with such people.

Hagbard came under some sharp criticism as well. But rather than go public with his own defense, he kept to himself and quietly entertained journalists' overtures. His price for an exclusive interview: 30,000 marks, or $15,000.

It looked as if Hagbard was beginning to straighten out his life. Before the news of the KGB activity broke, a friend had helped him get a job as a messenger in the Hannover office of the conservative Christian Democratic Union. It was menial, low-paying work, but by all accounts he was reliable and well liked by others in the office. Even after his spying came to light, along with his dependence on drugs, the CDU office kept him on, convinced that he deserved another chance. Some of Hagbard's friends viewed the CDU job as further proof that this erstwhile social democrat's political perspective had gone completely awry. Others saw the job as Hagbard's first small step toward folding himself back into society. His life, at least to outsiders, seemed more stable. After years of rootlessness, he was finally planning to move into an apartment of his own. And a recent embrace of conventional religion had probably added to his calm bearing. His fellow workers would hardly have supposed that this quiet young man had been in and out of psychiatric hospitals, battling his drug addiction.

In reality, Hagbard's life was still a mess. Contrary to appearances, he was still struggling to free himself from drugs. His rent in the new apartment was being paid by the West German authorities, engendering a strange dependence. And being in service to the West German authorities was probably more than someone so invested in paranoid fantasies could withstand. Hagbard was increasingly plagued by the notion that he was under constant surveillance by one or another real or imagined organization. And he was convinced that the officials were reading his mind and using his thoughts to their own end. Perhaps just as difficult for Hagbard was the realization that his commercial value was a direct function of his newsworthiness, which was beginning to wane. Within a month or two of the arrests of the others and the disclosure of the full extent of Hagbard's and Pengo's involvement, the story had lost its sheen and journalists had all but forgotten the young hacker with the unusual nickname.

On the morning of May 23, Hagbard set off in the CDU's Volkswagen station wagon to make a delivery to a government office in Hannover, and he didn't return. After lunch, friends went to look for him, and by 4:00 that afternoon the Verfassungsschutz had dispatched a search party as well. After a week, the friends abandoned their search.

Hagbard was missing for nine days.

▲ ▼ ▲

The first time farmer Ernst Borsum saw the VW Passat parked in an isolated patch of woods outside Ohof, a small village of six hundred people north of Hannover, he thought it belonged to a jogger. But for a few days in a row, each time the farmer went out to irrigate his fields he saw that the car was still there, leaves gathering on its hood. He called the police. Not far from the car they discovered a fully charred corpse. Next to the body were the remains of a gasoline can. All vegetation in the surrounding three or four meters was black. The driver, the police concluded, had taken the gasoline can, poured its contents over himself, soaked the surrounding earth as well and lit a match. He was consumed in flames immediately. Even if he had screamed out in pain, no one would have heard him. When he was found, he was lying facedown, one arm under his stomach, another over his head. That was a sign that he might have changed his mind at the last minute and dropped into a roll to try to put out the flames. Or, if someone had done this to him, he might have been trying to save himself. He was wearing no shoes.

It was, of course, Hagbard.

The news of Hagbard's death brought the case of the hacker spies once again into public view. West German magazines and newspapers carried lengthy, speculative articles about the death. Just how sensitive was the material that landed in the hands of the KGB? And just who might have had an interest in seeing Hagbard dead? Could such a hideously painful death really have been the result of suicide? Unlike a drug overdose or suicide by gunshot, death by burning usually leaves the cause more ambiguous. One complicated theory began to make the rounds that Hagbard was murdered by left-wing terrorists who had commissioned him to hack his way into state computers to retrieve information and wanted to kill him before he could talk.

Hagbard's attorney, Johann Schwenn, regretted that he hadn't done more to help someone whose cries for help he had somehow failed to recognize. Schwenn didn't subscribe to the murder theories. But he was quick to lash out at the authorities, whose only concern was to build their case, not to care for the troubled young man. Others began to conclude that Hagbard *had* been murdered, or at least pushed to his suicide by pressure from journalists and from the West German authorities, all of whom wanted something from him and pursued it without regard for his condition. If Hagbard had been driven to suicide by his own paranoia, speculated one friend, he might have chosen self-immolation because he believed it would prevent his thoughts from being misused further. The fact that he died on the twenty-third day of the month was also no coincidence: In *The Illuminatus! Trilogy*, twenty-three is a number of high significance. "All the great anarchists died on the 23rd day of some month or other," explains one of the *Illuminatus!* characters to another.

▲ ▼ ▲

A group of Hagbard's friends placed a notice in the *Tageszeitung*.

> We are angry and sad about the death of our friend.
> We are certain that he would still be alive if the police and media hadn't, through their criminalization and unscrupulous sensationalism, driven him to his death.

Pengo was shaken by the news of Hagbard's death. It forced him to confront a horrifying reality. More than his first confession to the journalists in the noisy Berlin café, more than the interrogations by the West German authorities, and more than the bursting open of the story with the arrests three months earlier, Hagbard's death forced him to acknowledge the serious consequences of the case. Pengo knew, of course, that Hagbard was so unstable that his involvement in espionage probably had little to do with his suicide. Nonetheless, the pretense that none of this actually hurt anyone was now so obviously false. Moreover, Pengo sensed that this could put more pressure on him as the sole witness in the case. Turning state's evidence in July had not yet given him full protection because his case in the espionage incident had been separated from the others'. It could still result in a trial against him.

Kohlhaas, the prosecutor in Karlsruhe, had different cause for concern about what effect Hagbard's death might have on the case. Hagbard's statements had been sufficiently tenuous, with enough assertions contradicting the others, to make Kohlhaas wonder if he could use Hagbard's testimony at all. Kohlhaas had little choice but to base his indictment on what Pengo and Hagbard—and the three defendants themselves—said. The gathering of matching testimony was of utmost significance.

The prosecution had to establish the connection between Markus Hess and the laboratory in Berkeley. Kohlhaas's most promising witness was Cliff Stoll. In fact, if it hadn't been for the efforts of the woolly scientist, there would have been just the barest of threads holding the case together. Sergei's telephone number had been found in encoded form inside Carl's small pocket computer, but it wasn't enough on which to build an espionage case. Stoll's most valuable contribution had been his girlfriend's invention of the SDInet material, which had not only cinched the telephone trace but also caused the appearance of the Laszlo Balogh letter. In fact, Kohlhaas viewed the Balogh letter as the most solid piece of evidence his office had to establish a link between Hess and the KGB. Someone in Moscow must have dispatched Balogh to seek the SDInet information, he reasoned. And since Stoll and the LBL hacker were the only two who could see the SDInet file, Hess must have taken the information and passed it to Carl, who in turn gave it to Sergei Markov. As far as Kohlhaas was concerned, there was simply no other explanation.

Stoll was a witness of special importance because Kohlhaas had no physical evidence of Hess's hacking. When Hess's apartment was searched on the day of his arrest, the police found no printouts that implicated Hess as having been inside American computers in general and LBL in particular. In fact, the police found none but the most remote signs that pointed to Hess's hacking. It was the Bundespost's work in completing the trace to Hess's apartment that proved Hess was in the LBL computers.

In June, the West German federal police summoned Cliff Stoll to Meckenheim for two full days of questioning. The police investigators were polite and deferential toward their American guest during the painstaking process of translating Cliff's account first into German that a judge unschooled in technical terms might understand, then into official police German. For the most part, it was a matter of matching Stoll's sitings of the hacker with the dates and times of the Bundespost's telephone traces. They were aided in their efforts by a meter-high stack of printouts from the dozens of hacking sessions Stoll had monitored. Stoll was then taken to Karlsruhe to meet the prosecutor and repeat what he had said in Meckenheim before a magistrate. Stoll's deposition went smoothly. Kohlhaas was encouraged when, at one point, Stoll said that he saw data flowing over his monitor, then two weeks later saw the hacker break into another computer using complex commands he must have copied down or stored from the earlier session. That indicated to Kohlhaas that the hacker had been storing information.

But there was one crucial point at which the prosecutor misconstrued Cliff Stoll's words. When Stoll was explaining what he saw happening on the computer screen at LBL, Kohlhaas concluded that Stoll had a piece of equipment at hand to prove that the hacker was actually transferring and saving the information he was seeing. "I saw him printing out screens full of military data" was one typical assertion of Stoll's, when what the Berkeley scientist really meant was that he saw screens filled with such data scroll by on his screen at the laboratory, with no way of knowing what was happening to the information on the other end. What Kohlhaas, with his limited computer knowledge, failed to understand was that it was quite possible that the hacker was doing nothing at all with the data he saw. It was in fact likely that the hacker disregarded the bulk of it, that he neither went through the time-consuming task of transferring it from Stoll's computer to his nor stored it on a floppy disk. But with Stoll using such words as *print* and *download* and *copy* interchangeably, Kohlhaas's mistake was understandable.

The prosecutor concluded that he was pursuing a solid case. From what he could tell, the Berkeley computer manager had just provided the prosecution with ample proof not only that Hess was the one hacking into LBL's computers, but that Hess's intrusions into LBL and other U.S. computers played an integral role in the dealings with the KGB. So Kohlhaas decided to call Stoll to the stand as the prosecution's star witness.

But Stoll was no expert in military matters. Kohlhaas's next step was to find a witness, preferably from the U.S. military, to corroborate Stoll's claims that what Hess had taken from LBL and other computers was of a highly sensitive nature. The names of the places whose computers Hess had compromised sounded sensitive enough. All Kohlhaas needed was for someone more qualified than Stoll to peruse the pile of printouts and offer an assessment. He contacted the Verfassungsschutz and asked that someone from that agency seek a U.S. official who would be willing to travel to West Germany and take the stand as an expert.

It turned out to be a far more difficult request to fill than Kohlhaas had imagined. Although the Verfassungsschutz official assured Kohlhaas that he got in touch with the appropriate U.S. officials promptly, to the prosecutor's bewilderment no one from the U.S. stepped forward to lend some expertise to the case. The National Security Agency had made its own assessment of the damage to national security that might have resulted from the West German hackers. It came in the form of a brief memo from Robert Morris, an agency scientist and the father of Robert Tappan Morris, the Cornell student who had released the computer worm throughout the Internet the previous year. Morris's summary: "Looks like the Russians got rooked." But even Morris's opinion wasn't sent to Karlsruhe.

Still, with the confessions of the three defendants themselves, and the promise of provocative testimony from Stoll, by late July Kohlhaas felt confident enough to issue a seventy-three-page written indictment of Carl, Hess and Dob.

Officials at the U.S. Justice Department were quietly mulling their own plans for a possible prosecution. They couldn't charge the hackers with espionage, since the information that was sold didn't appear to be classified. Instead, the government wanted to charge the West Germans with unauthorized access to U.S. computers. Mark Rasch, the same prosecutor who was working on the Robert Morris case, was considering ways to lure the West Germans onto American soil, make an arrest and try them in U.S. courts. With their frequent electronic travels inside U.S. computer systems, Hess and Pengo were the Justice Department's main targets, and Pengo was the only one who came close to ensnarement. With plenty of time on his hands while he waited for his case to be decided, he toyed with the idea of going to the States in the summer of 1989 to visit some friends he had made over the network, but his lawyer pointed out the possibility of an arrest in the United States and he stayed home.

▲ ▼ ▲

By the time the case came to trial, it was a new era in Europe. For nearly thirty years, Berlin had been two cities divided by a wall. Two months before the trial, in one spontaneous burst in November 1989, the border dissolved overnight. Thousands of jubilant Germans clambered atop the wall and danced. That night marked the beginning of German reunification, a process that was to be surprisingly swift and peaceful. Democracy was breaking out all over Eastern Europe, and the toppling of the Berlin Wall was the most dramatic event signaling the end of the Cold War. The long-standing restrictions on the export of technology were diminishing quickly.

But the prosecution didn't intend to allow world events to influence its case against the three hackers. As Kohlhaas saw it, if the defendants had been working for Stasi, the East German secret service, a trial would have been of dubious relevance by now. But these young men were accused of getting their pay from the Soviet KGB, which was still very much in business. The trial would take place at the beginning of January, with a score of witnesses.

Celle, a small, conservative city just north of Hannover and one of the dozens of picturesque towns that dot the West German landscape, is the seat of the Higher Regional Court for Lower Saxony. All half-timbered houses, bakeries and pedestrian malls lined with shops selling knickknacks, Celle seems an unlikely judicial terminus. The trial took place inside a courtroom that the Federal Republic spent $8 million to build for trying terrorists. Despite its inviting blond wood paneling, pea-green carpeting and comfortable seats, the Celle courtroom is designed to be highly secure.

A panel of five judges, all middle-aged men, none possessing so much as a rudimentary acquaintance with computers, was appointed to sit in judgment of the three young computer hackers. The presiding judge, Leopold Spiller, a soft-spoken man in his middle years, had a tenuous grip at best on the technology that had played such a crucial role in the case. Sitting on the evidence shelf behind the judges in the space that usually held more traditional weapons were the spoils of the previous two years' house searches: Hess's Atari and Apple II computers, Dob's battered Rainbow, Carl's miniature Casio containing Sergei's phone number.

On the first day of the trial, January 11, 1990, curious spectators and at least a dozen reporters sat among stony-faced officials from the police and the Verfassungsschutz. As the three men stood trial for the same crime, they were seated together with their lawyers before the judges' bench. Dob was the first to give his statement. First, the presiding judge wanted a short lesson in computers. He asked a reluctant Dob to explain what *hacking* meant.

"As I understand it," the judge ventured, "hacking is an attempt to get into a computer system. How does it work?"

During the months Dob had spent in jail awaiting trial, a legal form of suspended animation in Germany known as *Untersuchungshaft,* or pretrial confinement, his mood had plunged so markedly that prison guards began to worry they might have a suicide attempt on their hands. He had let his trim goatee grow to an unkempt mass of beard, and his walk had degenerated to a shuffle. The time in jail—much of it spent in solitary confinement—had only heightened Dob's natural antipathy for West German authority.

From Dob's slouched bearing in the courtroom, it was evident that he didn't feel up to giving a lesson in computers. "Hacking is the modern art of telephoning," he said.

The judge was puzzled. "How does it work?"

"Well, you have to have a telephone line in any case," he answered.

The judge wouldn't be deterred from his attempt to understand this strange new world. "What if I want to get into the University of Hannover and read a book in the library?"

"I was never interested in the contents," Dob answered. "Just in the computer itself."

And so it went, each of Dob's answers more listless than the one before. The other four judges grew restless. One played with his watch, another began shuffling papers. A third began to leaf through a copy of *Hacker für Moskau*, a colorful, breathless account of the espionage story written by journalist Ammann and his colleagues and published shortly after the March arrests. The book was to become a major reference for the confused judges as they struggled to understand some of the computerese that dominated the trial.

Dob's truculence came as a bit of a surprise. Despite his low mood, Dob had offered more cooperation to the prosecutors than either of his fellow defendants at the prison. He had served as a technical consultant and helped the investigators restore the Rainbow computer Pengo had damaged and retrieve the Digital *Securepack* security program from the hard disk.

"Okay," said the judge. "Let's say we're in a hacker's apartment. What happens now? How long does one of these sessions last?"

No answer.

Despite his apparent indifference, Dob didn't waste any time before trotting out what was to become known as the "world peace" defense. The group's motive, he said, was to equal the balance between the world powers. Hence the code name Equalizer for the operation. They adopted as their role model Manfred von Ardenne, a leading German physicist who immigrated to the Soviet Union after the Second World War and worked on nuclear fusion. In his testimony, Dob perpetuated a popular myth that Ardenne was an atomic spy who wanted to bring world peace through espionage. With technology transfer, the hackers wanted to do something similar, Dob explained. Better that the Soviet Union should have more reliable Western software controlling its missiles than Soviet-made software that could inadvertently trigger a nuclear war. What he and his friends were doing was one more step toward world peace.

It was an inventive line of reasoning, to be sure, a twist on the code name Equalizer.

When his turn came, Peter Carl continued with the world peace defense, but it got a cold reception from the judge. "World peace?" the judge asked, looking a bit annoyed now. "Nowhere in your one-hundred-ninety-three-page testimony to the police did you mention anything about world peace."

The judge instructed Carl to describe to the courtroom his trip in the late summer of 1986 to the trade mission in East Berlin, his first conversation with Sergei and his subsequent role as courier.

"So you were responsible for marketing?" asked prosecutor Kohlhaas.

"Did you ask for someone from the KGB specifically?" asked the judge.

"No," Carl answered. "It was all the same for me. KGB. Trade mission. Ambassador. It was understood, but not said."

If nothing else, the defendants' ostensible concern for the world's military stability proved a successful exercise in manipulating newspaper headlines: "Hackers Wanted to Assure World Peace" was the headline in the Berlin *Tageszeitung* the following morning.

On the trial's second day, it was Markus Hess's turn to tell his story. In contrast to Dob's phlegmatic demeanor, and Carl's occasional coarseness, Hess presented himself as a bright, articulate, wholesome young man. Whereas each time Carl was asked a question he leaned toward his lawyer as a shy toddler might retreat from a roomful of strangers into his mother's skirt, Hess took command of his own defense. His lawyer was there, it seemed, as a mere formality.

What Hess's testimony did more than anything was remind everyone that this was an espionage trial. Nobody had been accused of illegal entry into computer systems. No one, that is, was on trial as a hacker. In fact, Hess seemed only too willing to describe his exploits inside the LBL computer. Hagbard had first introduced him to hacking, and in August of 1986, when Hess discovered LBL, which he described as being "open like a barn door," he began his most intensive search for military information. It was "a bit titillating" to go into military computers; mostly, he liked them because other hackers tended to stay away from them. He explained that he sometimes got passwords by guessing them, sometimes by transferring a file of encrypted passwords out of a computer and comparing it with words in a standard dictionary using a standard decoding technique.

Hess rattled off the list of computers he had been in. In his hundreds of attempts, he got into about thirty computers, with full privileges on at least half a dozen of those. And he acknowledged his hacking as something he had been irretrievably drawn to.

"How is that?" the judge asked.

"Hacking is an addiction, Your Honor," Hess said. "And as long as you stay out of trouble, the addiction is difficult to shake." When he did get into trouble in June 1987, Hess testified, he stopped hacking entirely.

Hess's principal defense was that his hacking had never had anything to do with the espionage. Yes, he had made copies of innocuous software, software that anyone could buy, none of it on the COCOM list of technology forbidden to cross into the East Bloc. And he had given that software to Peter Carl with the knowledge that a KGB officer was buying it on the other end. But that business arrangement was a separate matter altogether from his hacking adventures. The only material to have gone from him to Carl to Sergei, Hess insisted, was the harmless public-domain software he had copied.

In a way, it was a brilliant defense strategy. From the day of his arrest, once confronted with the testimony of others, Hess knew he couldn't deny that he had been involved in espionage, but there was no proof that he had sold any of the fruits of his electronic wanderings to the Soviets. Hess had long since decided that the Laszlo Balogh letter was a flimsy piece of evidence at best. He didn't express this theory in court, but Hess was convinced that the frizzy-haired Stoll must really be an FBI operative who had worked together with the agency to fabricate the SDInet material, then feed it to Balogh directly and frame Hess. Even with the Balogh letter in hand, Hess would argue, he didn't see how the prosecution could establish a strong connection between Balogh and himself.

Even Kohlhaas was impressed by Hess's self-confidence, but he also believed Hess was lying. Kohlhaas had his own careful plan for proving that Hess did transfer information out of LBL and other computers. Cliff Stoll would be the witness to establish that. As far as Kohlhaas was concerned, if Stoll simply repeated the story he had told during his interrogations, especially the part about watching data travel from his computer to Hess's, the case was open-and-shut. It didn't take much imagination to conclude that Hess was storing the information for a reason.

▲ ▼ ▲

Pengo was to testify in Celle on January 19. He had received ample warning from his lawyer that he should tell the absolute truth, and he planned to comply, in his own way. But rather than get nervous, he decided to approach his time on the witness stand as pure theater. So he gathered a small group of friends—most of them from Berlin's punk scene—and drove from Berlin to Hannover. Once the group arrived at another friend's place in Hannover, an all-night party ensued, with the consumption of copious hash. Pengo and his entourage were in some disrepair when they arrived at the Celle courtroom. Dressed in his standard black jeans, black turtleneck and heavy black boots, Pengo strode to the witness table and began to answer questions.

"How old are you?" asked the presiding judge.

"Uh, twenty-one," Pengo replied haltingly.

"Your profession?"

"Programmer." Since 1985, Pengo said, he had been working as a programmer and studying computer science at the technical university in Berlin.

"What semester are you in?" the judge asked.

At a loss, Pengo reached into his pocket and pulled out his student identification. "Eighth," he pronounced after consulting the document, and laughter rippled through the courtroom.

Pengo might have figured that he had little to lose by being cheeky in the courtroom. He was there strictly as a witness, and not as a defendant. In his own investigation it looked as if he would soon be in the clear, the charges against him dropped. Besides, he was happy to put on something of a show for his friends. One of them, a young woman with hair dyed violet, was the object of a budding romantic interest. Pengo's refusal to yield in the slightest to those in authority, especially in a nation so conscious of rules and regulations, carried with it a certain perverse appeal: Pengo was someone who wasn't frightened to speak his mind, to shrug off a rebuke.

Those authorities born and raised in West Germany who were involved in the case—Kohlhaas, Judge Spiller, even Pengo's own relatively conservative lawyer—might have explained away Pengo as a product of the chaotic sensibilities that dominated West Berlin. But they could just as easily have rationalized his behavior by saying he was part of a generation that had managed to divest itself of the exhausting and burdensome thing known as the German past. If postwar Germans hated the Nazis and their children hated the Soviets, many of those in Pengo's generation could only be described as indifferent.

Pengo recounted his first meeting with Hagbard at the Chaos gathering at the end of 1985. "For Hagbard, hacking was a skill with political relevance," Pengo said. "His belief was that you have just one chance in your life to influence world politics, and with all this hacking knowledge in hand, this was it."

Pengo also told of the seeds of the idea, of his trip with Peter Carl to East Berlin and of the meeting with Sergei. "We went to a restaurant afterward and talked about God and the world," Pengo testified.

After that first trip, Pengo testified, he waited for Sergei's feedback. To his great disappointment, none came, and by the beginning of 1987 Pengo's role in the enterprise had fizzled. Carl continued to drop by "for coffee" and ask him for source code, but Pengo couldn't deliver them. "The whole thing was more hot air than anything else," he declared. After a while, Carl stopped calling. By that summer, Pengo said, he and Dob were no longer speaking, and he had never had much contact with Hess. Hagbard, he said, was completely "outgespaced," talking of nothing but conspiracies and having religious hallucinations. Once, Pengo recounted, when he and Hagbard were on a train together that was traveling 180 kilometers per hour, Hagbard wanted to open the train doors and jump out.

Pengo turned himself in to the authorities, he said, not because his conscience weighed on him or because of a recognition of a moral breach, but because he "had to find a way to get myself out of this mess."

It was the first time Pengo had had any direct contact with his erstwhile friends Carl and Dob in nearly two years. Both defendants sat next to their lawyers and frowned. Dob barely moved, only occasionally reaching up to play absentmindedly with his beard.

The judge's questioning turned to Pengo's hacking career. "Where have you hacked?"

"I've been all over the world," Pengo answered.

"Which companies have you hacked?"

"I don't know," Pengo said. "The computers don't tell me what companies they're from."

"Did you share the passwords you got with others?"

"No, I always kept the passwords I had for myself."

"How many did you have?"

Pengo took a few moments to consider this one. "In the course of my active hacking time? Maybe fifty or a hundred."

Pengo then decided to bring up the science fiction that had inspired him in this esoteric enterprise. He described John Brunner's *Shockwave Rider* and William Gibson's *Neuromancer*, and their cyberpunk anti-heroes.

"Did the people in the books get a lot of money?" the judge asked, clearly missing whatever point Pengo was trying to make. Pengo's friends in the back of the courtroom giggled.

"The chief thing for me was the adventure, suddenly being inside a movie."

Then it came time to ask about the *Securepack* software that Pengo had stolen from Digital's Singapore computer. A witness from Digital had already testified that it was difficult to put a market price on the security program, as it was restricted to Digital's internal use, but it had cost $375,000 to develop. Digital had also sent a lawyer from its Munich office to sit at the trial every day. She was particularly interested in Pengo's testimony.

There seemed to be some confusion on the judges' part over whether Pengo had copied the software once in late 1986 and a second time a few months later. The computer kept track of the last time the software had been worked with and not the time it was copied out of Singapore. So when *Securepack* was found inside the reconstituted Rainbow, February of 1987 referred to the last time Pengo had looked at the software and nothing more. But he couldn't seem to make that clear to the judge.

"Did you copy *Securepack* twice?" asked the judge.

Like Dob, Pengo was short on patience when it came to explaining the fine points of modifying software. "Why should I copy it twice from Singapore when I have a complete version in Berlin?"

Pengo thought he had done pretty well. His friends thought so, too. In fact, he seemed to be emerging barely scratched from the espionage ordeal. It was an ordeal that had landed two of the others in jail for nearly a year, plunging Dob into irretrievable depression and causing Peter Carl's ex-wife to cut off what little remaining contact he had with her and their daughter. A third had taken his own life. Markus Hess, the wily hacker of LBL, was the only other one whose life had proceeded on a relatively steady course. Though Hess viewed Pengo as ice cold, and Pengo wanted little to do with someone of Hess's conservative nature, they were in some ways similar. Their middle-class upbringings and concerned parents had given them a sense that they could negotiate their way out of tight spots. Their styles were different, but they were both able to enlist sympathy and help. Hess had had the more traditional upbringing, but both knew better than the others how to manipulate the system to their own ends.

Cliff Stoll was to testify on January 30, 1990. It was Stoll's testimony that Markus Hess at once dreaded and awaited eagerly. Depending on what Stoll might say, the American's testimony could incriminate Hess more than Hess's own testimony. On the other hand, Hess planned to engage Stoll in a match of technical wits.

By late 1989 Stoll was in some disfavor among his colleagues in the U.S. Some of his peers questioned his talent as a computer scientist, which shouldn't have been surprising, since he had never claimed to be one. Stoll was by no means the first system manager to have successfully beaten a hacker at his own game. And he certainly wasn't the first to impart a small but invaluable warning to the computer community at large: when choosing passwords, don't choose something obvious such as your spouse's name or your dog's name, or even an English word. Other security experts had been saying that for years. But Stoll somehow struck a nerve. His unorthodox personality had made him a perfect media icon. He had written a breezy first-person book about his zealous hunt for the odious hacker that became a best-seller and for several weeks he was a fixture on television talk shows. There was no avoiding his message: the nation's computers were at risk.

When Stoll arrived in Celle, he was flanked on one side by an FBI special agent, whom the West German spectators assumed was a body-guard, and on the other by an American public television crew filming a documentary on the amazing hacker catcher from Berkeley.

First, the judge asked Stoll simply to tell the story of what had happened at the Berkeley laboratory in late 1986. Stoll had already spent the better part of a year refining his story and was happy to tell it once again, starting with the seventy-five-cent discrepancy and ending with the news from West Germany in June 1987 that the hacker had been caught.

Gesturing wildly and speaking through an interpreter, Stoll relived for the judges, the prosecutor and the spectators his alarm and concern as he saw the hacker break into military computers and search for specific military information. As soon as Stoll began to tick off the military sites the hacker had been in, the judges started scribbling notes: the Jet Propulsion Laboratory, Mitre Corporation, Anniston Army Depot, the

Redstone Army Missile Base in Alabama and an Army computer at the Pentagon. The hacker searched for such hair-raising words and phrases as *nuclear, Strategic Air Command* and *stealth bomber.*

"Excuse me," interrupted the judge. "What does *stealth* mean?"

"I consider myself lucky that I don't know what it means," answered Stoll. And just as he had done the previous summer before the police in Meckenheim and the magistrate in Karlsruhe, in court in Celle Stoll used such words as *copy* and *print out* to describe what he saw happening on his computer screen in Berkeley. "As an astronomer, I wasn't all that familiar with military stuff," Stoll said. "So it surprised me to see space shuttle information being copied out." Once he had planted the false SDInet inside the LBL computer, Stoll testified, the hacker "spent several hours copying it back to his own home."

This was the very point on which Hess and his lawyer were poised to attack Stoll's credibility. Hess's lawyer spoke up. "We seem to be having some trouble with the language," the defense attorney said. "Herr Stoll, what is really meant with *print* and *copy?*"

Stoll began to bounce in his seat. "Excellent question! Excellent! Excellent!" he exclaimed, as if he were in front of a freshman computer science class, encouraging a clever student. "*To print* means 'to look at, to list, to see.' "

Once that point had been set straight, it was clear that what Stoll had said in his testimony during the summer, and what he was repeating in the Celle courtroom, was scant proof that the hacker was downloading and storing anything at all. It could well have been months of browsing, mere electronic joyriding and nothing more. Prosecutor Kohlhaas saw his case weaken.

Stoll's testimony lasted nearly three days. Most of that time Stoll spent reciting page after page of his logbook. Exact dates and times of the break-ins were of particular importance, as German telephone officials were expected to come to court immediately after Stoll's testimony with corroborating evidence matching the dates and times of their telephone traces to Markus Hess with the dates and times in Stoll's logbook.

On the final day of Stoll's testimony, it was the defendants' turn to interrogate the witness. Neither Dob nor Carl had much to ask Stoll. Hess, on the other hand, was ready for some confrontation, a battle of technical prowess. Hess first asked Stoll if it wasn't true that he had little way of knowing that it was the same hacker inside the computer each time. Stoll agreed that there were some break-ins that occurred on VMS computers, not UNIX computers, and that with the exception of the traces made directly to Hannover, he had no way of knowing that it was the same hacker each time. Hess also wanted to know if Stoll's SDInet file was his and Martha's own invention entirely. Hess's motive, apparently, was to provoke Stoll into admitting that he was in league with the FBI or the CIA. Hess would maintain to the end that the FBI had assigned Laszlo Balogh to ask for the information, in order to speed up the pace of the investigation.

Stoll's answer was curt. "It was fictitious," Stoll responded, "but I tried to make it look as realistic as possible."

Hess wouldn't give up. He questioned Stoll's "typing rhythms" experiment. How could this prove anything given the great distance? Any unique characteristics would certainly be lost, he suggested, in the labyrinth of networks between Germany and California. Stoll countered that he had believed the validity of his experiment.

The interrogation at times required that Hess and Stoll gather in front of the bench. Together they bent over computer printouts to examine the details of their arguments. It was odd to see the hunter and the hunted analyzing documents like two academic colleagues. One morning before court was in session, Stoll and Hess happened to arrive before anyone else. It was an awkward but cordial encounter. Hess suggested that the two get together for a beer sometime. Stoll, an outspoken nondrinker, nodded politely and excused himself.

By the time Stoll's testimony in Celle had ended, it was unclear whether he had helped or hurt the prosecution. Ask Stoll what he considered to be the ultimate value of his sleuthing and he would say, "Chasing the bastards down." For his part, Kohlhaas was grateful for the existence of Stoll's logbook, and pleased that Stoll had kept such careful note of the times the hacker had been inside LBL. Otherwise, the Bundespost's telephone traces would have been difficult to present in court as incontrovertible evidence. Then again, Kohlhaas had to admit that establishing Hess as the LBL hacker wasn't as relevant to the espionage charge as proving that the information he had seen had gone to the Soviets. And no one would know until the verdict was delivered how the testimony was viewed by the judges. For the judges to impose stiff sentences, the prosecution had to demonstrate that sensitive military information had changed hands. Prosecutor Kohlhaas's efforts to bring a U.S. military expert to Celle to assess the sensitivity of the material had ended in frustration. And in the end, Hagbard's testimony to the police, in which he had claimed that hundreds of logins to sensitive U.S. and

European computers went to the East, was stricken from the record for lack of credibility.

At the time Markus Hess was caught, the media were making him and his friends out to be half the KGB, when in reality Hess was a fairly conventional young man who claimed to take inspiration from a movie. As notorious as the hacker spies became, the judges would have to admit that the defendants really did very little damage to Western security. In the end the only person who was damaged irretrievably was Hagbard.

So it wasn't a great surprise when the sentences turned out to be light. Peter Carl, whom the court viewed as the one most actively engaged in espionage itself, and the one who exhibited the most "criminal energy," got two years and a 3,000-mark ($1,500) fine. Hess was sentenced to a year and eight months and ordered to pay 10,000 marks; Dob got a year and two months and a 5,000-mark fine. But instead of having to serve prison time, all three defendants were put on probation. At the time of their actions, the judges remarked, Dob and Carl were operating in such a drug-filtered haze that they were in no position to recognize the gravity of their deed.

In his conclusions to the court, presiding judge Spiller said he believed the hackers had indeed sold information out of military computers to the KGB, and that the KGB had probably found the information very interesting. But, he added, Sergei couldn't have seen it as terribly valuable because he didn't yield to the hackers' demands for a million marks. In the end, all that hacker know-how went unappreciated, even by the Soviets.

PART THREE

RTM

Phil Lapsley, an engineering student at the University of California at Berkeley, was puzzled. No sooner had he logged in to a Sun Microsystems workstation than it was clear something was amiss.

Computers such as the Sun run dozens of programs at once, so it is routine for people like Lapsley who maintain them to peek periodically to see which programs are currently active. But on November 2, 1988 he saw, hidden among dozens of routine tasks, a small program controlled by an unusual user named *daemon*. *Daemon* is not the name of any particular human, but an apt label conventionally used for the utility programs that scurry around in the background and perform useful tasks. But this program was not one that Lapsley recognized.

"Is anyone running a job as *daemon?*" he asked the others in the "fishbowl," room 199B at the Berkeley's Experimental Computing Facility. People shook their heads. Then somebody else in the room pointed to one of the screens, where a program that monitored the status of various other computers in the department was displayed. Lapsley looked more closely and discovered that a number of people appeared to be trying to log in to other Berkeley computers. He decided it must be an attempted break-in. At least once a year, someone tried to break into the computers in Cory Hall, which houses the school's prestigious electrical engineering department. The school year wouldn't be complete otherwise.

Whoever this intruder was, he was apparently quite intent on getting in, trying time after time to log in to Berkeley's computers. So Lapsley started to jot down the names of the machines from which the break-in attempts were coming. But he was startled to see that they were scrolling by faster than he could write them down. In fact, they were coming so rapidly they were scrolling straight off the screen before he could even read them. At that point, Lapsley realized it wasn't a person at all who was trying to break in. It was a program. When it wasn't running as *daemon,* it was running under the names of other users.

The program kept pounding at Berkeley's electronic doors. Worse, when Lapsley tried to control the break-in attempts, he found that they came faster than he could kill them. And by this point, Berkeley machines being attacked were slowing down as the demonic intruder devoured more and more computer processing time. They were being overwhelmed. Computers started to crash or become catatonic. They would just sit there stalled, accepting no input. And even though the workstations were programmed to start running again automatically after crashing, as soon as they were up and running they were invaded again. The university was under attack by a computer virus.

Lapsley called Mike Karels, a programmer a hundred yards away in Evans Hall, an imposing concrete tower and home to the school's computer science faculty. As the principal programmer at the Computer Systems Research Group, Karels was the scientist most knowledgeable about Berkeley UNIX, the operating system widely adopted by universities and research institutions everywhere. If anyone would have good advice, it would be Karels.

All Lapsley got from Karels was a short, stiff laugh, then, "So you've got it too, huh?"

After another thirty minutes of puzzling over the enigmatic intruder, Lapsley and others in the fishbowl discovered that the program was expanding beyond Berkeley. Peter Yee, another undergraduate working with Lapsley, logged in to a computer at NASA's Ames Research Center fifty miles to the south and saw it there. And when Lapsley logged in to a computer at Berkeley's sister campus in San Diego, he saw it there, too. By the time a call came from a system manager at Lawrence Livermore National Laboratory to say it was on his machines, there was no doubt that this was no local problem. It was all over the nationwide network known as the Internet.

▲ ▼ ▲

The people who care for the networks of computers used on college campuses and scientific research centers had spent many years preparing themselves for various eventualities. And for years, computer scientists had spoken theoretically of the possibility of a program running loose in the network. But no one was prepared to cope with the massive assault on November 2, 1988.

Within minutes of each other, computers all over the nation felt the presence of the rogue program. Shortly before 6:30 P.M., computer managers at the Rand Corporation in Santa Monica, a famous think tank where Daniel Ellsberg once photocopied the Pentagon Papers, noticed that their computers were unusually sluggish. There appeared to be a program running that was robbing the computers of speed and slowing them to a near standstill. Fifty-five minutes later, across the country in Cambridge, Massachusetts, computers at the MIT Artificial Intelligence Lab were under attack. Almost immediately after penetrating MIT, the program struck Lawrence Livermore National Laboratory and a computer at the University of Maryland. Then it struck Stanford, Princeton and the Los Alamos National Laboratory in New Mexico. Once inside a computer, the program propagated to other computers much like a biological virus. On some computers there were hundreds of copies of the program running, slowing the machines to a halt. Even when its attempts to get into a new computer were unsuccessful, this electronic virus's repeated knocks on the door were often enough to cripple the machine. And even after it was killed, it would reappear almost immediately. Moreover, once it entered a workstation, the program had a mysterious way of finding other computers to attack. Throughout the night it hopped back and forth through the network, setting off havoc wherever it touched down.

System managers around the country, responding to frantic calls from night operators, were racing to their offices at 2:00 A.M., 3:00 A.M., and 4:00 A.M. to wrest back control of their computers. Others noticed it when they had trouble logging in to their institutions' computers from home. Still others wouldn't learn about the program until they arrived at work on Thursday morning to find their computers besieged. Programmers at the University of Illinois at Champaign-Urbana were convinced they were going to have to rebuild the software for their entire campus computer system from the ground up.

Worse than what could be observed about the program was the fear that it might be a Trojan horse program—apparently innocent, but carrying a string of code instructing the computer to carry out a specific damaging instruction at some later time. System administrators at an aerospace company in San Diego got so frightened by the threat of a malevolent string of code that they pulled everything off their computers and installed their most recent set of backup tapes.

When the program started entering computers shortly before midnight at the Army's Ballistic Research Laboratory in Maryland, system managers feared invasion by a foreign power. And since the program came in over the network, they were afraid it might also be taking Army data out over the network. Assuming the worst, at 10:00 on Thursday morning Mike Muuss, the chief Ballistic Research Laboratory system programmer, did what dozens of other managers across the Internet had already done: he disconnected his computers from the network. The laboratory would stay off the network for nearly a week.

Taking computers off the network stopped the program from coming in or leaving, but it had the unfortunate side effect of cutting off communications among people accustomed to staying in touch with electronic messages. Few people thought to pick up the telephone, and those who did were at a loss: the electronic network had become the sole form of communication for most computer experts, who seldom bothered to give out their telephone numbers.

▲ ▼ ▲

If any place could grapple with such a bizarre and troubling situation, Berkeley could. The university was the birthplace of the very version of the UNIX operating system that the rogue program was targeting when it broke into computers on the network. During the evening it became apparent that the intended targets of the mysterious program were computers made by Sun Microsystems and the Digital Equipment Corporation, two of the most common machines on the Internet.

Fifteen minutes after Lapsley first noticed it, the program had broken into at least thirty workstations on the Berkeley campus. From the way it was acting, it appeared to be a selective beast, setting its sights on machines that were connected to as many other systems as possible. It used simple, quick and powerful methods to break in immediately. Two Berkeley computers were especially attractive targets. One, called CSGW, which was a gateway to local area networks on the Berkeley campus, crashed after dozens of copies of the virus arrived. So did UCBVAX, a major gateway to the Internet. It seemed that infecting such a vital organ had strategic value for the program, increasing many-fold the number of computers it could reach in just a single hop. Thus, UCBVAX was under constant attack. Still, the team of Berkeley defenders decided against pulling their computers off the network. That would have been admitting defeat. The challenge, they decided, was to stay connected to the network and still kill off the program.

One of Berkeley's first tasks was to capture a snapshot of the program as it was running, in effect to catch it in freeze-frame, and then to analyze it. From there, they could examine strings of code and try to figure out what the program was doing. But most of it was encrypted, as if whoever wrote it knew someone would take such a snapshot. The Berkeley programmers found that the coding scheme used to obscure the program was a simple one—unscrambling the data was much like the child's game of translating words from pig latin. The programmers quickly uncovered the original instructions. The snapshot of code also told the programmers that the program was trying to crack passwords using what was known as a dictionary attack, comparing encrypted passwords to an on-line dictionary that had been encrypted. The snapshot also showed them that the program was using cracked passwords to get onto one system, then go from there to other computers by taking advantage of the fact that a password validation on one computer often grants access to other computers across the network.

Other things quickly became obvious. The Berkeley programmers soon figured out that the program was exploiting a subtle flaw—or bug—in a communications program called sendmail, which it used to send messages and data between computers over the network. The flaw in sendmail arose from the subtle concatenation of two features of the program, much as a binary poison gas is deadly only when two inert gases are combined. One feature made it possible for someone at a remote location to embed a program in a message. Instead of being handled as an electronic letter, the message fooled the computer into running it as a program.

The second feature allowed those programmers who needed to "debug" or maintain the mail program to examine mail connections over the network. This "debug" feature made it possible to switch on the first feature from a remote location. Once the first feature had been switched on, a program embedded in electronic mail could be sent to run on another computer immediately. The combination of the two features, known to only a few, proved to be a glaring loophole in the mail program.

Whoever had written the rogue program made use of this obscure flaw to send a small "scout" or "grappling hook" program across the network. This program in turn immediately called back and brought over the main body of the virus. Having taken hold of each new computer, the process would repeat itself indefinitely. That much, at least, was obvious.

Their first look at the program suggested to the Berkeley group that the invader had no intention of destroying data. Apparently, it examined actual information inside computers only in order to find ways of breaking into other systems. But they realized that was just a superficial impression. The possibility of a Trojan horse still lingered. The only way to determine what it was actually doing would be to pick it apart line by line, a painstaking task that could take days or weeks. Until the program had been thoroughly analyzed "with microscope and tweezers," as programmers at the Massachusetts Institute of Technology titled a later paper on the virus, there was no knowing what kind of dangers lurked inside.

For the next three hours, programmers at both the Experimental Computing Facility in Cory Hall and the Computer Systems Research Group (CSRG) on the fourth floor of Evans Hall worked simultaneously at shaking the program out of their systems and building a clearer understanding of how it worked. If it were a playful hoax, they reasoned, wouldn't it have come with a set of instructions on how to get rid of it? But there were no such instructions, and all potentially useful details hidden inside were encoded, sheltered from prying eyes. The program tried to remain hidden by giving itself the name of an innocuous command that its author obviously hoped would avoid scrutiny. Apparently, the idea was that to anyone taking casual stock of the computer's activities, nothing would appear out of the ordinary. And to further elude detection, like a chameleon the program constantly changed its identifying number, taking on new aliases to make itself less conspicuous.

By 11:00 P.M., most of the Berkeley programming staff had congregated in one of the two computer labs. Keith Bostic, a twenty-eight-year-old programmer at CSRG, was in his office at Evans Hall, working with Mike Karels. As two of the principal software engineers behind Berkeley UNIX, they had good reason to take the attack personally. Bostic had seen his machines hacked and crashed by outsiders before, but this episode was on an entirely new scale. Meanwhile, Lapsley and a group of others gathered in the fishbowl at Cory Hall. At 11:30 P.M., Peter Yee sent a message from Berkeley to an electronic mailing list on the Internet: "We are currently under attack from an Internet virus," the message began. "It has hit UC Berkeley, UC San Diego, Lawrence Livermore, Stanford and NASA Ames."

Fueled by adrenaline, sugar and caffeine, the Berkeley group was meeting the break-in head-on. A software invader that brought scores, perhaps hundreds, of computers to their knees was just the sort of nightmare that every computer manager feared. At the same time, it was as if someone had just handed the group at Berkeley an imposing crossword puzzle to solve—the ultimate challenge. And the possibility that the program could contain virulent code infused the evening with tension. Someone made a sign that read, "Center for Disease Control," and taped it to the door of the Experimental Computing Facility.

Sometime after midnight, Lapsley walked back to a machine room containing most of Cory Hall's largest computers and began the arduous task of going from machine to machine, plugging the holes the invader was using and killing off all copies of the foreign program. He reconfigured each system with a patch that blocked the *sendmail* loophole. And in Evans there were dozens more. One by one, the Berkeley computers were immunized from the attacker. Berkeley had survived the plague.

By 3:00 A.M. Thursday, the tired programmers knew enough about the program to issue a broad alert to other computer sites. Bostic sent messages to several electronic mailing lists describing how to fix systems in a way that would stop the program. His message reached some parts of the network, but, unfortunately, not the centers that had already cut themselves away from the network and were working in solitude to stop the rogue program. The Internet sites that stayed connected found their messages bogged down by a form of electronic gridlock; the program had clogged some of the network's crucial mail machines. In some cases it took messages hours, even days, to travel routes that normally took a few minutes.

Exhausted, Bostic went home to sleep. Lapsley stayed in the fishbowl. He knew the program was somehow using other methods to break in, but he hadn't yet been able to figure out exactly what they were. At 8:30 A.M. Thursday, Lapsley finally went home, too.

It had been one of the most harrowing nights that anyone in the computer science community had ever faced. For others, though, it was just the beginning. The following morning, news of the program spread around the country. Word of the previous night's invasion had circulated not just among computer scientists but in the national press as well.

During the day, Lawrence Livermore National Laboratory, one of the nation's leading weapons laboratories, held a press conference to describe the attack in detail.

▲ ▼ ▲

The anonymous caller to *The New York Times* on Thursday afternoon made it clear that he didn't want to disclose who had written the Internet virus. He just wanted to let the *Times* know that the person who had written it was a well-intentioned soul who had made a terrible mistake in the code.

The switchboard first routed the call to the paper's national news desk.

"Uh, I know something about the virus that's going around," said the caller.

"What virus?" The editor sounded confused.

"The computer virus that's crashing computers all over the country."

"Give me your number and someone will call you back," said the editor.

The editor gave the message and a telephone number to John Markoff, the paper's computer reporter. Markoff had already heard about the incident. He had received a call at 10:00 that morning from Cliff Stoll, the Berkeley astronomer who had gumshoed his way to the bottom of the West German hacker case a year earlier. Stoll, who was now working at the Harvard-Smithsonian Center for Astrophysics, told Markoff he had been up the night battling the program, which had swamped fifty of the center's machines. The reporter then spent the morning calling universities and research centers to see if they, too, had been infected. One of his calls was to an occasional contact at the National Security Agency. Markoff had called the NSA in the past on security-related stories, and he thought his contact there might tell him something about what was going on. But his contact wasn't there and his call wasn't returned.

Nobody Markoff spoke with at universities, corporations or military sites seemed to have any inkling of the program's origin. Theories ranged from prankster to foreign agent. So the anonymous call to the *Times* was intriguing. When Markoff returned the call to a number in the Boston area, it was immediately clear to the reporter that the caller, who would identify himself only as Paul, knew a great deal about the program and how it was written. The excited-sounding young man said he was a friend of the program's "brilliant" author. The author, Paul said, had meant to write a harmless virus, but had made a small error that caused the program to multiply around the network.

By Friday, Paul and Markoff had talked on the phone several times. Paul referred to the author only as Mr. X. The two went back and forth about what kind of trouble the program's author might be in. By this time, news of the program, which by now was being described alternatingly as a "virus" or a "worm," had been on the front pages of newspapers and on the nightly television news around the nation. It was the first wholesale assault ever on the nation's computer systems. More disturbingly, military computers had been infiltrated. The program had been contained, but there was still no full assessment of the damage it had done.

Then Paul made a mistake. During one conversation, instead of saying "Mr. X," he slipped and referred to his friend by the initials *rtm*. Markoff was close enough to the computer scene to recognize *rtm*, in lower case, as a likely computer login. After hanging up, he phoned Cliff Stoll, with whom he had been trading information all morning. From his computer at home, Stoll used a network "white pages" directory and *finger*, a utility program that acts as a computerized directory-assistance tool giving users limited information about others on the network. When Stoll fingered *rtm* on the Harvard University computers, he retrieved the name Robert Tappan Morris, identified as a graduate student at Cornell University. A phone number and an address were included. Stoll called the *Times*.

Now Markoff had a name, but he still didn't have a story. Without an independent confirmation, he couldn't be sure if Paul was telling the truth, or if the *rtm* the caller was referring to was the same person as the *rtm* Stoll had just found on the Internet directory. Markoff called the Cornell telephone number. Nobody answered.

Late that afternoon, the NSA finally rang back. The caller was Bob Morris, a computer security expert who was the chief scientist at the agency's National Computer Security Center.

"I think I know the name of the person who wrote the virus," Markoff said.

"Who is it?" Morris shot back.

Markoff bridled. "I'm not going to tell you. You're the computer police."

A strained conversation followed and before long it became clear that

Morris knew exactly who had written the program. In fact, the man from the security agency appeared to know more about the event and its perpetrator than the reporter did.

Finally Markoff said, "I think the program was written by Robert Tappan Morris."

"You're right," Morris answered, giving the paper the confirmation it needed. They talked for a while longer. It had been a chaotic day of sifting through dozens of sometimes contradictory reports on the event. Markoff still wasn't certain what to do with this new information. If the program had been written by a Cornell graduate student, how was it that computer security experts at the National Security Agency already knew that? What was going on, anyway? Just as he was about to hang up, he had a sudden thought. "Isn't that a funny coincidence," he said. "You both have the same name."

Without missing a beat, Morris replied, "That's no coincidence. He's my son."

▲ ▼ ▲

Bob Morris entered Harvard in 1950 as a chemistry major. His father had been a salesman for an etching and engraving company and Bob thought he too might end up a salesman, perhaps at a place like Du Pont. He interrupted his studies to spend two years in the Army, and when he returned to Harvard he decided his options would be broader with a degree in mathematics, so he switched majors in his senior year and completed all the remaining math requirements in time to graduate that same year. He earned his master's in 1958 and embarked on his Ph.D., with plans to write his dissertation on number theory. In 1960, he took a summer job at Bell Labs.

By the time Bob Morris got there, Bell Labs had as many scientists with Ph.D.'s as did most universities. Since its inception in 1925, Bell Labs has been a monument to the degree of innovation that can spring from within the walls of a monopoly. From its considerable investment in both basic and applied research AT&T has seen ample reward. Bell Labs scientists hold nine Nobel prizes. There are few other industrial research institutions like it.

Some areas of research were of uncertain commercial value, but it was rare that a project would be blocked simply because its immediate practical benefit to telephony was unclear. AT&T managers were astute enough to recognize that significant breakthroughs are born not of rules or plans but of people, and those in the upper echelons at Bell Labs governed their hiring practices accordingly. Bob's summer job was to stretch into two summers and eventually into a twenty-six-year career. He started out in telephony, working on data transmission, but he was a frequent visitor to the mathematics department, some of whose scientists were developing computer software. Morris made the acquaintance of Doug McIlroy, a mathematician who was to become his boss and close friend for many years. After a year or so Bob transferred to the mathematics department, took over the job of someone who left and became so entrenched that there didn't seem to be much point in returning to his Ph.D. work. Besides, he was pushing thirty, an age at which almost all of the best mathematicians have already done their major work.

It didn't take long for him to migrate into computer research. By then, in the sixties, the field of computer science had begun to burst open with new discoveries. Computers were appearing everywhere within the scientific community. Practically every mathematician who came near a computer had a chance of doing something original. None of those people had been brought up on computers, of course, but all were captivated by them. Mathematical solutions to problems were more and more often complemented by a computer's ability to calculate rapidly. The notion of programming, of being able to instruct a machine to do almost anything, of inventing artificial worlds on a computer, was a source of absolute fascination for McIlroy, Morris and their colleagues. Morris, in particular, seemed to have an uncanny understanding of computers. Whenever a convoluted computer problem presented itself, bringing it to Morris's attention was almost certain to produce a creative answer.

Morris established himself early as a programming wizard. One of his first displays of such mastery came when he joined in on a simple game called *Darwin*. *Darwin* was the 1962 invention of McIlroy and a colleague named Victor Vyssotsky. Vyssotsky thought it would be fun to have a computer game in which the program played against other programs rather than against people. The idea was to create a program that tried to kill opposing programs. At the end of each round, the winning design would be shared with the group. A concept well ahead of its time, *Darwin* was a predecessor to a later program called *Core Wars*, a simple computer game that became popular after the advent of the personal computer. *Core Wars* came with its own simple programming language. Players designed tiny software "warrior" programs, then turned them loose on an imaginary playing field in the computer's memory. The winner was the program that disabled the opposing program and still remained functioning at the end.

The three young scientists had been playing around with *Darwin* for a week when Bob came in one day with the toughest survivor of all. Bob's program was composed of just thirty instructions, and its power lay in the fact that it was adaptive: it learned how opponents were protecting themselves and devised its attacks accordingly. Bob's program was unfailingly lethal, and the game ended.

Even in an environment where quirkiness abounded, Bob was regarded by his colleagues as an original. Shortly after arriving at Bell Labs, he grew a beard, which was to remain aggressively unkempt for the next three decades. An iconoclast by nature, Bob had a habit of challenging others' assumptions and would go to any length to make his point. But he carried out his challenges playfully, never dogmatically. In answer to a colleague's assertion that all the equipment in the computer room was fireproof, he took a lit match to a computer tape's write-protect ring (an attachment that prevents computer information from being erased) and tossed it in the wastebasket, setting off smoke alarms and causing pandemonium throughout the lab. And so acute was his sense of a system's vulnerable points that as his colleagues buzzed proudly around the computer lab on the day a new operating system called Multics was first unveiled, Bob strode in and typed two specific characters he suspected would confound the system. They did. The computer crashed.

It was such stunts that earned Bob his reputation as a scientist whose biggest strength was his capacity for offbeat thinking. When he wrote a program designed to spot typographical errors, the program contained no dictionary and knew no English. Instead it was based on statistical probability; it sifted through a document searching for uncommon sequences of characters. And it found many of the mistyped words.

Bob loved having inside information, and he enjoyed possessing insight into arcana that others were only vaguely familiar with. He often subjected his colleagues to intellectual popquizzes. If he discovered that someone down the hall had a passing interest in, say, relativity, he would learn all that he could on the subject and start asking questions. It was less a show of intellectual bravado than a sign of Morris's constant curiosity.

In a report for the twenty-fifth reunion of his Harvard class in 1979, Morris wrote, "A long time ago I promised myself that I would learn to read Greek, learn in some detail how the planets move in their orbits and how to decipher secret codes. I have gone a long way toward keeping all three promises." In his thirties, he had taught himself ancient Greek. And one project that occupied him for nearly a year at Bell Labs was an astronomy program for predicting planetary orbits. But the promise with the most relevance to his work at Bell Labs was the third one.

▲ ▼ ▲

In 1964, Bob was one of the first people at Bell Labs to have a terminal in his home. His modem carried data back and forth at an excruciatingly slow rate of 135 bits per second, about a tenth the speed of the slowest of today's most common modems. Retrieving or sending even a small amount of information was a process that left plenty of time to go get a cup of coffee while the modem churned away.

The terminal itself, called an IBM 2741, looked like an oversize IBM typewriter with an IBM type-ball mechanism. The typewriter perched atop a pedestal and inside the pedestal was a mass of electronics. Later came a slightly faster terminal, the Teletype Model 37, a cumbersome affair that was roughly half the size of a standard desk. The Teletype terminal had a mechanical encoding grid of rods under the keys that converted keystrokes into binary signals, which in turn made their way over a modem into the Bell Labs central computer. Any Bell Labs scientist with a terminal got to know the repairman pretty well. Every repair visit finished with an oiling of the encoding mechanism; the next time the terminal was used, fresh oil often dripped onto the user's legs.

In the early 1960s computer security wasn't a problem. Locked doors sufficed. It first became an issue with time-shared computers. The original idea of time-sharing was simply that everybody could apparently have his own computer when he needed it, with the cost shared among many people. This was the first time people had thought about sharing the power of a computer. With the development of time-sharing, there arose the need for accounting and security mechanisms of some kind, because more than one person at a time could use a computer.

Multics was one of the first time-sharing systems that paid real attention to security as an explicit design goal. The main goal of Multics, a joint research project of MIT, Bell Labs and General Electric, was to make time-sharing commercially available. The dream was that Multics would be a computer utility with capabilities far beyond those of existing commercial time-sharing systems. It had to permit cooperation among users who wanted it while guaranteeing privacy to others.

The earliest self-described computer hackers, those at MIT who abhorred computer security, or anything else that would inhibit the sharing of information and free access to computers, had it in for Multics from the start. MIT hackers often tried to bring the system to its knees, and occasionally they succeeded.

But ultimately, Multics developed to the point of becoming too unwieldy. As Morris would describe it many years later, continuing to support its development was "like kicking a dead whale down a beach." Bell Labs pulled out of the project in early 1969, after which Multics was adopted by Honeywell as a secure operating system to run on military computers. But the "tiger teams" of the 1970s—groups of people who were authorized to probe the security of Defense Department computers by trying to break into them—put Multics computers through rigorous tests and eventually got in. The teams even managed to confound the system's meticulous audit trail, modifying it so there would be no trace of a penetration of the computer.

Breaking into computers in order to improve security was an important tactic used by people who worked in the field of computer security. Members of a tiger team were allowed to have at least limited access to a target computer. That was one thing. But what about those would-be invaders with no legal access at all? People like Bob Morris and Ken Thompson, another talented computer scientist, thought about such problems extensively. The first step, of course, would be for intruders to find out what telephone numbers dialed in to a computer, perhaps by using a scanning program that could dial every possible telephone number sequentially. Ten years later, it would be common for twelve-year-old computer hackers to write programs similar to those they saw depicted in the movie WarGames. If a hacker's modem detected another computer, signaled by a high-pitched tone, his next step was to log in and identify himself to the computer's satisfaction by supplying an account name and a password. Unless the perpetrator already had inside information of some kind, coming up with the correct password could be the most difficult step. But once he had logged in, it was possible for him, depending on the level of privileges achieved, to enter other computers over a network illicitly.

Bob's interest in computer security grew with the development of UNIX, the successor to Multics. UNIX was a play on the name *Multics*. Where Multics was complex and its name referred to computing in multiples, UNIX signified simplicity and uniformity. UNIX began as a backlash to the Multics system; it was developed for a small computer, and programmers grew to like it for two primary reasons: its flexibility let them tailor it to suit the needs of whatever program they were working on, and it was designed to be "portable," meaning it could be made to work on computers of many different brands. Future versions of the system grew slightly in complexity as new capabilities were added, but each new edition of UNIX remained faithful to the principles of simplicity. UNIX would bring fame to a few Bell Labs programmers and become a fixture at universities and research institutions around the world.

The UNIX development team consisted of two principals—Ken Thompson and Dennis Ritchie—and a peripheral group that made smaller contributions. Bob's work on UNIX involved the mathematical functions of the software. Something as simple as asking for the time, for instance, involved a calculation. But his main interest lay in writing the encoding algorithm used in UNIX, the procedure that transformed the uncoded, plain text in a file into encrypted text.

When Bob wrote the *crypt* program, his fascination with ciphers intensified. He was a mathematician, and his deepest interest lay in number theory, which typically involves the study of prime numbers and creative uses of randomness. Cryptology is a natural extension of number theory since it requires turning a message of clear text into a code through manipulation of numbers. Cryptology is more than a mathematical discipline; it requires linguistic skills as well. To do exceptional work in cryptology requires remarkable intuition and leaps of imagination. Morris had that. He also had an ability to see the security holes where others saw protection.

In the mid-1970s, Morris was working on a method of cracking the encryption machines developed during the 1930s by a Swedish cryptologist named Boris Hagelin. The machine, known as the M-209, was an updated version of the earlier German Enigma machine used by the Nazis in World War II, which was decoded in 1939 by British cryptanalysts, including the famous mathematician Alan Turing. The M-209, which looked like a cash register with letters as its keys, coded messages in such a way that each letter was turned into one of more than a hundred million possible substitutes. Morris devised an elegant method for taking a passage of text encoded by the M-209 and transforming it into clear, readable English without relying on machines. At the same time, Jim Reeds, then a mathematician at the University of California at Berkeley, came up with a different method for breaking the code that could be done with a computer program. Reeds and Morris learned of each other's work and, with the help of Dennis Ritchie, created a program that would read encoded text and generate a clear translation. The trio then wrote a joint paper describing their feat, and submitted it to the academic journal *Cryptologia*. At the same time, however, as a courtesy Bob sent a preprint of the paper to the National Security Agency, whose mission —indeed, whose very existence—was largely unknown to the general public at the time. The NSA spreads a far-flung net for gathering communications intelligence in every corner of the world. For example, when a Korean airliner strayed off course in 1983 and was shot down by a Soviet interceptor, NSA monitors captured the radio conversation between the Russian pilot and his flight controllers. And in 1989, when the United States accused a German company of selling materials that enabled Libyans to build a poison gas plant, intelligence on the matter was gathered by a massive and permanent NSA communications surveillance operation in Europe.

The mission of the NSA, which was classified until recently, also required that the agency maintain the world's best cryptographic capabilities. The NSA maintained that it was not in its interest for the most advanced cryptographic research to be widely disseminated to the public. So it was perhaps not surprising that shortly after submitting their paper to the agency for review, the three computer scientists received a visit from a retired Virginia gentleman to discuss the impending publication of the paper. He was, in fact, a former intelligence officer and still had close ties with his former employers.

The agency was divided, he told them. Some didn't see a problem with the article, but one conservative group was opposed to any publication of information that would advance the public knowledge of cryptography. The initial contact over lunch at Bell Labs led to other meetings. The researchers traveled to visit agency officials several times. In the end, the Bell Labs scientists decided to withdraw the paper.

As Ritchie remembers the incident, it was at this time that Bob Morris's flirtation with the NSA began. What went on inside America's most secret intelligence agency held a certain fascination for all of them, and for Morris in particular. Already, the NSA was a customer for UNIX and the accompanying C programming language that the Bell Labs group had designed. Morris was offered a summer appointment at the Institute for Defense Analyses, the NSA's classified think tank. But at this point all three still felt that if they took security clearances, it would mean sacrificing much of the freedom they enjoyed as outsiders. They decided to keep their contact with the computer spooks informal.

▲ ▼ ▲

Anne Burr Farlow came from a long line of New Englanders. Moonfaced and slightly plump, Anne was a music graduate fresh out of Bryn Mawr College in 1959 when she moved to Cambridge to work as an office assistant in the geology department at MIT. On occasion, a Harvard graduate student in mathematics named Bob Morris would stop by Anne's apartment to visit her roommate, but Anne didn't take much notice of him until one day when he asked the roommate to a concert. When the young woman declined the invitation, the serious young student turned straight to Anne and asked if she would be interested. She accepted. Their two-year courtship consisted of frequent ski trips in the winter and long sailing trips in the summer. In June of 1962, Anne and Bob married.

When Bob decided to settle permanently into the Bell Labs job, the young couple went house hunting. Bob, who had grown up in farming country north of Hartford in the Connecticut Valley, wanted ample privacy. They settled on a farmhouse in the small town of Millington, New Jersey. The house dated back to 1740; it had few modern amenities and abutted a steep wooded hillside, a wildlife preserve that served as a woodlot for the family. The house sat on a nine-acre triangle of land on a dead-end road. A two-acre field lay between the house and the Passaic River. Unless the temperature dipped below twenty degrees, the nine-room house was heated entirely by a wood stove and a fireplace.

Meredith, the first child, was born two years before Bob and Anne moved into the Millington house. Three weeks after the family moved in, Robert was born, in November 1965. Ben arrived two years later. On the property, farm animals were accumulating: sheep, chickens and geese. The family had three large dogs, and at least a dozen cats, "working cats" as Bob called them, roamed freely. When Meredith asked for a horse one year, Bob compromised and got her a pig. The entire household later joined Meredith in her hobby of training Seeing Eye dogs. A large vegetable garden provided much of the family's fresh produce, and within a few years nearly half the family food came from the animals and the earth. Every lamb they owned was named Lambchop, lest the children lose sight of its fate.

Anne would always describe marriage to Bob as "complex." He was completely lacking in conventional traits. For stretches at a time, in fact, he was missing in action. He kept odd hours; for years his pattern was to work half the night and sleep until eleven the next morning. He believed firmly that if he was going to work hard, it should be at something he enjoyed doing. The children occasionally had trouble understanding that their father wasn't going to be like the fathers of their friends, conventional workaday men who would spin through the kitchen at 8:00 A.M., briefcase in hand, returning promptly at 6:00 P.M. Bob's erratic schedule was dictated by the nature of his work.

But once Bob finished a major project at work, he would spend several weeks at home working on a domestic job that he found equally absorbing. Bob had a natural inclination to integrate his practical skills with his broader cultural knowledge. One of his more ambitious projects was to design and build a sheep pen. Dissatisfied with the conventional designs he found in home construction books, he turned to the February leaf of a famed fifteenth-century illuminated manuscript, the Très riches heures du duc de Berry, a wintry scene of peasants working on a farm, depicted in exacting miniature in blues and golds and whites. A centerpiece of the picture is a simple yet elegant sheep pen. The sheep pen on this page of the fabled Book of Hours proved the most practical and aesthetically pleasing design. Bob built its precise twentieth-century replica in rural New Jersey.

Bob's good salary at Bell Labs enabled Anne to do what was important to her rather than work just to supplement the family income. For the first few years, that meant raising the children. Then she involved herself in local and state environmental work; eventually she became executive director of the Association of New Jersey Environmental Commissions, a statewide umbrella organization for municipal environmental commissions. For his part, Bob became chairman of the local planning board.

Sending the children to the best schools possible was all part of the plan, too. Bob and Anne considered the quality of the local public schools inadequate to the task of educating their children, so they sent them all to private schools. Not only was it a financial burden, but for Anne it meant she would have to drive them to school every day for fourteen years. In order to pay the hefty school bills, Bob and Anne remained frugal. The house was furnished with pieces both inherited and discovered. Seldom was a new household appliance purchased. Washing machines and other large appliances were mostly other people's discards that were old but still functioning. Bob kept a stockpile of appliances in various stages of disrepair in the barn, with at least one at the ready should the one on duty fail.

Bob and Anne Morris provided their children with an idyllic, if somewhat eccentric life. Their range of options and their exposure to life in general were far broader than those of the vast majority of children their age. And the family was unusually close-knit. They played in orchestras together, sang in choirs together and took regular trips to Manhattan. And when the family went on vacations together, it wasn't to island resorts, but to Iceland for a month, or to England for canal boating. Bob Morris's work was on the cutting edge of a discipline that was defining the future; yet when friends came to visit from more suburban communities, they felt as if they had stepped into a time warp. And to complete the picture, the computer terminal resided in the basement next to an enormous eighteenth-century beehive oven.

Bob left the task of child rearing to Anne. She believed that children should be exposed to a full range of experiences and should be able to draw on that broad spectrum when choosing the direction of their adult lives. Ben would eventually pursue a life outdoors, working as a tree surgeon in Millington. Meredith would choose liberal arts and become a researcher at the Library of Congress. And from an early age, Robert seemed destined to follow his father into science.

It was an admirable approach, but it also required a disciplined household. Anne instilled in her children a strict work ethic: each had morning chores to do outside, animals to feed, eggs to collect, wood to gather. For the most part, the three children discharged their chores without complaint. Some of the work was also a great deal of fun. Gathering wood in winter, for instance, meant cutting a path up the icebound river with shovels and skates, and carrying the wood back down on sleds. Anne also made it clear that they were wholly responsible for their assigned work around the house. If Robert neglected to feed the sheep in the morning, he would return from school in the afternoon to a chorus of bleats.

Through the years, changes were made to the house that reflected the Morris way of life. Bookshelves went up everywhere. In time, the family library included six thousand volumes, their subject matter ranging from theology to natural history to sailing and navigation. Every book had been read by at least one member of the family. One day, Bob brought home one of the original Enigma cryptographic machines. On one of what had become regular visits to Fort Meade, Bob had simply walked out the front door of the NSA, accompanied by the agency's deputy director, with the machine stuffed into a brown paper bag. Eventually it became yet another Morris household curio.

The children weren't given an allowance. Instead, they were paid for work they did around the house outside of their normal obligations, such as digging drainage ditches and building fences. Anne was always careful not to pay them very much, in order to get them to see that they could earn more money by working elsewhere. Other kids always had more money than the Morris children, and later, when others had the use of their parents' cars, Bob and Anne told their children that if they wanted to drive, they had better figure out a way to buy their own cars. Robert and Meredith accepted the arrangement perhaps more easily than Ben, who groused mildly at the restriction.

Like many children in rural settings, the Morris children grew up without a gang of neighborhood kids to run with, a nearby shopping mall or a video arcade. When the children were very small, the family had no television set. But when it turned out that six-year-old Meredith was a mass-culture "illiterate," as Anne described it years later, the family bought a tiny black-and-white set specifically to view "Sesame Street." An upgrade in size came only because all three children had trouble seeing the screen at the same time. Still, television wasn't so much prohibited as quietly discouraged. The black-and-white set with poor reception competed for space in the living room with the computer terminal, which had migrated upstairs. When Anne voiced complaints about having a computer terminal in the middle of the living room, Bob gently reminded her that he could have put it where some of his colleagues had theirs—in the bedroom.

The young Morrises were early and voracious readers. Meredith was already reading at age four. By the third grade, Robert had read Tolkien's Lord of the Rings trilogy and memorized its many poems. By age nine he was devouring back issues of Scientific American, and by the time he was in his early teens his reading list had expanded to include the classics, history and copious science fiction.

Robert's intelligence was especially apparent from an early age. As a preschooler, he built working-scale models of cars out of whatever tools he found lying around—paper clips, cardboard and file folders. Before long, following his father's example, he was pulling electronic equipment apart and piecing it back together.

Anne saw that Robert sensed that he was different from others his age. But he recognized only that he was different, not why he was different. In fact, he once confided to his mother that he thought he was "weird." She occasionally tried to inquire just enough to test whether he understood that his abnormality lay in his intelligence. But even though it was clear to his parents that Robert was more intelligent than his schoolmates, he seemed only confused, and occasionally frustrated, by the difference between him and his peers.

Robert and Ben started out at the Country Day School in Far Hills. Robert was easily bored and his performance suffered accordingly. When Robert got to the fifth grade, Bob took matters into his own hands: he went to the headmaster and suggested that Robert skip into sixth grade. The headmaster refused, citing school policy. Bob's response was to keep Robert home for four days. The headmaster relented, put Robert in sixth grade, and his grades improved immediately. Nonetheless, unhappy with the direction of the school's curriculum, Anne and Bob pulled the boys out and enrolled them in the Peck School, twelves miles away in Morristown.

Following the Peck School's more traditionally rigorous curriculum, Robert improved his scholastic performance dramatically. Still, he was well beyond his classmates on most subjects. By the time he was in seventh grade, Robert was reading science fiction at a clip of two or three volumes a day. Bob and Ben were avid science fiction fans, too, but Robert was the one who seldom went anywhere without a science fiction novel tucked under his arm. When Anne went to parents' day at the Peck School one day, she saw her son seated in the front row of his math class, his nose buried in a science fiction book. When called upon, Robert simply looked up from his book, recited the correct answer and returned to his reading. It was clear to Anne that this wasn't cheek on her son's part. It seemed a perfectly suitable arrangement between the teacher and a student who could read his books and still stay a step ahead of the class.

High school meant yet another private school. Delbarton was an exclusive boys' school, also in Morristown, run by Benedictine monks. Delbarton was known for its excellent music department, and after Robert's third week there he came home one day and announced that he planned to learn the violin. Once Robert started, Ben took up the viola, Anne played the bassoon, Meredith started on the French horn and Bob dabbled, starting out on the oboe, then turning to the cello. Anne and Bob made music a family focal point. Each child was introduced to grand opera at age ten, with a trip to New York. For years, Bob's annual Christmas present to the family was *Hänsel und Gretel* at the Metropolitan.

Ben and Meredith liked using the computer well enough. They logged on mostly to play games. But of the three children, Robert was the one to fasten onto computers most earnestly. When Bob stepped away from the terminal it was only a matter of minutes before Robert logged on. In front of the terminal was a cavernous and comfortable old armchair, its back facing the rest of the living room. Whoever sat in the chair was enveloped by it, and the young Robert nearly disappeared.

Then there were the electronic friendships computers created. To give the children a sense of what was possible with computers and communications, some of the parents gave their children their own accounts on the computer at Bell Labs. Aside from some fairly strict ground rules about behavior on the network, the kids were allowed, and even encouraged, to explore the world of computers firsthand. Ken Thompson's son Corey was a regular on the network. At times, there were up to twenty-five kids using the Bell Labs computers and communicating with each other. Many of them, in fact, developed strong electronic friendships before they ever met in person.

Robert's was the first generation to grow up with ubiquitous computer networks. Using the computer gave the fourteen-year-old Robert his first taste of the power of instantaneous communications, and the social equality that computers made possible. Tapping the computational power of a machine ten miles away presented an irresistible lure. Robert became a regular, making friends on-line and exchanging homemade computer adventure games. Not only were they cleverly programmed, but the kids' games also required a fairly sophisticated knowledge of data communications. They were similar to early adventure games such as Zork and Adventure. These games were really vast puzzle-solving exercises played at a computer terminal. They were interactive, permitting the player to explore by typing commands at the keyboard. The games revolved around treasure hunts and magic words. One of the teenagers wrote a game called t4c (The Four Corners), complete with underground passageways. The best thing about the game was its interactive, multiuser nature. Players ran into each other while playing.

Robert then wrote a game called Run-Me, an enhanced spinoff of t4c. In t4c, characters could only talk to one another. Run-Me players could also hug, kiss, hit and tickle. With Run-Me, Robert established himself as the games master of his group.

Not only were the teenagers learning about computers, but they were learning the rules of the computer community. For some of them, the Bell Labs computer was a telephone and television rolled into one, fulfilling their social needs and their need for entertainment.

One of Robert's best friends on the network was the unusually bright daughter of a Bell Labs scientist, one of the few girls on the network. Robert set up some features of the *Run-Me* game specifically for her. For example, the altar in the church would shimmer when her character entered. One of her most impressive achievements was her own wardrobe program, which told her what to wear each day. Her automated decision was a function of the articles of clothing in her drawers and closet. Each morning when she called up the program, it would tell her which pants and shirts hadn't been worn recently and would select several possible combinations for her. Though they lived just eight miles apart, she and Robert carried on an electronic courtship for a year before actually meeting.

Only rarely did the children of the network overstep their bounds. One day Bob arrived at work and stormed into an office where some colleagues were sitting. He announced to the group in his trademark summary manner that all the kids' accounts had to be taken away immediately. Deciding there must be a story behind this sudden decision, the others prodded him into telling them that one of the kids had been operating as a superuser on the computer.

"Well, then, just take away that kid's account," suggested one in the group.

Bob shook his head.

After more probing, Bob broke down and said it had been his own kid.

"How did he manage to get the root password?" someone asked.

"He didn't."

"Well how did he get in there?"

It finally surfaced that Bob had absentmindedly walked away from the terminal at a point where he had access to everything on the Bell Labs computer, leaving Robert the run of the system. Robert had just walked up to the machine and started using it.

Robert was clearly interested in more than just playing games on the computer. By the time he entered junior high, his father had introduced him to UNIX and he was already finding holes in it. He was soon writing his own UNIX "shell," a sophisticated program for carrying out user commands. As soon as the UNIX source code was on line, Robert started to study it with a special zeal. In his mid-teens, Robert was showing his best friend, Doug McIlroy's son Peter, how it was possible to get superuser privileges on one computer, then parlay those privileges into a tour of various computers at the lab. Robert even modified a few files before alerting his father's colleagues at Bell Labs to the security hole he had found. If researchers at Bell Labs were amused, or grateful to a teenager for pointing out weaknesses in their own handiwork, they didn't let on. He was told to stop and that was that.

Even as a ninth grader, Robert was more his father's colleague than his disciple. Bob was careful never to sit Robert down and say, "Here, I'm going to give you a lecture." For weeks at a time, father and son could be steeped in an ongoing discussion of a technical problem. A conversation could last for hours or for days, and while they were talking about whatever it was—it could be a discussion of a security flaw in UNIX or of building an electronic circuit together—they remained oblivious to the rest of the family. As the one with more knowledge to impart, Bob could occasionally be hard on Robert, and extremely challenging. In overhearing some of these exchanges, Anne could tell from the tone of Bob's questions and Robert's quiet responses that Bob was pushing Robert. But that was Bob's way. He was accustomed to quizzing everyone anyway, asking his questions in short, clipped phrases that might seem abrupt and impatient to an outsider but were unthreatening, at times even playful, to those who knew him. Mostly, the part of their relationship that involved computers centered on theoretical questions. Yet Bob always encouraged Robert to refine his practical programming skills.

To outsiders, it seemed that Bob might even be encouraging Robert to break into computers. In 1982, Gina Kolata, a writer for *Science* magazine working on a story about computer crime for *Smithsonian* magazine, went to interview Bob Morris about security. He told her about tiger teams and smugly predicted that after a few minutes of looking in her wallet he would know enough about her to guess her computer password. When she asked him if he knew of any young hackers she could interview, he suggested she speak to his son on an anonymous basis and invited her to the house. The sixteen-year-old Robert struck Kolata as unusually shy, almost intimidated by the reporter. Anne Morris supervised the interview, and while Anne appeared to be protective of Robert, Kolata got the impression that father and son were a duo, egging each other on. Young Robert told her that yes, he had read private computer mail and had broken into computers that were linked together in networks. "I never told myself that there was nothing wrong with what I was doing," he told her. But, he said, he continued to do it for its challenge and excitement. In an ironic coda, that year Robert placed eleventh in a state high school physics competition. His prize was a subscription to *Smithsonian*, and the first issue he received was the one with Kolata's article in it.

The *Smithsonian* article came out at a time when awareness of computer security was growing gradually. By the 1980s, hundreds and then thousands of personal computers were linked together via networks and one user, one machine was the new computing philosophy. But then another idea began to form: why not create a computing system that wasn't found in a single computer but was spread throughout a network of computers? Could the system itself be so intelligent that when a particular computing task needed to be done, it could be distributed automatically to the geographic point that had the best available resources? A computer revolution that is still only partially realized was under way.

▲ ▼ ▲

When Robert was growing up, networks were for the most part private laboratories used by computer scientists who were experimenting with new ways of using computers. The things he observed his father do, and the research he heard and learned about, served only to reinforce that perception. But the world was changing rapidly, and the most powerful instrument of change over the next decade was the Arpanet. Its name derives from ARPA, the Pentagon's Advanced Research Projects Agency, which was renamed Defense Advanced Research Projects Agency during the 1970s. This agency was run by scientists rather than soldiers and it was charged with exploring high-risk ideas. For American computer science in university and corporate research centers, DARPA created an entirely new world. During the 1960s and the 1970s DARPA funding was crucial to the most significant advances in computer science. Personal computers, networks, artificial intelligence and voice recognition all in one way or another were the fruit of DARPA-funded experiments.

The Arpanet network in turn was the brainchild of a community of computer scientists who, during the late 1960s, were among the first to envision permitting scientists and engineers to share computers and expensive resources instantly and easily no matter where they were. That a computer network could serve as both a means for instantaneous communication among researchers and an experimental communications laboratory was a revolutionary notion.

At the beginning of the 1960s, Paul Baran, a scientist at the Rand

Corporation, was searching for ways to make telephone networks more reliable in the event of nuclear war. Out of his research came the idea of breaking digitized messages up into "packets of numbers." Each packet would carry an electronic address, and each could be routed by the most efficient route. Packet switching dramatically lowered the cost of data communications, making low-cost computer networks possible. The notion of actually linking computers to share these networks came from J. C. R. Licklider, a psychologist who went on to become the first director of DARPA's information processing and technology office.

The original Arpanet was built around separate message-passing computers known as Interface Message Processors, or IMPs, which were the backbone of the network. Later small computers, known as TIPs (terminal interface processors), which handled connections with slow dial-up terminals, were added. Each of the IMPs would be connected to another IMP on a leased phone line and was capable of sending and receiving at what then seemed like an extremely high speed. Contemporary networks routinely carry data at twenty times that speed and network designers are working to build a "national data highway" that would increase the speed of today's fastest commercial links by up to seven hundred times.

The first Arpanet node was installed at the University of California at Los Angeles in late 1969 and the next three nodes were placed at the University of California at Santa Barbara; Stanford Research Institute, a California think tank; and the University of Utah. The next year three more nodes were added on the East Coast: at the Massachusetts Institute of Technology; Bolt, Beranek and Newman, the Cambridge, Massachusetts, think tank that designed the Arpanet network; and Harvard University.

Other research projects had linked computers experimentally, but the Arpanet was to grow into the first nationwide computer network. The Arpanet connected research centers, military sites and universities. Initially, virtually all of the computers on it were identical (almost all were Digital PDP-10s), and virtually all of the people at those sites were government-funded computer science researchers. By 1973, the Arpanet consisted of twenty-five machines.

To be at a site connected to the Arpanet was to be among an elite. So coveted was a connection to the network that academic job offers were sometimes accepted or turned down on the basis of promised access to the network. For some computer scientists, access to the network was a requirement for doing their jobs. For these scientists, going to a university without a network connection would have been like a research microbiologist accepting a job at a school with no microscopes.

In its early days and even into its middle years, the Arpanet had the feel of a private club. "Are you on the net?" was a question heard among the most elite computer scientists. Getting into the club wasn't easy, but once you were in, you were given free rein. There was no concept whatsoever of security. Anyone anywhere could read a file anywhere in the network. At Carnegie-Mellon University, for example, every file on every computer, save those that were explicitly protected, was available for examination or copying by anybody on the Arpanet. Graduate students at those places spent many happy hours cruising around through the files on outside computers to see if there was anything worth reading.

In 1975 the operation of the network was turned over to the Defense Communications Agency, a Pentagon organization that is responsible for military voice and data traffic. By then, there were more than sixty sites on the network, and the amount of data traffic carried by Arpanet had increased dramatically.

Part of the clubbishness that defined the early Arpanet grew up around the technical limitations of the network. Through the 1970s, the Arpanet could only support 256 computers. But by 1982, a new network-addressing scheme was developed to allow for exponential growth of the network. By the mid-1980s, the Arpanet had become the seed for a complex of networks called the Internet, which touched down in more than fifty countries. It was no longer just an engineering experiment. Computer centers used the network for technical support, researchers sent papers back and forth in an instant and software of all kinds flowed around the globe. Commercial enterprises adopted the technology of the network to create their own private versions of the network based on the same set of communication protocols. These corporations also used the Internet itself to stay in contact with operations spread around the world. The Internet in turn was connected through gateway computers to hundreds or thousands of other networks. Some began to speak of an even broader concept of interconnected networks. They referred to it as the Matrix, taking the name from the all-encompassing computer network in William Gibson's *Neuromancer*.

It was with some indignation that the Arpanet pioneers watched their network be appropriated by society at large. Whereas in the early days it could cost as much as $250,000 a year to maintain a connection to the network, the base of support had since grown to the point where the cost was minimal. Universities and corporate research centers still composed most of the links, but by 1988 the function of the network had broadened considerably. The Arpanet was supposed to be primarily a computing laboratory, but mostly it was used for sending electronic mail about every topic imaginable. The network pioneers were puzzled and not a little miffed to see newspaper reporters, of all people, with accounts on machines linked to the net. That was nearly as preposterous as the notion of walking into the campus chemistry lab and seeing a bunch of reporters wielding Bunsen burners and pipettes.

Gradually, over a period of years the original Arpanet network links were supplanted by faster data paths, and by 1990 the Arpanet ceased to exist as a separate entity, having been absorbed into the Internet. By one current estimate, several hundred thousand different computers are currently on the network, from supercomputers to personal computers. The best guess is that there are more than four million Internet users. This data highway already carries the work of scientists, students, soldiers and businessmen, and many now argue that connecting it to millions of American homes and businesses will revolutionize the country with new business, educational and entertainment services.

But at first, the reason for the existence of networks was to carry out experiments that explored the reach and power of the networks themselves. In 1971, Bob Thomas, a scientist at Bolt, Beranek and Newman, was working on distributed computing software. His group designed an air traffic control simulation that was intended to model different airports on different computers. The idea was to be able to move control of an airplane from one computer to another and tell all the other computers so that they would know the location of a particular aircraft had been changed. To do this, Thomas wrote a clever program whose mission was to crawl through the network and pop up on each screen, leaving the message, "I'm creeper! Catch me if you can!" Some time later, as word of the program grew within the early network community, other hackers wrote similar programs—some of which multiplied as they worked their way around the net (*Creeper* didn't reproduce itself, it simply moved), and others of which included "reaper" programs that sought out and destroyed creepers. Writing such programs became a minor fad for a few months and then died out.

In the early 1980s, two computer researchers at Xerox's Palo Alto Research Center started experimenting with programs they called "worms" that were able to run on many computers in a local-area network. (The Arpanet was a wide-area network, connecting computers over long distances.) The term *worm* was taken from the book *The*

Shockwave Rider, a science-fiction classic written by John Brunner in 1975. It describes an authoritarian government that exercises power through an omnipotent computer network until a rebel programmer infests the network with a program called a "tapeworm." In order to kill the worm, the government has to turn off the network, losing its power in the process.

Brunner became a cult figure, as the book swept through the world-wide community of science fiction readers. It had a strong influence on an emerging American computer underground—a loose affiliation of phone phreaks and computer hackers in places like Silicon Valley and Cambridge who appeared simultaneously with the development of the personal computer. John Shoch and Jon Hepp, the Xerox researchers, were looking for a way to make shared computing power more widely and easily accessible over a local area network. They came up with five or six useful worms. One was called a "town crier worm." Its job was to travel through the network posting announcements as it went. Another was a "diagnostic worm." It was intended to hop from machine to machine, constantly checking to see if anything was amiss. Certainly the most dramatic distributed program the two conceived was the "vampire worm." Such a program, they suggested, would take advantage of the almost limitless free processing power in a network of computer workstations. After all, these machines spent many idle hours that could be harnessed for useful work. The Xerox vampire worm automatically turned itself on at night when people had gone home, setting to work on complex problems that required vast amounts of computing power. In the morning, when the computers' human owners returned to reclaim their machines, the vampire program would temporarily store the partial solution computed so far and shrink back to wait for the next evening.

But early on, Shoch and Hepp were also to learn of the potential dangers of worms. One night a malfunctioning program went out of control on a local area network at the Palo Alto Research Center. In the morning, when scientists arrived, they found that computers throughout the building had crashed. They began to restart their systems, but soon found that each time they attempted to start a machine the defective worm caused it to crash again immediately. The problem was that many computers were behind locked doors and couldn't be reached. Finally they wrote a "vaccine" program that traveled through the network and electronically inoculated each computer in the network against the worm.

The Arpanet was also a resource to Bell Laboratories scientist Ken

Thompson, who used it for a computer security experiment. In the late 1970s, when Thompson was working on a paper about breaking password security, he used several network sites, such as Harvard, MIT, Carnegie-Mellon and Berkeley, on which password files were publicly accessible or on which he had accounts with access to these files. His password-cracking program was successful and he discovered that he had inadvertently captured passwords used by some of the Arpanet's key administrators, people with accounts on many machines throughout the network. He tried the passwords and discovered they worked. In the hands of those whom Thompson and Bob Morris thought of as network "bad guys," such a security flaw was dangerous. So Thompson sent mail to the people who owned the passwords to tell them about the problem.

▲ ▼ ▲

Robert leapfrogged entirely the process of learning computers in school. The Delbarton School had early Apple computers, but from age twelve Robert had access to a machine ten times more powerful. While the school was handing out computer achievement awards to other students for mastering the Apples, Robert was already writing complicated programs and technical papers.

Yet few of Robert's friends and teachers at Delbarton even suspected that the diffident sophomore had such a level of expertise in computers. Robert had launched his computer career entirely from home. During Robert's senior year, Bob's old friend Fred Grampp hired Robert part-time at Bell Labs. Robert at sixteen behaved like any of the dozens of college students who took part-time jobs and internships at the Labs. Unlike his father, he was inordinately quiet, but he shared his father's tremendous curiosity about the world around him.

Robert had already made something of a name for himself at Bell Labs with his earlier unsanctioned tours of Bell Labs computers. But he was a hard worker. His project there was his own idea: to write a more secure and efficient implementation of UUCP, the program used for copying files from UNIX machine to UNIX machine. The challenge was to write a UUCP implementation that could cope with the volume and variety of traffic that had evolved on the network over the years. It's not every high school student who can redesign a major piece of software. Despite a few problems, Robert's program was so good that it became the model for UUCP that Bell Labs eventually adopted. He even produced a technical paper on results of his work, titled "Another Try at UUCP."

Yet he wasn't single-minded in his devotion to computers. Robert distinguished himself early in other ways at Delbarton. He swam on the school team and sang with the chorus. Yet he remained shy and, as far as his parents could tell, still unaware of his intellectual gifts, so Anne took it upon herself to have a talk with the headmaster at Delbarton. She explained that she thought Robert might benefit from a boost from his superiors at the school. Apparently in agreement with this concerned parent, the headmaster went out of his way to praise Robert in the presence of other students. On the day the school received the results of that year's SAT exams, the headmaster greeted Robert in the front hall of the school and, in front of a dozen other students, told Robert that his scores—a perfect 800 in verbal and a 790 in math—were the highest in the school's history. From that point on, Robert's self-esteem seemed to soar. For college, he set his sights on Harvard. Not only had his father gone there, but for several generations back the Tappans and the Farlows on Anne's side of the family had as well. Robert applied for early acceptance and got in.

When Robert entered Harvard in the fall of 1983, he was still shy and socially awkward. But he knew of one place he could go where he would have a good chance at fitting in quickly: the Aiken Computation Laboratory. Most college campuses have a research computer center, distinct from where the university's own central data processing is performed. At Harvard it's Aiken, which takes care of the computing needs of the university's Division of Applied Sciences. The central computer center, with operators who don't need to know much more than how to feed the printer, is across the campus at the Office of Information Technology. Aiken is a little faster and looser with its computers, and hence a more interesting place to work. When Robert arrived at Harvard, there was in fact no computer science department per se at the school. Instead, there was a computer science faculty, a group of seventeen faculty members within Applied Sciences. Serious computer science students at Harvard gravitated, more often than not, toward Aiken, where computer science faculty spent their time.

An all-brick monument to an architectural aesthetic grounded in common sense, Aiken stands diagonally across from the magnificent law school building on the Law School Quadrangle. Inside the Aiken lobby, spanning an entire wall, stands Howard Aiken's fifty-one-foot-long, eight-foot-tall legacy to modern computing—the Mark I Automatic Sequence Controlled Calculator. During the thirties, the mathematics professor had a dream of building a large-scale calculator, a switchboard-mounted device that would do arithmetical operations without the intervention of an operator. In 1944, in collaboration with IBM, Aiken completed the Mark I, at a cost of $250,000. It was the world's first large-scale electric calculator. A typical problem that would have taken a team of four experts three weeks to solve occupied the machine for only nineteen hours. By the late 1980s, a problem that would have really challenged the Mark I could be done in a second or two on a $40 programmable hand-held calculator. Still, the Mark I was a revolutionary development in its time and Aiken's place in the history of technology was duly secured inside the building named for him.

Across from the Mark I is a glass-enclosed room of terminals and workstations, a place where students and Aiken staffers work. In 1983, the hard-core computer people at Aiken who didn't have their own offices spent most of their waking hours in the room and others drifted in and out. In the 1950s, Bob Morris had also spent time at the computation lab, helping to build the Mark IV, the fourth generation of Aiken's calculator.

Shortly after arriving at Harvard, the younger Morris walked into the Aiken administrator's office and asked for an account on the lab's computer. Eleanor Sacks, the Aiken administrator, patiently explained that freshmen weren't given accounts at Aiken, that Aiken was the exclusive province of faculty and more advanced students. She gently told him to join the other freshmen a few doors down at the Science Center. But Robert didn't especially want to join the masses in the basement of the Science Center, a sea of computer terminals and personal computers that resembled a word-processing pool more than a computer science lab. Not only was Aiken a more civilized place to sit and program, but there were more computer resources available there. But instead of trying to argue about it, Robert wandered back out of her office. A few days later, he took care of the problem himself by turning the Aiken VAX into a single-user machine, creating an account for himself and then returning the VAX to multiuser status. His login, which he had used since the days on the Bell Labs computer, was *rtm*. Shortly thereafter, a faculty member who knew Bob Morris saw to it that Robert received a legitimate account.

Nick Horton, the Aiken manager, knew little about UNIX, and before long Robert was a permanent fixture there. He wasn't one to learn a little about a lot of things. Like his father, Robert learned a lot about a lot of things. He could handle hardware emergencies as well as software problems. Once his expertise became know at Aiken, his services were in great demand.

The number of computer science majors at Harvard is relatively small —each year, about thirty students get their undergraduate degrees in computer science—but they pride themselves on being more well rounded than their counterparts down the street at MIT. Robert may have had an unusual aptitude, but he was certainly no freak. Everyone around him had a dozen other interests. The spirit of a place like Aiken was personified by the students who worked there as part-time staff programmers. One professor who needed a student to do some programming for him was amazed by what he saw: there the student sat, and as he waited for output from the computer he appeared to be reading two books at once, one in French and one in German, just to keep himself occupied.

Perhaps what made Robert stand out most was his impressive knowledge of UNIX. He could sit for hours just reading UNIX manuals. The UNIX documentation had grown to consist of more than two thousand pages, and on each page was a new set of minutiae concerning the workings of the operating system. Most people kept the manuals on hand purely for reference, but Robert appeared to enjoy simply reading them. Before long, he was regarded as the most knowledgeable UNIX technician on campus. And he had more than a theoretical understanding. He had a tremendous capacity for remembering details. If someone had a question about UNIX, it was often easier just to ask Robert than to look it up. While some people were in awe of his capacity for minutiae, others wondered if that was all he thought about. He was, in computer parlance, a systems hacker through and through.

Like his father, Robert was especially talented at putting together a quick program that would solve a pressing problem. On one occasion, when a professor got a new computer, he needed some software written for it. He asked Robert if he could do it, and without needing to pick up a pen to sketch an outline of the program first, Robert sat down at the computer and just typed. He was finished within a few hours, and the program, while not the most refined of code, certainly did the job.

By the end of his freshman year, Robert was spending almost all of his time at Aiken. He was doing odd programming jobs and technical support, all of it *gratis*, and he had become indispensable. When others asked him why he didn't apply for a job at Aiken so that he could at least get paid, Robert replied that his father had told him not to take a job right away so that he could concentrate on his schoolwork. This way, he could keep to his word but still have fun at Aiken. And since he wasn't on the payroll, he could work on projects of his own choosing.

Robert spent the summer of his freshman year living at home in Millington and working at Bell Labs. A second technical paper came from the summer's work, calling attention to a security hole in Berkeley UNIX. By this time, Robert's expertise was so well appreciated at Harvard that a special data line was set up between Harvard and the research machine at Bell Labs so that Robert could perform remote diagnostics and maintenance from New Jersey during the summer. Robert's notes were always terse and they almost always fixed the problem. Even when he wasn't asked directly for help, he offered it anyway. For example, while browsing around in the Aiken system he noticed that some hardware had been installed improperly. A message showed up in Nick Horton's mailbox one day: "Try swapping the two boards in Positions A and B." It was from Robert. The problem was fixed.

When it came time to declare his major, Robert started out in math, but soon switched to computer science. In the first semester of his sophomore year at Harvard he was hired as a staff programmer. He wasn't actually doing much more than he had done when he worked for no pay, but now he spent even more time at Aiken, to the exclusion of nearly all else, including his coursework. His academic performance fell to the point where the college ordered him to take 1985 off. Rather than tell his parents right away that he was in trouble with Harvard, Robert lined up a full-time job as a programmer at Convex, a hot new computer company in Dallas, and presented the news of academic probation to his parents as a problem for which he already had the solution. Once again, while in Dallas, Robert was a remote diagnostician and consultant for Aiken. When Nick Horton asked him a technical question, he would send back not just a paragraph of explanation, but a lengthy example and tutorial.

While at Convex, Robert helped run the company's time-sharing systems, and wrote software that would analyze and simulate the performance of Convex hardware. He was also flown out to customer sites as the company's troubleshooting whiz kid. It was a pretty lonely time for someone so young and so shy. When people gave him projects to do, he invariably finished them early and waited for something else to do. Outside of work he learned rock climbing and scuba diving, and he played *Photon*, a high-tech version of Capture the Flag. When he went to visit his friends at Aiken, he said he was looking forward to coming back. One of the conditions of Robert's return to Harvard in early 1986 was that he not work again on the Aiken staff, at least not right away.

So although he continued to spend time there, he was no longer a formal employee.

The Aiken staff came and went, but in early 1986 the esprit de corps was especially strong and the group especially diverse. There were Nick Horton, a psychology major and social activist; Andy Sudduth, a tall, red-haired Olympic rower; Steve Kaufer, captain of Harvard's fencing team and in the midst of starting a software company; Karen Beausey, on her way to law school; and David Hendler, one of Robert's closest friends, a linguist and history of science major. It was a group that somehow clicked together particularly well, doing things outside the lab, taking trips to museums, going on ski vacations and eating dinner together. They all knew a lot about computers but they also knew a lot about other things. David was a gourmet cook who found some of his best recipes on the USENET Cookbook, a network recipe exchange. Nick Horton was also an avid subscriber to the recipe exchange and, as a Christmas present one year, Nick printed out the archives of the cookbook, bound it and sent it to everyone at the lab. Around his friends at Aiken, Robert's shyness melted away. Engaged in a technical discussion, Robert was quite animated. In his element, he could be positively expansive.

Working at Aiken meant keeping irregular hours and adjusting to a hectic and occasionally demanding environment. Aiken staffers did so uncomplainingly. Dozens of computer start-ups were eager to use places like Aiken as a test-bed for their new hardware and software. There was also a push among the Aiken staff and some of the faculty to procure equipment that was new and interesting, if not entirely reliable. The popular sentiment was that if it didn't work, then it could be made to work, so Aiken staffers spent a lot of their time trying to fix things. If there was a problem in an area that no one knew anything about, someone would volunteer to become an expert in the course of an evening.

Robert always managed to find time for some harmless pranks. Exploiting people's tendency to type "mial" by mistake when asking for their electronic mail, Robert wrote a program so that each time someone made the error, instead of mail a Dungeons & Dragons–like adventure game appeared on the screen. He excluded senior faculty members from the prank; when they made the typing error, the system simply said it did not recognize the command. The "mial" prank was clever and harmless, but after a while people became annoyed with the game and Robert was told to remove it from the system. Then, as an April Fool's joke, Robert wrote a program that made it appear to anyone who logged in that Harvard had gone back in time ten years and was using a long-obsolete operating system on equally obsolete hardware. Whenever Robert was asked if he was the source of a prank, he would look down with a shy smirk.

Then there was the Oracle. Anyone logging on to the computer was told to ask any question of the Oracle. But before the question could be asked, a question from the Oracle had to be answered first. Some questions tested one's knowledge of technical trivia; others were just silly ("Why do we have 8:30 a.m. classes?"). It took everyone a while to figure out that it wasn't the computer itself generating questions but others using the system. Whenever someone logged on, he or she fulfilled the computer's request to ask a question, which was sent on to the next person to log on. That person's answer was mailed to the user who posed the question, and so on. The Oracle's cleverness lay in making it look as if the computer were doing everything, when in reality people were both asking and answering questions and the computer was just mailing messages back and forth.

Those who knew Robert well were aware that he had a special interest in computer security. He wasn't one to boast about his computer security expertise, but it helped to explain his preoccupation with studying UNIX line by line. Careful examination of the code itself was the best way to unearth security flaws. But he didn't flaunt his detailed knowledge of the operating system, and he certainly didn't announce plans to follow a career in security. Nonetheless, one of his favorite refrains was how many holes there were in Berkeley UNIX.

At the same time, Robert had a sense for where to draw the line when probing security. Once, he and David Hendler were discussing a particular way of logging in to machines around the network. Taken with the notion, David considered logging in to Brian Reid's computer at Digital Equipment's research laboratory in Palo Alto, but Robert advised him strongly against doing that. David knew Reid for his network recipe exchange, but Robert knew him to be an especially conscientious network sentinel who would notice something amiss immediately if someone logged on to his computer. Robert made a practice of breaking into only the computers of people he knew wouldn't mind.

▲ ▼ ▲

Paul Graham, a hyperactive and pink-cheeked computer science graduate student, had always considered himself more intelligent than virtually everyone else. In his twenty-one years he hadn't seen much
evidence to indicate otherwise. Then Paul heard from a friend about
someone who, the friend said, was on another plane altogether.

One day at an Aiken party shortly after Robert's return from Dallas,
someone pointed out the brilliant young Morris.

Paul went up to him. "Hey, aren't you Robert Morris?"

The young man lowered his head and grinned, then pointed across
the room at someone else and said, "No, that's him."

It wasn't until several days later that Paul learned that he had been
duped, if only because the same person who had disowned the name
Robert Morris was always at Aiken Lab, always working until at least
3:00 A.M. and always working on something that seemed complex.
When Paul started spending time at Aiken, Robert Morris was writing a
program called a ray tracer for a graduate-level computer graphics course.
A ray tracer produces images of three-dimensional scenes. Given a model
of the scene in geometric shapes, the ray tracer follows the path of
individual rays of light from their source as they bounce off objects in
the scene and eventually enter an observer's eye. The most impressive
thing to Paul was that even though the course had ended and Robert
had already received his grade, he was still working on perfecting the
program for the sheer intellectual challenge of it. In fact, Robert's program was so interesting that it caught the interest of his roommate, Greg
Kuperberg, a math student, who helped him with some of the more
complex mathematics needed for constructing solid shapes.

Ray tracing requires vast numbers of computing cycles and Robert
took them wherever he could find them. But he was scrupulous about
not affecting the other users on the system. So he wrote a program a bit
like John Shoch's vampire worm for making use of perfectly good cycles
on computers around the lab that would have otherwise gone to waste.
When a user sat down at a workstation and began to type, the computer
stopped doing Robert's work and went to work for its rightful owner.
Robert extended his clever redistribution of cycle wealth so that other
students could use it, too.

Paul came to call him by his login: *rtm.* There was no limit, it seemed,
to rtm's knowledge. He not only knew about the workings of the VAX,
but he also knew about graphics, and he had read all of the UNIX source
code. In Paul's view, rtm was no single-minded geek. This guy had read

all of the Norse Sagas. And he liked to go to the opera, of all things. He was nothing like Paul's suburban contemporaries, who had grown up addicted to video games, television and junk food. When Paul left Monroeville, a Pittsburgh suburb known for its immense shopping mall (it served as the set for the cult film *Dawn of the Dead*), his years in front of the television haunted him through college. When he got to college, he tried to make up for lost time by going cold turkey. One glance at a television screen could well turn into a week-long binge. But here was someone with no interest in that electronic drug, nor in video games. Paul felt that, compared to Robert's, his childhood had been wasted. He envied Robert's upbringing: the rustic environment, the private-school education, the adventurous vacations, the prominent father. Paul was in awe. He was a reluctant computer scientist who would rather have studied painting and looked upon others in the graduate program as irredeemably narrow-minded digit-heads. Meeting rtm was the best thing that had happened to him all year.

Paul knew that he and rtm were going to be good friends when he discovered one thing they had in common: neither of them liked to sit in classrooms, and if a course failed to challenge them, both were inclined to skip class frequently. One day, Paul was sitting outside on the steps of Aiken reading a book when he was supposed to be sitting inside absorbing a lecture on artificial intelligence. When rtm walked up to him and took a look at Paul's book, the historian Jacob Burkhardt's history of the Italian Renaissance, he smiled. Both agreed that reading Burkhardt was a far better way to spend one's time.

Others at Aiken considered Paul too unrestrained, but Robert was more willing to be his friend. He once took Paul along to a relative's house on an island off the Maine coast. As they were in a boat headed for the barren island, which had no electricity or telephones, Robert said, "You're going to like this. From now on, things are done right." It was Robert's appreciation for things that had nothing to do with material possessions that impressed Paul.

But Paul was concerned that his friend rtm didn't have girlfriends. "If you like someone, rtm, you've got to say something to her," Paul would insist. "You can't expect her to read your mind."

"But what are shy people supposed to do?" Robert would reply.

More than once, Robert and his friends got the itch to make a killing from their specialized knowledge. After completing their much-praised ray tracer, Robert and Kuperberg gave brief thought to launching a computer graphics firm. And with David Hendler, Robert mulled over the idea of a computerized method for predicting the commodities market. Robert hatched his most farfetched get-rich plan with Paul Graham, when the two decided they could make a bundle predicting the horse races at Suffolk Downs. Paul kept copies of the *Racing Form* locked in his desk drawer and the two spent hours entering reams of data about past races into the computer. But after two depressing afternoons of mingling with the crowds of desperate middle-aged men as they walked from the subway station to the racetrack, they decided that it wasn't worth the effort.

The summer after his junior year in 1987, Robert worked at Digital for the second year in a row, this time in Palo Alto. The previous summer he had spent at Digital's engineering facility in Nashua, New Hampshire—the same facility Kevin Mitnick would later break into electronically—working on routine programming tasks, which he found only moderately interesting. But the Palo Alto summer was wonderful. While there he worked on graphics programs and programming languages, trying things that had never been done. The work was extremely challenging and Robert thrived.

The Morris family, in the meantime, was uprooting itself from New Jersey and Bell Labs after twenty-six years. Bob had gotten frustrated at the labs. He had been waiting for months to be appointed to a new position that would oversee the creation of a secure version of UNIX, a version with no security flaws. The job got stalled in the bureaucracy, and while Bob's patience was wearing thin, the National Security Agency came to him with an offer he couldn't refuse: to be chief scientist at the National Computer Security Center, the unclassified component of the NSA. The center had been established to improve the security of computers within the military services but was later given a broader mandate that encompassed establishing computer security standards in the commercial world as well. The job was particularly attractive to Bob because, while most of his work revolved around the more public center, there was a classified aspect as well. Part of the time he worked in the arcane intelligence-gathering world of the NSA.

Bob and Anne sold the old house in Millington and moved to Arnold, a small Maryland suburb along the banks of the Severn River. Bob had crossed the line from theoretical research into a real game with real players. Anne was sad to give up her job as director of the Association of New Jersey Environmental Commissions, but she knew this was precisely the career boost Bob wanted. She eventually took a job with an environmental group in Washington, which required a long commute each day.

Meanwhile, Robert's senior year was another period of both intense work and good fun at Aiken. Schoolwork was shunted aside yet again— one geometry course that Robert found excruciatingly dull he scarcely attended at all. Despite a lot of cramming for the final, he failed the course. He spent little time in his room at Dunster House (the Harvard dormitory where his father had also once lived) preferring to sleep on the couch at the group house where David Hendler was living. Many evenings were spent cooking elaborate dinners and baking cookies to send to out-of-town friends. During his spring break, at his father's suggestion, Robert gave a talk at Bob's division of the NSA on everything he knew about UNIX security. The following day, he repeated the lecture to a group at the Naval Research Laboratory.

▲ ▼ ▲

When Robert applied to graduate school in computer science, Stanford was at the top of his list, followed by Cornell and Harvard. Stanford has the most rigorous program; while Harvard fosters a more nurturing atmosphere for its students, it isn't uncommon for first-year graduate students at Stanford and Carnegie-Mellon to fail their qualifying exams. Stanford's program is also by far the most difficult to get into. Of 1,000 applicants each year, the graduate program admits just 30. Cornell, which ranks among the nation's top ten graduate programs, is also very difficult to get into. Of 550 or so applicants, the school admits just 40 into each entering class.

Robert gathered letters of recommendation from some of the most respected figures in computer science. Doug McIlroy from Bell Labs wrote one. And Mark Manasse, for whom Robert had worked at Digital's Palo Alto Research Center, wrote an effusive letter. "I fully believe that Robert will succeed at almost anything he undertakes," Manasse wrote. Nonetheless, Stanford rejected him, partly because of his spotty academic record, possibly because his score on the math section of the standard Graduate Record Exam, though high, wasn't a standout in the fiercely competitive Stanford applicant pool.

But both Harvard and Cornell accepted him. His thesis adviser recommended against his staying at Harvard for graduate school. Perhaps it was time for a change, and Cornell was a renowned center of computer science theory. If Robert was to be faulted for anything, it was his tendency to allow himself to be seduced by the machines themselves, at the expense of a theoretical understanding. Robert's father, on the other hand, had such a strong mathematical foundation that he would instinctively bring mathematics to bear on problems that didn't appear mathematical at first. Cornell would be a perfect place for Robert to gain a better theoretical foothold on computer science. So Robert decided to go to Cornell.

He spent his last summer in Cambridge working at a plum job. At the recommendation of Jamie Frankel, the same adjunct professor who had recommended Robert for one of the summer jobs at Digital, Robert spent the summer of 1988 at Thinking Machines Corporation in Cambridge. One of the most interesting companies to work for, Thinking Machines had developed a supercomputer based on "massive parallelism," applying thousands of small processors—rather than one or a handful of processors running at tremendous speed—to divide up the burden of particularly numerically intensive tasks. The "Connection Machine" was being used for such applications as picking out ground structures from satellite photographs, predicting the behavior of molecules and making three-dimensional maps. The company was doing so many interesting projects that any job there promised to be fun.

Robert's principal project at Thinking Machines was to refine one of the sophisticated languages that took special advantage of the Connection Machine. On the side, he wrote a crossword puzzle generator, which took a blank pattern and a list of about fifty thousand words and filled in the grid. The only manual labor was in writing clues for the words. It was a perfect use of the Connection Machine's ability to try out millions of combinations of words very quickly. By the end of the summer, he had a working puzzle generator. He was pleased enough with his work to send one of the puzzles to the crossword editor of The New York Times. To Robert's disappointment, the puzzle was rejected.

▲ ▼ ▲

Robert enrolled at Cornell in the last week of August 1988.

Cornell is among the most isolated of major universities. Its campus is in Ithaca, a small city with a population of twenty-nine thousand at the southern tip of Cayuga Lake, one of the five large Finger Lakes in upstate New York's lush farming region. In the first week he was there, Robert skipped most of the computer orientation talks given by Dean Krafft, the campus computer facilities manager. It seemed unnecessary to attend basic lectures on logging in to the system and sending electronic mail. Krafft had handed everyone a copy of the computer science department's computer use policy, which prohibited the "use of . . . computer facilities for browsing through private computer files, decrypting encrypted material, or obtaining unauthorized user privileges." While Krafft was giving his talks, Robert was already logged in to the computer.

He didn't make many friends at first. He moved into an old house about a mile from campus with two other graduate students, but the students kept to themselves. It was nothing like the easy communal atmosphere at the house in Cambridge.

Upson Hall at Cornell didn't seem to foster the camaraderie and closeness of Aiken at Harvard. It was larger and more anonymous. Robert shared an office with seven other graduate students on the building's fourth floor. The office had just two terminals. From his desk, Robert had a southerly view that looked out on Cascadilla Creek gorge and on Ithaca College across the hills on the other side of the valley. Interesting science was happening all around him. One floor above in the building's newer wing, the plasma physics group of the electrical engineering department was working on space plasma physics using data from the space shuttle program. And in Cornell's computer science department there were major research efforts in physical modeling and simulation, robotics, and computer vision, as well as in reliable distributed computing, which looks at ways of building systems to survive the failure of individual pieces.

Robert started out taking basic graduate courses. In one, a small class on microprocessor design, the professor noticed that Robert had an unusual curiosity about how things work. He seemed less interested in concentrating on his assigned piece of the project of building a microprocessor than in the bigger problem of chip design. If something didn't captivate him right away, Robert was blunt about it. When another professor gave him a paper to read, he returned it, saying it hadn't interested him. He had spent enough years staring out the window when something bored him. Now he spoke his mind freely.

Robert was lonely and slightly distracted. He was late in submitting one of his first mathematics papers, and received only a fair grade. He was spending a lot of time at the computer but he wasn't necessarily concentrating on his schoolwork. He made friends with one of his officemates, Dawson Dean. Dawson had gone to MIT and was just the kind of one-dimensional digit-head Paul Graham so often complained about, but Robert was also quick to give someone the benefit of the doubt, and he thought Dawson was an okay guy. He enjoyed having technical discussions as much as Robert did. One night while both were working late at Upson Hall, Robert and Dawson started talking about network security. Robert pointed out that he had figured out several ways to bypass security on local area networks.

"Are you one of those people who breaks into computers for fun and then gets hired?" Dawson asked.

Robert smiled and nodded. He told Dawson that he had given lectures on security at the National Security Agency and the Naval Research Laboratory. But, he told Dawson, he wasn't particularly interested in making a career of computer security. "It's too boring," he told his officemate.

Robert exchanged lots of electronic mail with his old friends from Cambridge, most of whom had scattered to other places. A great deal of mail came from David Hendler, as well as from Janet Abbate, a housemate from the previous summer in whom Robert had taken a romantic interest. A graduate student at the University of Pennsylvania, Janet was getting ready to return to Philadelphia. She checked in regularly with Robert, sending him warm and high-spirited notes over the network and baked goods by mail. Eleanor Sacks, the former manager at Aiken, sent him a note saying she hoped Cornell would give him *rtm*, his beloved login. He still hadn't bothered to change the *morris* login Cornell had assigned him. Nick Horton, who had moved to Oregon, forwarded Robert a half-dozen recipes for Thai dishes from the USENET recipe exchange.

Robert quickly developed a reputation as a talented programmer who kept to himself. He wasn't aloof. He was just quiet. He sat apart from others in classes and declined invitations to join professors and fellow students for beer at a local pub on Friday evenings. Nonetheless, he took advantage of many other things Cornell had to offer. He signed up for a rock-climbing course, joined the computer science department's intramural ice-hockey team and started singing with the Sage Chapel choir.

▲ ▼ ▲

While the first personal computer virus probably emerged on the Apple II computer in the early 1980s, it wasn't until 1987 that viruses exploded into the public consciousness. A computer virus that struck at the campus of Lehigh University in Pennsylvania drew national attention that year. A year later, viruses made the covers of *Time* and *Business Week*. The programs were captivating because they were so mysterious and because they offered such a clear analogy to their biological namesakes. Computers became "infected." "Vaccines" were possible. Soon people were drawing comparisons between computer viruses and the plague of AIDS.

The American public assumed that all viruses were malicious and that they invariably destroyed data. But those who knew more about computing understood that there was no rule that such programs had to be harmful. It would be much more interesting, in fact, to write a program that was at once subtle and benign but still capable of spreading.

Writing a virus that would spread to as many computers as possible was an idea that just seemed to come to Robert, and the recent examples emboldened him. He loved the thought of an invisible piece of software that could propel itself through an electronic universe of thousands of computers, spreading slowly and imperceptibly, achieving immortality by protecting itself against anyone who might want to destroy it. And there were certain flaws in Berkeley UNIX that he had known about for at least two years, perhaps collecting them with the intention of someday putting them to use. By early October, Robert was thinking in earnest about writing the program. The goal of the program was simply to see how many computers he could reach. On October 15, he produced a wish list, a set of two dozen or so goals for his program, including these:

```
The goal is to infect three machines per
Ethernet (local network).
Only work if all users are idle.
Try to avoid slow machines.
Look through host table for the other
interfaces of known gateways, then find
hosts on that net.
Steal his password file, break a password,
and rexec.
```

In Robert's mind, it was a perfectly harmless plan for probing security of the network. It was something his father, in the early, clubbier days of computer science, might have dreamed up to earn the respect of his colleagues. It probably didn't occur to Robert that this was the kind of thing computer saboteurs might generate in order to bring down an entire international computer network.

▲ ▼ ▲

Robert got a ride to Cambridge for the long weekend over fall break with Dawson Dean. David Hendler was out of town on a long trip to Europe, so Robert spent most of his time with Paul Graham. Andy Sudduth, who was rowing that weekend in the Head of the Charles race, was around too.

It was like old times. Robert sat upstairs, glued to one of the workstations. Paul was sitting downstairs in the office of David Mumford, an eminent mathematician on the Harvard faculty, whose office Paul occasionally used when Mumford wasn't around. Early Saturday evening, Robert walked into Mumford's office, wearing his trademark smirk. Paul knew something was up. Robert was pacing furiously. He announced that he had been reading UNIX source code and had found a big bug in *ftp*, the file transfer program that enables users to copy files from machine to machine over a computer network. The hole enabled someone to read or write any file on the target computer.

From the level of Robert's excitement, Paul got the impression that he had just discovered this hole a few minutes earlier and was bursting to tell someone. Robert's pacing in the small office picked up.

Encountering Mumford's desk at the end of one of his paces, instead of turning around again, Robert walked straight ahead and paced atop the desk. It was a sure sign that he was completely lost in his discovery.

"rtm! You're on Mumford's desk," Paul cried as he watched his friend's sneakers shuffle the papers on Mumford's desk.

"Oh," Robert replied, and descended from the desk.

At first, Paul wasn't sure what Robert was driving at. It sounded like just another way of breaking into UNIX. "Well that's an amusing hole, but what's the point?" he asked.

"I could use this hole to write a virus," Robert explained. He said that for much of the fall term at Cornell he had been thinking about writing a virus, one that would spread slowly over the Internet. As Robert described it, the virus wasn't going to do anything malicious and certainly would not destroy data. In the end, it should do nothing at all except spread to as many machines as possible.

Paul was immediately intrigued. He had been badgering Robert all semester to make more friends at Cornell and improve his social life, but when he heard that such efforts had been deferred in favor of something as intriguing as this computer virus, he was delighted and envious.

"That's really great!" Paul was getting just as worked up as Robert. "You should do this for your dissertation!"

Paul was, in some ways, the perfect friend. When he got excited about an idea, his support was all the encouragement one needed to keep going. He could pull others into his excitement just by virtue of his own enthusiasm. Especially when it came to his friend and role model rtm, Paul was a one-man cheerleading squad.

When Robert began to talk about the virus he was planning, Paul's enthusiasm tripped into high gear. It must have had its effect even on the usually placid and low-key Robert. Robert had meant to keep his virus plans to himself. By telling Paul, of all people, Robert must have gotten both a validation for his idea and a stronger sense of urgency. Had he told anyone else about it, such as Nick or Andy or David, they might have been less encouraging. They would probably have suggested he simulate the experiment first, perhaps by running it on a smaller network that had been disconnected from the Internet at large. Safety measures like that would keep the virus from affecting the entire network in case it contained an error. But a controlled environment would have made it a more modest and less scientifically interesting experiment. Robert wanted a large proving ground.

Paul and Robert went to meet Andy for dinner that night at Legal Sea Foods, a restaurant across the street from MIT frequented by programmers and engineers. While Paul and Robert were standing outside waiting for Andy to arrive, the subject of the virus came up again. Since neither Paul nor Robert knew of anything like this having been done before, thinking about it required a lot of creativity. Both of them thought the idea sounded like the kind of "great hack" that was often dreamed up in the computer world. Robert started thinking out loud, describing to Paul some of the more important features such a program would require. First, of course, it would spread through the network, secretly planting itself in many different machines. An important goal was to make the virus as inconspicuous as possible, so as not to arouse suspicion from system managers. Once it had taken up residence it would need a means of knowing whether or not another copy was already present. And it would have to regulate itself in order to limit the number of copies in each computer. But a difficult question was still unresolved: how to limit the growth without halting it completely.

While they waited, Robert sketched out his ideas. The virus would enter a computer through the UNIX loopholes he had found and look around the system for any possible copies. If it found one, the two would "talk" to each other and decide what to do. Ideally, one would automatically stop running in order to limit growth. But what if someone discovered the virus and tried to trick the incoming virus into believing that a copy was already running on the machine it approached? A programmer could design a decoy to fool the invader into thinking that a copy already existed on a computer. Such a program, easy to write, could prevent the virus from spreading—serving the same purpose as a biological vaccine. Thinking like chess players, Robert and Paul decided that there would have to be a countermeasure against potential defensive programs. How could they fool the decoy?

Why, randomization, of course! They had taken a graduate course together in efficient algorithms taught by Michael Rabin, a prominent mathematician and cryptologist. The concept of randomization was big with Rabin, who told his students again and again that if a problem seemed impossible to solve they should reduce it to a simpler one and apply randomization. (This was the philosophy behind Bob Morris's probabilistic typo checker.) Rabin had discussed randomization as it applied to abstract problems, such as prime-number searches. But it occurred to Paul and Robert that they could use the concept in the virus program. When the program entered and detected another copy, it would toss an electronic coin to decide which one should stop running.

Another way of insuring the virus's survival occurred to Robert. One in N times the virus should enter a computer, forget about the electronic coin toss and simply command itself never to stop running. But then came a new question: what should N be? Five? A thousand? Ten thousand?

Just as they were beginning to ponder this, Andy walked up. Although he was a close friend, his job as system manager at Aiken would place him in an awkward position, to say the least, if he were suddenly privy to discussions of huge security holes in UNIX. Andy thought the conversation had ended abruptly because his friends were discussing a woman in whom both Andy and Robert were interested.

Robert was still thinking about the virus. And he couldn't contain his enthusiasm for his discovery of the *ftp* bug. The next day, he walked into Andy's office at Aiken and casually told him about it. Just as casually, he told him not to spread it around.

Andy didn't waste much time before attempting to verify the bug. When he couldn't, Robert had to give him a more explicit description of how it worked.

▲ ▼ ▲

The Wednesday after Robert returned to Cornell, Paul sent him some electronic mail: "Any news on the brilliant project?" Two days later,

Robert sent a message back: "No news. I'm buried under legitimate work." Paul took that to mean schoolwork. But the virus project was still alive. One of the most time-consuming things Robert had done during his four days up at Harvard was to decode a collection of encrypted password files he had taken from various machines around the country.

While directly decrypting a password may not be possible, guessing often works. It's impossible to decode a password by reversing the process that created the coded version. However, nothing prevents the decoder from guessing at the original password by coding, say, an entire dictionary in the same manner and matching the results against the coded password. Since many passwords are ordinary English words, a dictionary method yields a surprisingly high number of matches. The faster the computer, the more computers used, the less time it takes.

On that Friday evening, Dawson Dean walked into the computer terminal room at Upson Hall, where Robert was seated at a Sun workstation. Dawson asked Robert what he was up to. When Robert showed Dawson what he had on his screen, Dawson's eyes widened. It was a long list of passwords in plain, unencrypted form. Robert scrolled down the list to reveal passwords of dozens of Cornell students and professors. Dexter Kozen, the graduate adviser, was on the list. His password? *tomato*. Keshav Pingali, who was teaching the course in microprocessor design, had chosen *snoopy*.

"Wow!" Dawson exclaimed. "Is my password on there?"

His password didn't show up because it wasn't a dictionary word.

"How about Aitken?" Bill Aitken was another graduate student whom Dawson thought to be "really obnoxious." Robert scrolled down and found Aitken's password, *subway*.

"Isn't it kind of dangerous to have a decrypted password list lying around in your account?" Dawson asked. The very tone of the conversation—Dawson's excited questions, as if he were getting a vicarious kick out of the illicitness of what Robert was doing, and Robert's careful answers—implied that there were some real taboos being broken here.

"Well," Robert replied, "you encrypt what you can. For the rest you basically take your chances."

Dawson's curiosity had gotten the better of him. "Could you do this all over the place and build a nationwide data base of passwords?"

Robert told Dawson that he had other ways of getting into machines without relying on dictionary encryption.

Dawson pressed Robert to tell him about the other ways he had of breaking into machines. Robert hesitated, but Dawson wouldn't give up. Finally Robert said that from reading UNIX source code, he had discovered several bugs. One was the back door in the *sendmail* program. Another was a bug in *finger* that would also allow him to run a program on another machine without logging in. Robert said he had known about the two bugs for a year and didn't think anyone else knew about them.

Dawson Dean wanted to skip the bugs talk and hear more about specific computers Robert could get into. He asked about a specific private company. Robert shook his head. "You could get in there, too," he said, "but you really want to stay with machines that are owned by universities. Universities are less strict about their security in general." Further, Robert said, it wasn't really a good idea to access machines across state lines.

Dawson then asked about one machine at the MIT Media Lab, a center at MIT that was studying technology and communications. Within a few minutes, Robert had logged on to the computer.

Dawson was amazed. "What's the account you're logged in as?"

Robert typed a command that would tell him the account name he was using. The computer replied with "nobody." Dawson was impressed. Obviously, Robert had fooled the machine and was logged on illicitly.

But Robert didn't tell Dawson Dean he had any plans do anything with the bugs. And Dawson Dean didn't ask. In fact, aside from Paul, Robert hadn't told anyone of the program he had been planning all semester, and on which he had been working in earnest since just before the trip to Cambridge. By this time, in fact, Robert had been programming the virus on and off for a little more than two weeks.

A week later, on November 2, Robert was dismayed to see a posting on the network: Keith Bostic, who worked on Berkeley UNIX, had posted a fix to the flaw in *ftp*. Since it had been just the week before that Robert had told Andy about the bug, this couldn't be a coincidence. Robert immediately suspected that Andy had alerted someone at Berkeley about the *ftp* bug. He fired off a note to Andy and asked him if he had leaked the secret. No reply. This meant that Robert could no longer use the *ftp* bug for his virus. But the flaws in *sendmail* and *finger* remained.

Robert spent that afternoon and early evening putting the final touches on the virus. He finished the work at 7:30 P.M., Eastern Standard Time. An hour later, having logged on to a computer at the MIT Artificial Intelligence Lab, he typed in a few commands to execute the program. He went to get some dinner.

In the time it took Robert to put on his jacket after pressing Return, the program began to spread. Within a few minutes it was already fanning out over the network. Computers started infecting one another like toddlers in a day-care center. Any VAX or Sun machine linked to any other VAX or Sun was instantly vulnerable. While Robert was eating his dinner, dozens of copies of the virus were already swarming around inside machines, vying for computer time. Machines had begun to slow down and then crash.

Robert had planned to go home after dinner, but he couldn't resist returning to Upson Hall to check up on the program's progress. When he logged in, the computer wouldn't respond. Something seemed to be going wrong. The virus was replicating out of control.

▲ ▼ ▲

Later that night, at about 11:00 P.M., Paul and Andy had just returned to Aiken from a late dinner. As Paul was pulling his keys out of his office door, the phone rang. Andy answered it for him. It was Robert, who asked to speak with Paul. Andy put Paul on the line and went back to his office.

Robert sounded miserable. "I think I've really fucked up," he said. It was the first time Paul had heard rtm utter such strong language. From the tone in his friend's voice, which was much softer than usual, Paul knew rtm was very upset. Paul's first thought was that it must have something to do with a woman.

"What do you mean?" Paul asked. "What'd you do?"

"I started a virus and it isn't working at all the way it's supposed to," Robert replied. "I got one of the numbers wrong on how it should propagate."

"What number did you use?" Paul asked, referring to the question of how frequently the virus should infect a machine even if there was already a copy present.

"One in seven."

"One in seven?! rtm, you jerk! Why seven?" At that instant, it was clear to Paul that the number should have been higher by a factor of a thousand or more.

But Robert wasn't eager to sit around evaluating his error in judgment. He told Paul that all the Suns and VAXes at Cornell were messed up, crashing every few minutes. And if all the Suns and VAXes at Cornell were messed up, it was reasonable to assume that lots of other computers around the country were, too. He said he had launched the virus earlier that evening from a computer at the MIT Artificial Intelligence Lab. He had gone out for dinner, and when he returned to check on the virus's progress, he saw that it was stalling machines everywhere he was able to check.

They discussed ways to fix the virus. Paul's idea was to send another program after the virus to kill it, a *Pac-Man*-like program that would run after it and eat it up. As Paul got more excited about his *Pac-Man* idea, Robert just grew more morose. If he had already made a mess of one program, what reason was there to believe that he wouldn't screw up on a second one?

The next idea was to get Andy involved. Paul went up to Andy's office and peered in the door. Andy was still there, working late to install new hardware on the lab's computers. "I think you'd better get in touch with rtm," he said. "There's something really big going on. But I can't tell you what it is." Paul was pacing back and forth in front of Andy's desk.

Andy was ever skeptical of anything Paul might consider really big. "What's going on?" he asked.

"You'd better talk to him about it," Paul said. "He told me not to tell you."

"Why don't you just tell me?" Andy was losing his patience.

That was all the coaxing Paul needed. "Well, don't let him know I told you, but he's written this computer program and it's taken over the whole country. It's out of control now! It's incredible!"

Andy was still skeptical. But then he remembered receiving an uncharacteristically conspiratorial message from Robert that afternoon, asking if he had told anyone about the *ftp* bug. Andy had in fact mentioned it to a few people; he had even demonstrated it to them. And he had used it to get full privileges on Nick Horton's machine in Portland and then later told Nick about it. He figured that someone had eventually told the Berkeley people about it. It wasn't until Robert registered his concern that Andy remembered Robert's having asked him not to spread the word. Andy hadn't responded to Robert's earlier note, but now he typed a message back: "Sorry about betraying other trusts," he wrote. "Tell me what's going on." Andy was, above all, concerned about what could happen to the Harvard machines.

An hour or so later, Andy got a call from Robert, who sounded unusually subdued. Robert told Andy there was a virus out in the network that seemed to be bringing down a lot of machines. He didn't say he had written it, and Andy didn't need to ask. Andy wanted to know whether the computers at Harvard would be affected. No, Robert replied, because Harvard had already patched the holes the virus was using to get in. An hour later Robert called back and asked Andy to send an anonymous note out to the network with directions on how to fix the virus. Robert told him the points he wanted to make and Andy composed the following message:

```
A Possible virus report:

There may be a virus loose on the internet.
Here is the gist of a message I got:

I'm sorry.

Here are some steps to prevent further
transmission:

1) don't run finger, or fix it to not
overrun its stack when reading arguments.
2) recompile sendmail w/o DEBUG defined
3) don't run rexecd

Hope this helps, but more, I hope it is a
hoax.
```

After Robert had dictated his brief apology and cure, Andy told him that he would make sure it was sent from a remote machine that wouldn't be traced to Robert or Andy. Andy decided that he didn't want to be the one to say Robert had done it. He thought Robert should be the one to decide when and if he was going to admit to having done it. Andy told Robert to be prepared to lie to people. If anyone asked him about the virus, he said, Robert should try not to smirk.

After he hung up, Andy thought about the best way to send the message. He knew theoretically how to send out anonymous electronic mail messages, or at least how to pretend to be another computer delivering mail. The message had to look as if it had come from some place other than Harvard, and certainly not from Cornell. He decided to post the message to a network discussion group on a computer at SRI. He realized that if he sent it directly to the computer at SRI that fed the network, it might get traced straight back to Harvard. So he created a fictitious origin for his message—*foo@bar.DARPA*—and routed the message through a computer at Brown University, expecting to see it get sent to SRI within a few hours.

As it turned out, Andy's message got bogged down on the first leg of its trip. The Brown computers were besieged by the virus. Worse, Andy had not indicated the subject at the top of the message, making it likely that it would be ignored, or given a low priority, once it finally did arrive at SRI.

Andy also tried to make a few phone calls to Berkeley to tell the Berkeley UNIX people there about the virus, but he wasn't sure just whom to call, or even how to find their numbers. Calls to the main number on the Berkeley campus brought no answer, and Andy decided the whole thing probably didn't justify rousting people from bed in California, where it was already midnight.

Andy knew that anyone, even as gifted a programmer as Robert, could make a mistake. Andy had once inadvertently brought down two hundred computers at Harvard because of a small error in a computer network routing command. The blunder had been a breach of Harvard's official policies regarding computer use, but because university officials recognized it as an honest mistake, they didn't censure Andy. This virus didn't seem to be as disastrous as Robert and Paul were claiming. If Robert had indeed brought down a bunch of Cornell computers, then perhaps some people would be upset. But it couldn't be too terrible. Finally, satisfied that he had done what he could for his friend, Andy went home at 4:00 A.M.

That Robert had committed a grave transgression—something *illegal*, in fact—didn't occur to Robert, Andy or Paul. Robert's worst fear was that people in the computer community would be beside themselves with anger. He hoped he wouldn't get in trouble with Cornell. Seeing what he had already done to the Internet was very upsetting. And his little program was probably still ricocheting out of control. He could only hope that Andy's message would help stem the damage.

But news of the virus that ate the Internet was all over Aiken by the time Andy got to work on Thursday morning. And Robert Morris's name, it seemed, was on the tip of many a tongue. After all, Robert's reputation at Harvard was that of a security expert, Internet habitué and occasional prankster. The only aspect of this incident that wasn't in keeping with Robert was the apparent malice behind a move that would crash computers all over the network.

It was unclear to Andy whether his message had reached people. Reports were filtering in from Berkeley and MIT telling everyone on the network how to get rid of the virus. No one mentioned an anonymous message that was going around, but the instructions given were exactly those Robert had dictated to Andy.

It was with some difficulty that Andy told professors around Aiken that he didn't know a thing about the incident. Paul appeared to be having an easier time with the deception. When another computer science graduate student asked Paul if Robert had anything to do with this virus he was hearing about, Paul looked him squarely in the eye and said no. During the day, a subdued-sounding Robert called Andy to ask if he had sent Robert's message out. Andy assured him he had.

▲ ▼ ▲

When Keith Bostic of UC Berkeley arrived at work at 6:00 A.M. Thursday, after three hours' sleep, the phone was already ringing. Calls were coming in from angry computer managers around the country, demanding to know what to do about the program that had infested their systems. Bostic had already anticipated some of the wrath. Did he know about these holes in Berkeley UNIX? No, Bostic responded, he didn't. Computer managers at the Defense Department, one of the largest customers for Berkeley UNIX, were particularly irate. Did Bostic have any idea who had committed this heinous act? Had he been aware of holes in Berkeley UNIX? Could he assure them that there were no Trojan horses lurking in the program? Did Berkeley plan to disassemble the virus code layer by layer?

One of Bostic's first tasks that day was to send out Virus Posting #2, an amendment to his first patch for the *sendmail* bug, providing a more complete fix to the program. That message went out at about 8:00 A.M.

Bostic and the others had already discussed the question of disassembling the virus. It would be a time-consuming and arduous task, but it was the only way to determine beyond a doubt whether there was any destructive code hidden somewhere inside.

The work at Berkeley was being duplicated in Cambridge, Massachusetts, by a group of MIT programmers who had also stayed up most of the night. In the middle of the day, a message from MIT arrived in Berkeley with news of a second method of attack the virus was using. Exploiting a hole in the small UNIX *finger* program, the virus was able to crash *finger* by sending it more characters than it could handle. Once it had overflowed the storage space, the invader was able to start a small program that called back across the network and brought the entire body of the virus into the target computer.

Bostic was skeptical about what MIT was telling him—*finger*, after all, was such a trivial little program. He couldn't imagine that a program only fifty lines long could contain significant bugs. But he was wrong: to prove its point, the MIT crew sent him a sample program that demonstrated the hole in *finger*. Later that day, Bostic sent out Virus Posting #3, a fix to the *finger* program.

Seeing the *finger* attack was enough to prove to Bostic that the only way to find out if other dangers remained was to pore over it line by line. The program would have to be laboriously decompiled.

Decompiling programs is something of an arcane art that entails translating a program from ones and zeros, which a computer reads as "on" and "off" instructions, into something a human programmer would write and understand. Decompiling a computer program is like taking a book that has already been translated from the original English into, say, French, then translating it back into English, all without seeing the original version. The new English version may not use the same words, but good translators can convey the book's proper meaning. When a program is decompiled, the language itself may be slightly different, but its behavior should be identical to the original.

Programs are normally compiled, not decompiled. That is, once a program is translated from the original source code into machine-executable code, there is seldom any reason to do things the other way around. In fact, many commercial software licenses prohibit disassembling programs precisely because the people who do so might want to break copy protection or modify the software. But having a program's original source code is invaluable because it is a window into the author's intentions. And since the author of this particular program had apparently gone out of his way to hide his program, there was no choice but to decompile the program, an arduous task that few programmers have much practice with.

Berkeley, as it turned out, was ideally suited to the job. Not only was the Berkeley version of UNIX created and still maintained there, the Berkeley campus was the site that week of an annual gathering of UNIX experts from around the world. A year earlier, during the same UNIX conference, the stock market had crashed. This year it was the Internet. Bostic even had Chris Torek, one of the nation's leading UNIX experts, staying at his home. In addition to Torek, there was a compiler expert from the University of Utah named Don Seeley at the conference. That much talent alone would probably have been enough to decompile the program. But Phil Lapsley and Peter Yee knew of yet another decompilation ace.

Dave Paré became expert at decompiling programs as an undergraduate at the University of California at San Diego in 1985, when he got miffed at the author of a computer game called *Empire* who refused to distribute his source code. The twenty-two-year-old Paré put his mind to decompiling *Empire* in its entirety. It took him the better part of two years. Now he was in Silicon Valley, fifty miles south of Berkeley, working for a software developer. Not only was Paré proficient at decompiling code, but he had also written his own disassembler, a program that tried to make decompilation as easy as possible by automating some of the more mechanical steps. So Peter Yee called him Thursday afternoon to say there was something they needed his help on.

This was the first Paré had heard of the virus. "Where's Phil? Can't he handle this?" he asked.

"Phil's asleep. He was up all night," Peter replied.

That was enough to convince Paré that this was something big. He had never known Phil to stay up all night.

Paré got in his car and made the hour's drive up to Berkeley. When he got to Evans Hall, he sat down with Chris Torek at one workstation, while Bostic and Don Seeley sat at another one across the room. The office was transformed into a disassembly line. The job of the Paré-Torek team was turning the raw ones and zeros of each of the program's routines into assembly code, then into rough code in the C programming language. Once they had each routine in hand, they gave it to Bostic and Seeley, who tried to make sense of the code's precise purpose. The UNIX conference on campus was alive with talk of the virus that had taken hold of the network the previous evening. Some who had planned to arrive in Berkeley Thursday morning were forced to stay home and battle the invader. For those who were there, discussions of the siege overshadowed the workshops on subjects such as "UNIX with NPROC = 3000" and "Kernelization of MACH." Two of the attendees had already been pulled from the conference to help in the decompilation effort, and during breaks others wandered over to Evans Hall to see how things were going. When the programmers got hungry, they ordered calzones from a nearby pizzeria and kept working as they ate.

Teams of programmers on both coasts continued pulling the code apart. But behind the spirit of cooperation, an element of competition crept into the exercise. Each school was privately hoping to finish first.

Besides, once each group had settled into its own rhythm, it was far easier just to do the work alone than to adapt to someone else's method.

When Keith Bostic wasn't busy taking panicked telephone calls or responding to electronic mail, he helped with the effort. At least once an hour he got an anxious call from various branches of the Defense Department, asking if the Berkeley team had finished disassembling the code.

The suspense came from uncertainty over what instructions the rogue program might contain. At one point, Paré got nervous when he saw that there was code that appeared to have a timer on it.

"Hey, guys," Paré called to the others in the room. "After twelve hours it does something."

"What?" came the chorus of replies.

"It calls a routine called H_Clean."

H_Clean? Could that mean Host Clean? And if it did mean Host Clean, did it intend to clean out the files of the host it was running on?

Bostic rushed across the room to peer over Paré's shoulder. Anything that was timed was not a good sign. They had no idea what would happen once the timing mechanism was tripped. With an edge of panic in his voice, Bostic said, "Dave. Time out. Do that routine. Now."

So Paré set to work on the H_Clean routine as the others watched. As it turned out, the routine was designed to erase the virus's own list of the hosts it had infected during the previous twelve hours. It was nothing to worry about.

There also appeared to be a piece of code in the virus designed to send a little bit of information—a signal at regular intervals—to Ernie CoVAX, a computer in Cory Hall that computer science graduate students used for sending and receiving mail. It appeared that this part of the program was supposed to act as a foil to throw pursuers off the scent by making it seem that the program was coming from Berkeley, but there was an error in the routine that was supposed to send data to Ernie, so nothing was ever sent. The Berkeley team came upon other errors in the virus's code as well. They appeared to be careless mistakes. For example, whoever wrote the program forgot, at one point, to assign a value to a variable; the author also misdirected a message to another program. The most mystifying thing to Dave Paré was the inconsistent quality of the code. Parts of it were extremely well written while other parts were so sloppily executed they appeared to be the work of someone else entirely.

In Cambridge, the MIT group found a more significant flaw: the dialogue between a newly arrived virus and one that was already established would necessarily end in disaster because the listening program would not always listen long enough to the new infection to acknowledge its arrival. Therefore, because each program thought it was alone on the computer, the "electronic coin toss," which in most cases was supposed to result in the self-destruction of one of the copies, would never take place. This was a major error. As it was, the author had guaranteed that the virus would clog the network because one in seven times neither the listening program nor the new arrival would destroy itself; this newly discovered flaw meant that a logjam would have resulted even if the author had chosen a frequency of one in a hundred thousand.

By 4:00 A.M. the next day the main structure of the program had been reconstructed. By now, it was clear the virus was basically harmless. So early Friday morning, Bostic sent out his fourth and final virus posting to the network. It was a list of fixes to the virus itself. This final posting was a bit of a joke because the Berkeley team was wagging a finger at the author of this clever but in some ways careless program. After that, Bostic went home to sleep soundly for the first time in two days. As soon as they were finished with the disassembly, the Berkeley team sent a copy to the anxious officials at the Defense Department.

But no sooner was the program decompiled than a controversy erupted over whether decompiled versions should be posted on the network. Bostic and others at Berkeley were against it, arguing that they didn't want some high school student to take it and try it out. Bostic's detractors argued against adopting this patronizing, "father-knows-best" attitude. Bostic stood his ground. As far as he was concerned, sending out the source code would be "the electronic equivalent of scattering guns through the network." At the same time, he said, Berkeley wasn't trying to hold back anything about what the program did. Moreover, although the Defense Department made no specific request of Berkeley, officials there told Bostic they were pleased that he wouldn't be sending out the disassembled code.

▲ ▼ ▲

Robert didn't return to Upson Hall on Thursday morning. He stayed home for much of the day, trying to concentrate on schoolwork. He went to choir practice that evening and on his way back from the chapel he stopped by Upson to log in and read his mail. When he logged in, he saw that most of the computers were working fine. In his mailbox were notices from the Cornell staff that said there was a virus loose but that Cornell had it under control. There were also some bulletin board notices from Berkeley about patching the holes the virus had used to get in. And there was a message from Paul asking Robert to call him.

Andy and Paul had dinner that night with David Hendler, who had just returned from his long trip.

"Well, have you heard?" Paul asked David.

"Heard what?" David asked.

"The virus that's been going around the Internet!" Paul could hardly contain himself. "Andy was up all night! It was out of control."

"Oh." David smiled. "Did Robert do it?"

Silence.

The trio went back to work at Aiken, and when David walked into Paul's office shortly after 11:00 P.M., Paul was on the phone with Robert, telling him what a huge media event the virus had become. Robert didn't have a television and he was shocked to hear that it was one of the top stories on all the networks. Paul was also trying to cheer him up by reading aloud from *The Oxford Book of Light Verse*. Robert asked to speak with David. David had expected an effusive welcome after his absence, but Robert was monosyllabic. "This thing is mine," Robert mumbled. The tone in Robert's voice was not only that of a programmer who was upset to have erred, but of someone who was aghast at himself for having erred so visibly.

David wasn't surprised, but was still of a mind to joke about it. "Do you want to meet in Montreal?" It was, after all, the city closest to Ithaca that was outside the U.S. When Robert didn't laugh, David knew his friend was extremely upset. So he got practical. "What are you going to do?"

"I don't have any idea," Robert replied.

Ten minutes later Robert called back. He had called his father and was going to leave Ithaca the next day. He didn't say where he was going.

▲ ▼ ▲

Bob and Anne went out to dinner Thursday night. They discussed the computer virus that had been going around the Internet. Cliff Stoll had called Bob that morning and told him about it, but Bob had been too busy with other things to spend too much time thinking about its origins.

At 11:30 P.M., the phone rang. Bob was already in bed. Anne answered. She was surprised to hear Robert's voice. Ben frequently called this late, but not Robert.

"Can I talk to Dad?" he asked.

"He's asleep," Anne told him. "Is it important?" From the tone in Robert's voice, she already knew that it was.

"Well, I'd really like to speak to him," came the reply. He was as insistent as she had ever heard him.

She called Bob to the phone.

It was a short conversation between father and son. When he heard what Robert had done, Bob was perturbed but not angry. Robert told his father he already had a plane ticket to Philadelphia for the following day; he had been planning to spend the weekend with his friend Janet. Bob told him to use the ticket, not to talk to anyone and not to tell anyone where he was going. It was likely he would need legal advice.

When Anne went to work the following day, the staff were milling in the coffee room, talking about the computer virus. They barely knew what Anne's husband did for a living, much less her children. Tables were strewn with newspapers, all with news of the virus prominently displayed. Queasy and unable to concentrate, Anne left work early. By late that afternoon, *The New York Times* had figured out that Robert was the author of the virus and was planning to run the story in the paper the following morning. Anne and Bob spent the afternoon looking for an attorney. By the end of the day, they had several names. If Paul hadn't kept talking to *The New York Times*, there would have been more time for the family to figure out what to do. But Paul's careless slip about Robert's login had accelerated the pace of events.

"AUTHOR OF COMPUTER 'VIRUS' IS SON OF NSA EXPERT ON DATA SECURITY" was Saturday's front-page headline in *The New York Times*. The paper hadn't been able to get a photograph of Robert in time, but the following day, photographs of both father and son appeared. Bob looked the very picture of an eccentric scientist. His long, untrimmed, graying beard complemented his eyebrows, whose arch resembled birds in flight. Hair all but covered his craggy face.

Even with the presidential election coming up the following Tuesday, the media had an insatiable appetite for the story of the young computer whiz, son of a computer security expert, who loosed a rogue program on a nationwide network and brought it to its knees. By Sunday morning, a crowd of television and newspaper reporters had settled at the end of the Morris driveway, where they would remain on and off through the weekend and into the following week. The telephone inside rang ceaselessly with calls from the press. Bob's sense of humor stayed sharp. When a friend called and opened the conversation with "This is *not* a press call," Bob responded, "Oh, then you must have the wrong number."

It seemed wise for Robert to remain in Philadelphia and steer clear of his parents' house and the pack of journalists, so on Sunday, Bob and Anne drove to Philadelphia to retrieve him. On their way back, they pulled off the highway for gas and Bob got out of the car. Just then, a red sports car pulled up alongside the gas pumps and the driver glanced at Bob. A wide grin spread across his face as he took a longer look. "Hey!" he called out. "You a computer scientist?"

The nation's press corps fastened onto the story first as an incident that had disrupted a network of military computers, then, as the identity of the culprit emerged, as the story of a remarkable family, and of intellectual pranksterism gone awry. For several days in a row, it was front-page news in the nation's newspapers and one of the top stories in television newscasts. By Monday, every newspaper in the country, it seemed, was already writing editorials. Mike Royko, the acerbic *Chicago Tribune* columnist, called for a stiff prison sentence. Journalists tapped every computer security expert and computer industry executive they could find. "The MacNeil/Lehrer NewsHour" interviewed Ken Olsen, president of Digital Equipment Corporation. Even though Digital's computers had been a target not just of this incident but of hackers with more felonious intent, Olsen urged the computer science community not to respond by placing increased security on computer networks. "The worst thing that could happen," Olsen said, "is that we clamp down on the free flow of academic information, because that should be preserved at all costs."

When Robert's name first came out, some of those who knew him well weren't terribly surprised. Bob's old Bell Labs friend Doug McIlroy got up early Saturday morning and read the news. The slumbering McIlroy family was awakened by a booming "Guess who did it!" There was already a story circulating among computer scientists that as the virus was knocking on the door at Bell Labs, some of the old UNIX hands were chuckling among themselves, saying, "Must be Morris's kid."

But others who knew Robert, even those who had been on the receiving end of some of his Harvard pranks, were incredulous, or at least willing to give him the benefit of the doubt. One Harvard faculty member who knew Robert well was hoping that he had released the virus on a small local area network at first, had gone home and had returned the next morning to find that it had somehow spread out over the Internet. Another faculty member, whose courses Robert had taken, shook his head in disbelief and asked, "Why didn't he simulate it first?"

Andy, Paul and David Hendler, in the meantime, sat in Andy's office at Aiken discussing ways to protect their friend. Cooking up a bit of a propaganda campaign, the group wanted to see the press describe Robert in the best possible light. Paul was enjoying his role as spin doctor on the story. He told the others that he had repeatedly described Robert to the *Times* reporter as brilliant, and the reporter had used that description in his story. As they were sitting there, Robert called. He didn't tell them where he was. He said he just wanted to check in.

"What are you doing?" David asked when he got on the line.

"Baking cookies to send to friends," Robert replied.

▲ ▼ ▲

FBI special agent Joe O'Brien had moved to Ithaca from New York City in 1984 with specific orders to lie low for a while. He had been involved in an organized-crime case; he was the principal agent responsible for placing highly sophisticated electronic surveillance equipment in the home of the late Mafia boss Paul Castellano, which eluded detection even by the experts Castellano hired to root out bugging devices. O'Brien's work had led to a series of trials and convictions of organized-crime figures, and the bureau suggested to O'Brien that he move to a nice quite place where nothing much happens. Ithaca, New York, was the ideal place. Compared with some of the other FBI field offices, the three-man outpost in Ithaca was one of the sleepiest. Background checks on graduates of Cornell University and Ithaca College who had applied for government jobs made up the bulk of O'Brien's caseload. The last thing he wanted was another high-profile case.

When O'Brien heard of a computer virus that was sweeping the nation, his only concern was whether it would affect his home computer, an old Apple II he was using to write a book about his undercover role in infiltrating the Gambino family. The next night, when the 10 o'clock news revealed that *The New York Times* was about to identify the culprit as a Cornell student, O'Brien realized he had much more to be concerned about. O'Brien knew next to nothing about computers. He switched off the news and called his neighbor, a Cornell scientist in charge of the school's supercomputer center.

Cornell had beaten O'Brien to the news by thirty minutes. At 9:30 P.M., the public affairs office received a call from *The Washington Post* asking about Robert Morris. The call set off a chain reaction: the public affairs officer called the vice-president in charge of computing, who called the president, the provost and the computer facilities manager. The chairman of the computer science department had already spoken with Bob Morris, and within an hour a small group was assembled in the chairman's office. Dean Krafft, the facilities manager, combed through Morris's computer files for telltale signs. There wasn't much to see in the student's active account, so Krafft went to the backup tapes. Within an hour he found enough to be virtually certain of Morris's authorship of the virus. Among recipes and invitations to hockey games he found two files, one called *try.out*, another called *Stanford*. Both were slightly hidden in a directory that wouldn't show up ordinarily. It was difficult to tell exactly what was going on because the bulk of the files were encrypted. He saw that the final version of the program was last modified at 7:26 P.M. on Wednesday, November 2. And among Morris's personal mail from the previous two days was a message from Greg Kuperberg, the math whiz who had helped Robert with the graphics program. Kuperberg had written Morris a message on November 3 warning him of a virus on the Internet.

▲ ▼ ▲

O'Brien could have lived without a computer crime case, and would have been perfectly happy to hand it over to Mike Gibbons, the FBI's sole computer crime expert. Gibbons had parlayed a computer retailing job into a position as the bureau's foremost authority on computer crime. He had been deeply involved in tracking down the West German hackers who had been plaguing Lawrence Berkeley Laboratory, uncovering what turned out to be a new form of espionage. Gibbons was the only FBI agent who knew how to write a search warrant tailored specifically to a computer crime case.

O'Brien half expected to see a special team of agents descend on Ithaca as soon as the case broke. It was routine for the bureau to dispatch specially trained units to handle hostage crises and difficult kidnappings. But there was no such unit at the ready for computer crimes—if this was in fact such a case. Gibbons was good, but there wasn't much his expertise could add to an investigation that, in the end, would be carried out much like any other, with agents talking to as many people as possible and gathering what evidence there was. So O'Brien didn't have much choice but to set aside all else to start getting to the bottom of what was turning into the nation's biggest computer break-in to date.

When O'Brien went to have a look at Robert's office on Saturday, there were at least a half-dozen reporters already milling around, scouring the contents of the student's desk: a squash racquet under a stack of computer books, a pile of pennies, a computer-generated crossword puzzle. A problem set done for a course on the design and analysis of algorithms was marked late, and had earned only six points out of twenty. But nothing seemed the least bit incriminating.

O'Brien told the reporters to leave. Robert's personal belongings were placed in several cardboard boxes and put aside as government evidence. It was indeed an investigation turned on its head. There was little doubt who had committed the act; the question was whether the act was a crime. But that wasn't O'Brien's territory. It was something for the Justice Department attorneys to worry about. O'Brien's task was to gather as much evidence as he could.

An obvious and crucial witness was the person who had placed the anonymous call to *The New York Times*. At first, O'Brien was sure it was one of Morris's officemates. Most of them, it turned out, knew little more than that Morris was quiet and kept to himself.

O'Brien's first helpful interview was with Dawson Dean. The nervous graduate student recounted his trip to Cambridge with the suspect. He told O'Brien about the passwords he had seen on Morris's screen shortly after their return, and how he had seen Morris seated at his terminal late Wednesday night, in the midst of a telephone conversation. Dean said he was surprised; it's difficult to carry on a telephone conversation and program at the same time. But it was apparent that Morris was conferring with someone about a program he was either running or still working on. Dean heard Morris mention Harvard, then MIT. As Dean turned to leave, Morris waved good-bye.

O'Brien asked Dean one last question: "Did you call *The New York Times?*"

"No!" Dean exclaimed.

One officemate remembered that Morris didn't seem to be around Upson Hall on Thursday afternoon. He also remembered there was a note in large letters scribbled on the office blackboard: "ROBERT: CALL PAUL IMMEDIATELY. VERY IMPORTANT."

The officemate didn't know who Paul was.

O'Brien's next interview was with Kevin Asplen, the graduate student who drove Morris home on Thursday morning. Asplen had arrived at Upson Hall at about 1:00 A.M., he told the agent, planning to spend just a few minutes sending an electronic mail message to someone. He saw Robert Morris sitting in his office and told him that if he would like a ride home, Asplen would be ready in five minutes. Morris said he would appreciate a ride and sat down to wait. Then Asplen saw that there seemed to be something wrong with the computer he was using. It was crashing frequently for no apparent reason. He couldn't get it to stay running long enough to get his mail out. He told Morris he was having problems with the computer and it might take him a while longer than he had thought. Morris didn't offer any advice. He just said he didn't mind waiting. At 2:00 A.M. Asplen finally got his message out and took Morris home. During the ride, Morris was his usual quiet self and didn't say a word about the virus.

▲ ▼ ▲

Tom Guidoboni arrived home in Arlington, Virginia, late Friday night, exhausted after a two-day trip to Texas, where he had been taking depositions in a new case. When the forty-one-year-old defense lawyer got up the next morning, his wife was absorbed in a newspaper story about a virus that had crippled computers around the nation. "This is fascinating," she said, pushing the article toward her husband.

It was the first Guidoboni had heard of the incident. "What's so fascinating about it?" He took a quick look and went back to his coffee.

"Well," explained his wife, also a lawyer, "whenever I call the court to ask for something they always say they can't answer because the computer is down. Or you get a computer on the telephone. Now thousands of computers are down because of this one thing." But her husband was already lost in another section of the paper.

Two hours later, the telephone in the Guidoboni house rang. It was a partner from his firm, wanting to know if Guidoboni would be interested in representing "the virus kid." An hour later Guidoboni was on the phone with Bob Morris, and they arranged an appointment for 10:00 A.M. Monday.

It wasn't clear to Guidoboni which law, if any, had been broken. Normally, if he got a weekend call on a criminal case, he knew instantly which section of which statute applied to the case. But this was new terrain for Guidoboni. He was familiar with basic wire-fraud law, which was the first thing to come to mind after the conversation with his partner. Even as a criminal lawyer, he was only vaguely familiar with the existing computer crime statute. The next day, Guidoboni went to his office to read the law carefully and prepare for the meeting.

The 1984 law, amended in 1986, was created in response to the public's demand for more computer crime protection. It was Congress's first attempt to criminalize computer trespass, making it a crime to gain unauthorized access to computer systems. The statute also made it a crime to modify, destroy or disclose information gained from the unauthorized entry into a computer. As of 1988, the amended law was still largely untested. Only one case had gone to trial, none to a jury. But while other areas of criminal law tended to address narrowly defined, well-understood conduct, the computer fraud statute cast its net over a wide range of computer activities. The law's critics claimed that it was too broad and too vague, that it failed to define adequately such words as *access* and *authorized*. But Guidoboni, new to the subject, understood only that he would have to meet his prospective client and hear the facts before he could consider how the law might be applied.

The first thing that struck Guidoboni when the Morris family entered his office was their unusual appearance. Robert, dressed in a navy jacket and narrow purple flowered tie, appeared to be in a state of shock. Pale and drawn, he looked as if he hadn't slept or eaten in several days. Bob's eccentricities were evident from his long and unkempt beard. And Anne, heavyset and with graying blond hair piled into a bun, came across as a forceful, educated woman concerned about her son. In contrast to Robert, the parents appeared to be relatively composed, although anxious about whether or not this stranger could help.

Bob and Anne made it clear that they were doing a little hasty attorney shopping. One prominent Washington attorney had already offered to take on the case *pro bono*, but the Morrises wanted to explore all their options. Guidoboni took down some general information and asked Bob to explain his obligations to the National Security Agency with regard to the incident. The elder Morris responded that he felt obligated to tell law-enforcement officials what he knew of the offense. Morris had already contacted the lawyers at NSA, and he had already spoken with the FBI. He said that he believed Robert should speak with the FBI as well. It is a common, but often misguided, impulse to go straight to the authorities with the whole truth, hoping for leniency in return, Guidoboni explained. He was quick to dissuade the elder Morris from that plan.

The Morris family then interviewed Guidoboni. He told them he had attended the University of Virginia Law School, that he specialized in white-collar crime, and that with the exception of one computer-related case in 1981, his background in computers was minimal. They discussed his fees.

The lawyer then excused both parents in order to have a private conversation with Robert. In some ways, Robert struck Guidoboni as the youngest twenty-two-year-old he had ever met. Once the lawyer assured Robert that federal marshals weren't about to swoop down and carry him away, he relaxed a bit. Guidoboni had spent years learning to gauge potential clients—from street criminals to congressmen to corporate executives—and he could see through the overwrought facade to a young man who was not only very smart but utterly without guile. Robert's main concern appeared to be whether he would get into trouble with Cornell.

Once Guidoboni established that there didn't appear to be any national security risk involved, and that Robert's program hadn't affected any financial institutions, Guidoboni asked the young man to tell him who else knew firsthand that he had written and released the virus, and what they knew. Robert said that Paul Graham and Andy Sudduth knew the most about it. Then there were others whom he had told afterward: David Hendler, Janet and, of course, his father. Paul was the only one who knew about it beforehand. Guidoboni also wanted to know what tracks he had left. Robert said he had "cleaned up" a lot of files on his Cornell account, but someone there had obviously gone into his old files and found incriminating evidence. Guidoboni said that he considered this an indefensible means of collecting evidence. Cornell's claim that it had a right to search Robert's files because it owned the computers was, in the lawyer's view, preposterous. Scouring someone's personal computer files is as much a privacy breach as opening someone's personal mail, or listening in on private telephone calls.

As the conversation wore on, Guidoboni saw that Robert was only gradually beginning to see that this was a serious criminal offense. Until now he hadn't even realized that anybody outside the computer community would pay much attention to the incident. It appeared likely to Guidoboni, in fact, that until Robert walked into the lawyer's office, he didn't know there was even a law to violate—and he certainly wasn't aware of the computer crime statute in particular.

Guidoboni wanted the case. He knew it would be interesting and challenging. It was a rare opportunity to take on a piece of the law that had never been interpreted. He liked Robert, who was clearly intelligent without being arrogant. And the publicity couldn't hurt. The next day, he got a call from Bob Morris asking him to proceed.

The session with the lawyer had taken its toll on Robert. When the family got back to the Metro station afterward, Robert collapsed. His anxious parents propped him up on a bench and waited for him to get his strength back before continuing the journey. When the Morrises returned home later that afternoon, a crowd of reporters, photographers and television crews was waiting at the end of the driveway. Robert put his head in his hands. At first it appeared to the journalists that the computer whiz might make a statement. Instead, he looked down at his feet, admitted in a barely audible voice that it had been a pretty rough day and was escorted by his mother into the house.

▲ ▼ ▲

Later in the week, Paul and Andy drove down from Cambridge to meet with Guidoboni. Robert was glad to see them, but at the lawyer's instructions he wasn't talking about any specifics surrounding the incident, in case they were ever called on to testify. Conversations had to remain general, and Robert stayed pretty quiet, yet he was naturally curious about the reaction among people he knew at Harvard, and he found it difficult to refrain from asking his friends about it. Telling him not to discuss the case was like telling him not to think about the color red.

It was clear to Robert's two friends that the incident and the attention being heaped on him had made him shier than ever. When he called Guidoboni's office one morning, he hung up after one unsuccessful attempt.

"What happened?" Andy asked.

"I guess he's not there," Robert responded.

"What did they say?"

"A secretary just said he's not there," Robert said.

"Of course she said that!" Andy said. "They're screening calls from the press. Did you tell her your name?"

Robert shook his head.

"Call back and say who you are. I'll bet he's there this time."

Andy was right.

▲ ▼ ▲

Among computer scientists, the incident set off a debate that was to last for months. Within days, analyses of the program were under way. Three lengthy scientific papers emerged. The Association for Computing Machinery's technical journal, Communications of the ACM, devoted an entire issue to an examination of the program. The opinions of critics ranged from "mediocre and sloppy" to "brilliant." Some people said it had caused damage that could reach into the millions of dollars. One industry group estimated the damage at $96 million. Others said it had done no damage at all, that its effect was just the opposite: it had alerted the industry to security flaws.

Another topic of debate was the question of how widely the rogue program had actually spread. Estimates from MIT put the total number of computers infected at about six thousand, or about 10 percent of the number of computers on the Internet in 1988. But other evidence suggested that the number may have been much higher. A conclusive number was never reached.

Among the semanticists, a debate erupted over whether the program was a virus or a worm. The general consensus was that even though Morris and his friends called the program a virus, the program more closely resembled a computer worm because worms can move under their own power while a virus piggybacks parasitically on another program. In the biological world, a virus is an agent of infection that can grow and reproduce only within a host cell. In strictly technical terms, so the argument went, since the program did not need to attach itself to another "host" program in order to propagate, nor did it alter or destroy any programs, it was more like a worm. A number of computer researchers disagreed. Mark Eichin and Jon Rochlis, who battled the program when it hit MIT, wrote a paper declaring that "virus" was a more accurate description of the program because it was a closer biological analogy to what the program actually did. But in the end, the "worm" label stuck.

One of the few computer scientists whose career was actually boosted by the worm incident was a Purdue University professor named Eugene Spafford. Before the worm, he was a software engineer. After the worm, he switched to computer ethics and began to travel around the country making speeches on the topic of computer viruses, computer security and the moral imperatives he perceived. He argued that there was no excuse for what Robert Morris had done, that it violated the spirit of a community built on trust and that trespassing is trespassing, whether it is done physically or electronically.

The ethical controversy wasn't confined to computer scientists. It also engaged people who knew nothing about computers but who were worried about how this new technology could be used for criminal ends. The fact is that virtually everyone has been victimized by a computer in some way or other, either because a bank deposits money in the wrong account, an insurance company loses a claim or an airline reservation is booked on the wrong flight. What people often don't take into account is that it isn't the computers making the mistake but the people who program them and work with them. Nonetheless, some argued, if a suitable punishment wasn't dispensed to Robert Morris, wouldn't computer criminals with far more malicious intent than Morris think they had an open invitation to set off similar programs that could topple banks, sabotage air-traffic control systems and perhaps even start wars?

Robert's motive became another hot topic of debate. Since he refused to speak with the press, journalists speculated as they pleased. Newspaper reports and magazine articles suggested that he got his inspiration from *The Shockwave Rider*, John Brunner's proto-cyberpunk novel, which Anne Morris told reporters was among the most frequently read books on his bookshelf. But chances are that Bob Morris was more taken with the story of the shockwave rider than his son was. Bob was an early admirer of Brunner, and when Andy and Paul went to Maryland, one of the first things Bob did was pull the book from a shelf with the words, "Let's see where this all started." Robert had read the book, of course. He had even read it more than once. But he didn't feel that it had influenced his life any more than did any other book he enjoyed. Moreover, in planning his program, he called it a virus not a worm.

Others imputed more sinister motives to Robert. The program, they argued, was a hostile act, intended to bring the network to its knees. Some even came to suspect that there was a deeper message buried in the program. Perhaps this was a conspiracy. After all, wasn't he the son of one of the nation's leading computer security experts, an official at the secretive National Security Agency, America's computer spy organization? Might this be some sort of NSA-inspired exercise gone awry? The truth, of course, was far less dramatic. Robert had never pried his way into commercial computer networks in search of power, money or state secrets. He was simply carrying on an intellectual tradition he had learned from his father.

▲ ▼ ▲

Slowly, with the help of Dean Krafft, the patient and helpful facilities manager, FBI agent O'Brien was catching on to the computer talk everyone at Cornell was throwing around. It gradually dawned on O'Brien that the real evidence in this case wasn't to be found among the suspect's personal effects. It was all on the computer. But one of the problems, he gathered, was that Morris had encrypted nearly all of his files—and not only had he encrypted them, but with few exceptions he had also compressed them first—he shrank them by running them through a special program. That made them even more difficult to decipher. By one estimate O'Brien heard, it would take scientists at the National Security Agency, the center of cryptography, two hundred years to decrypt Morris's files. The information O'Brien got from his sources turned out to be out of date. Krafft's first step was to get *Cryptbreaker's Workbench,* a program to break encryptions. With the help of the program, Krafft was able to break the encryption and give Cornell access to all versions of the Morris worm. The process took less than a day.

The matter of the anonymous call to *The New York Times* was also becoming clearer. The FBI had obtained Robert's long-distance telephone record, and it appeared that that night he had called someone at Harvard named Paul Graham.

Special agent O'Brien couldn't muster much sympathy for this Morris fellow, even if he was as brilliant as people said. All that talk of an innocent backpack-carrying kid who made a mistake didn't move O'Brien. At twenty-two, the agent figured, a man should be considered a responsible adult. At that age O'Brien was already a full-time parole officer putting himself through graduate school. And he didn't think much of Morris's Harvard friends, either. When he flew to Boston soon after the incident to interview Paul Graham and Andy Sudduth, he found Graham first, in his office at the Aiken computer center. When O'Brien introduced himself, Graham was unfriendly and uncooperative.

"I don't think I want to get involved in this," he told the agent.

"But you are involved, Paul," O'Brien responded. "You dropped a dime on your buddy."

Paul stood his ground. "Well, I'd rather not talk to you."

O'Brien nodded, reached into his pocket and produced a piece of paper. "Paul," he said, "would you mind telling me if I spelled your name right?" He handed Graham the piece of paper.

"What's that?" Graham asked as he took the paper from O'Brien. His eyes widened. It was a subpoena summoning Paul to Syracuse to testify before a federal grand jury. O'Brien had to smile at the young man's ignorance. Most people to whom he served subpoenas knew what was coming the instant O'Brien reached in his pocket. By reflex, they would put both hands over their heads and take a step backward to avoid being served.

"It's a subpoena, and we'll see you in Syracuse next week."

"And what if I don't come?" Graham sounded irritated.

"Then I'll be back with an arrest warrant."

Andy Sudduth was more gracious, but he too declined to be interviewed. He was served a subpoena as well.

On the morning of the grand jury testimony, Sudduth and Graham were late. The assistant U.S. attorney, Andy Baxter, began to show concern. "What time does their flight get in?" he asked O'Brien.

"What flight?" O'Brien asked. "I heard they're driving."

The prosecutor looked confused. "It's a six-hour drive! You didn't tell them that they can fly and the government reimburses them?"

"Gee, that must have slipped my mind," O'Brien said, barely concealing his lack of sympathy for these young men.

When the two witnesses did arrive, they had to spend some time warming up. The heater in Paul's car was broken. They had stopped several times along the way to warm their hands, and once to buy Andy a warmer jacket. When they heard that the government would have paid their airfare, both were furious.

In December, Robert agreed to give a proffer, a statement to the government detailing everything he had done. It was an unusual concession from someone suspected of committing a crime, but Robert and Guidoboni thought that full cooperation might soften the Justice Department. O'Brien and Gibbons were ready for him. One of the bureau's interrogation experts flew to Syracuse beforehand to coach the two agents on effective questioning techniques.

O'Brien had already seen a videotape of the lecture Robert Morris had given half a year earlier to the NSA on computer security. Part of the lecture had been called, "How Not to Get Caught."

"Isn't it kind of strange," O'Brien asked the suspect, "that you're giving a talk like that on how not to get caught and you're sitting here right now?"

Robert looked down at his hands and smiled. He had no answer.

Shortly before Christmas, Guidoboni called Alan Rubin, the defense attorney in Los Angeles representing Kevin Mitnick, who had been arrested earlier in the month for breaking into Digital computers and stealing proprietary software. The lawyers didn't know each other and their clients were clearly spun of different cloth. Yet there was so little existing case law for the computer crime statute that Guidoboni thought he and Rubin might be able to exchange a few ideas. As it turned out, the two attorneys could only commiserate and wish each other luck.

With just a handful of keystrokes, Robert not only had paralyzed thousands of computers but had brought his life to a standstill as well. Returning to Cornell, continuing to attend classes and starting work on his dissertation were, of course, out of the question. By Thanksgiving, Robert had withdrawn from Cornell, his career as a computer scientist on hold indefinitely. Between Christmas and New Year's, Bob and Anne took him to Cornell to pack up his things. He moved in with his parents and got a programming job at a private international development agency, accompanying his mother on her three-hour daily commute into Washington. Cornell, in the meantime, in a measure to guard against civil lawsuits from victims of the worm, embarked on an extensive inquiry. By February, the university had issued a forty-five-page report on the incident, which concluded that Robert Morris had violated the university's computer-use policy. As if that weren't enough, Robert had to go through an academic disciplinary hearing. Cornell formally suspended Robert. He would be eligible to reapply for admission in the fall of 1990.

In an aside during the academic hearing, Dean Krafft asked Robert about the keyword he had used to encrypt his files. "It's in the dictionary," Robert told him. Krafft ran the dictionary through the encryption mechanism, compared the output and found the word. It was *simple*.

▲ ▼ ▲

The U.S. attorney in Syracuse was apparently willing to send a recommendation to Washington that Robert be charged with a misdemeanor, but Guidoboni heard nothing from prosecutors directly. The Justice Department was maintaining an ominous silence in the matter, placing the case in a tense state of suspended animation. Here was someone who had supposedly committed the computer crime of the decade, and he had yet to be charged with it. Still more ominous for the defense, in the spring of 1989 the case was lifted from the Syracuse prosecutor's hands and given to prosecutors in Washington. The waiting game continued. Beside himself with boredom and loneliness in Maryland, Robert moved to Cambridge and started working at a small software company run by a couple of his old friends from Aiken, then switched to a programming job in Harvard's classics department. Robert had yet to utter one word in public about what he had done, about his reasons or his motives, yet in some circles he had acquired folk-hero status. One young hacker had Robert's photograph taped to his bedroom wall; others were choosing *rtm* as their password.

In June, seven months after the incident, Mark Rasch, the Justice Department's young computer crime expert, called Guidoboni to introduce himself and say that the Justice Department planned to indict Robert on one felony count. If Robert agreed to plead guilty, Rasch said, the Justice Department would consider granting some concessions on the sentencing.

In one last attempt to reconcile the matter outside of court, Guidoboni had a meeting with Edward Dennis, the assistant attorney general

in charge of the criminal division of the Justice Department. But Guidoboni's pitch—that his client had intended no fraud and should be
treated lightly—failed to convince the prosecutors. After the meeting
with Guidoboni, it still took Dennis some time to make the decision to
proceed with the indictment. Robert, who would have pleaded guilty to
a misdemeanor, decided he would rather take the case to trial than plead
guilty to a felony.

The whole family was affected by the long, tense wait. Exhibiting his
own strange way of coping, Bob decided to take up a hobby of his
daughter Meredith's that he had always found intriguing and signed up
for a church-tower bell-ringing class in Washington. The class met twice
a week and demanded that Bob spend even more time away from home
than he already did. It irked Anne, who had quit her job in order to
work from home and hold things together. She was particularly edgy,
still stinging from what she considered the Justice Department's targeting
of her son when there were so many true criminals to worry about. All
her children, and Robert in particular, had grown up understanding that
you simply don't hurt people. Those who portrayed Robert as a malicious
felon didn't understand that it was antithetical to Robert's moral makeup
to do anything harmful. For his part, Robert remained quiet, asking his
parents for very little. When Anne asked him if he wanted the family to
be at the trial, he said there was no need. Anne told him everyone
would be there anyway.

▲ ▼ ▲

The question of Robert's intent was of pivotal importance from the start
of the legal proceedings. With few exceptions, the computer science
community agreed that Robert had intended no harm. Even a cursory
look at his program was enough to show that he had designed the worm
to be as innocuous as possible while simply taking up residence inside as
many machines as possible. Moreover, he had included mechanisms in
the program to limit its growth. In Guidoboni's view, this was a strongly
mitigating point and could even win him the case. He argued that the
section of the statute under which Morris was being tried was for people
with malicious intent and not for someone like his client.

This was a cornerstone of the Morris defense. Guidoboni's strategy
was to paint Robert as a well-meaning innocent who got caught up in
what he intended as a harmless experiment that exploded in his hands.
No jury, if allowed to consider the intentions of this young man, could
conclude that Robert Morris had meant to cause any harm. As far as

Guidoboni was concerned, it might even be enough for the judge to dismiss the case entirely.

So Guidoboni filed a motion to dismiss the case, arguing that the law required proof of intent to prevent authorized use and cause damage. Obviously, he argued, there was no such proof. In October 1989, Guidoboni went to Syracuse and argued for dismissal in front of the judge, but the judge denied the motion. At that point, Guidoboni saw his case weaken. The trial was scheduled for late November.

Two weeks before the trial, the prosecutors surprised the defense with a list of a dozen or so extra witnesses whom they planned to call. Then came another surprise.

Just how the Justice Department came into possession of the videotape of Robert's NSA lecture wasn't precisely clear, but it appeared that the agency had alerted Justice Department attorneys to its existence. Robert had already told Guidoboni about the lecture. He explained that his father had originally been asked to give the talk but had recommended Robert instead. Robert told Guidoboni that the speech hadn't gone particularly well and he was embarrassed about it. The following day's lecture to the Navy, he told the lawyer, was much better.

The videotape showed an extremely shy Robert in a blue shirt and blue jeans, his left hand wedged into his pocket and his eyes focused on some middle distance. For more than an hour, he rambled on about some of UNIX's biggest weaknesses, system managers' carelessness, sloppy user habits. Desperate, it seemed, for something to do with his eyes, he referred frequently to handwritten notes, but the information seemed to be so familiar he clearly could have done without them. It would have taken someone with a particular interest in UNIX security to stay alert through this talk.

Under any other circumstances this would have been regarded as an endearing if painfully awkward speech by a budding computer scientist. But as soon as the same shy young man was the subject of a federal indictment, the speech became an open window into the defendant's frame of mind. Did this Harvard student think like a computer criminal? A careful look at the videotape and the meticulously rendered transcript that accompanied it revealed that there were indeed ways in which this twenty-two-year-old appeared to be able to get inside the mind of a computer criminal. Moreover, he had a clear knowledge that certain uses of a computer were illegal. In one particularly revealing moment, in which he discussed "how not to get caught," Robert started out by saying, "Close to the heart of every hacker out there has got to be how to stay out of jail." He went on to list some of a computer criminal's most useful tricks: covering his tracks, laundering calls and never returning to the scene of a break-in. This was clearly someone who knew whereof he spoke. And if there was doubt in anyone's mind about whether Robert Morris was aware of the number of computers connected to one another over the network, such doubt could be dispelled by the second line of his speech: "There are thousands and thousands of UNIX systems out there," he said.

It's likely that a screening of the entire hour-long videotape would have just caused the jurors to doze off. But any juror listening closely couldn't have helped but notice some of the language Robert chose. Far from distancing himself from computer criminals, he fairly took on a hacker's way of thinking. "If you have foresight," he said on the matter of outsmarting auditing mechanisms, "the right thing to do is to fill up the disk which the audit trails reside on before you go doing bad things, so that when programs go to log something, they find the disk is full and nothing gets audited anymore."

The next thing Guidoboni knew, the videotape was on the prosecution's list of exhibits. More unnerving, the prosecutors had selected to view only the "How Not to Get Caught" portion of the tape.

Guidoboni's first move was to call in Bob Morris and question him on the lecture.

"Whose idea was the lecture?" the lawyer asked.

"Mine," Morris answered.

"And who decided which topics should be covered?"

After a few moments of considering the question, Morris said he couldn't answer that without disclosing classified information.

Guidoboni tried another question. "Who attended the lecture?"

Morris shook his head. The answer to that one would reveal classified information, too.

"Were there any topics of particular interest to the audience?"

"I can't answer that one either," Morris said.

Guidoboni's hunch was that NSA did its share of breaking into computers, or at least learned the mechanisms by which break-ins are carried out. The defense attorney wanted the trial to reveal just who chose his client's speech topics, and why.

Guidoboni told the prosecutors that if they planned to show the tape, then they had better show the whole thing or he would file an objection, primarily on the grounds that the portion they showed had been lifted out of context. Further, if the tape was to become a piece of evidence, then Guidoboni was going to put Bob Morris on the stand and ask him to disclose classified information under oath.

The matter of the videotape remained unresolved until shortly before the trial, when, as quietly as it had appeared on the list of exhibits, it disappeared. Guidoboni could celebrate a small victory.

The trial was rescheduled for January 1990.

▲ ▼ ▲

Syracuse, a city of about 170,000 in upstate New York, is hardly a high-tech mecca. Its largest employers include a Carrier Corporation air-conditioner plant and Syracuse University. Prior to the Morris case, the crimes that made the news in Syracuse didn't involve computers. In December of 1989, local citizens were captivated by headlines of serial prostitute murders up the road in Rochester and an *In Cold Blood*–style murder in nearby Dryden. These came on the heels of the dramatic trial of former Syracuse mayor Lee Alexander, who was convicted in 1988 of accepting kickbacks. The Robert Morris trial might have played to a more suitable audience in Silicon Valley. On the other hand, Syracuse jurors would be a realistic cross section of average Americans whose lives had yet to be touched by technology any more directly than by super-market scanners and digitized directory assistance.

Even if the locals weren't riveted by the case, the press turned out in force. Local newspapers had already heralded the start of "the hacker trial." Television cameras were stationed outside the building, waiting for the family to appear. Sequestered with relatives in a Syracuse suburb, the Morris family guarded its whereabouts closely.

The press showed up because the case tapped America's ambivalent feelings about the power and reach of computers. It had somehow come to symbolize the nation's collective anxiety about computer hackers and the threat they might pose. And the case had family drama: the father and son who belonged to a computer science elite were both obsessed with exploring the subtle intricacies of the complex computers that had come to control much of society.

The case also marked the sudden awakening of a national recognition of the fragility of tens of thousands of interconnected computers. The havoc wrought by Morris's program symbolized as never before the na-tion's increasing dependence on computers, and the increasing vulnera-bility of those computers. As computers became more and more interwoven, each link interconnected and interdependent on hundreds of others, and as these computers became accessible to more and more people, it was inevitable that something like the Morris program would come along. Still, it was a surprise when it came.

The two attorneys representing the Justice Department discharged their duties with proficiency. Mark Rasch, a short man with sharp, handsome features who was now the Justice Department's star computer crime prosecutor, wasn't the slightest bit flustered. His command of the complex technology under discussion was so impressive that it gave him an air of extra authority. His partner, Ellen Meltzer, at thirty-seven a Justice Department veteran, was actually senior to Rasch, but that wasn't altogether clear in the courtroom. Meltzer deferred to Rasch on most technical points.

Even before the trial opened, Rasch and Meltzer filed a motion asking the judge to dismiss as irrelevant any evidence involving Morris's intent. "The evidence of lack of intent to cause loss is simply not relevant to any issue in this case," Rasch declared in his motion. According to the law, Rasch said, the prosecution needed only prove that Morris intended to break in and that damage resulted from the worm, not that Morris intended to cause damage.

Jury selection went swiftly enough. Both the prosecution and the defense were looking for one important criterion: none of the jurors should have so much as a fleeting knowledge of computers. All prospective jurors were asked if they owned a personal computer, if they knew anything about computers. "I don't know a thing," replied one woman, and she was promptly seated on the panel. Only three prospective jurors owned computers, and all three were rejected by either the prosecution or the defense. One man who knew nothing about computers himself but whose son worked at IBM was excused. In the end, two women who used computers in their jobs as clerks were seated on the panel.

▲ ▼ ▲

The first and most striking impression Robert Morris made in the court-room was that of an unlikely felon. Traditional white-collar criminals felled by their own greed or malice were usually much older than this defendant. He was pale and thin, and his new suit seemed a little too large. It was hard not to wonder whether the suit was an imperfect fit intentionally, a subtle message from his defense team: If he's not even old enough to know how to pick a suit, how much of a white-collar criminal could he be? His seeming lack of worldliness was what the defense attorneys wanted to convey to the jury of nine women and three men, many of them with children the defendant's age. In court, his attorneys referred to him as "Robert," a "kid."

Howard G. Munson, the federal district judge in the case, was a sixty-five-year-old Republican whose fourteen years on the federal bench had accrued a record of fairness and neutrality in criminal cases. A heavy smoker with a resounding cough and thunderous voice, Munson had a sharp memory for the details of a case; at the same time, he possessed a knack for extracting the essence from a witness's testimony.

In his opening statement to the jurors, prosecutor Mark Rasch told them they would be hearing some difficult technical language, but that shouldn't distract them from what the government intended to prove: that on November 2, 1988, there was a full-scale assault on computers throughout the United States launched by the defendant, Robert Tappan Morris. Rasch pronounced Robert's middle name "Tap-*in*." "This assault was deliberate, planned, calculated," Rasch said. "It was calculated to break into as many different computers as he possibly could to gain what the law will call 'unauthorized access.' "

Rasch went on: "What the government intends to prove is that early in the evening of November 2, 1988, computer scientists throughout the country started noticing that something was going drastically wrong with computers from California to Massachusetts to Florida. All of the computers were starting to slow down. These were computers not just at government sites, not just at military sites, but at commercial facilities, at private companies throughout the country, and many of the people we'll hear testimony from worked at different universities doing scientific research. Their research was interrupted. They couldn't do their work because of the actions of the defendant, Robert Tap-*in* Morris. Valuable computer time was lost. Valuable experiments were lost. People could not communicate with each other. They couldn't talk to each other. They couldn't find out what was going on. Their computers, and you will hear testimony about this, were crashing, stopped running, came to a grinding halt because of the actions of the defendant, Robert Tap-*in* Morris."

As Rasch described it, Morris had sat for hours at a computer terminal in Upson Hall at Cornell, planning "the break-in." "It was deliberate." Rasch repeated the word *deliberate* like a mantra. "As the time for release of this worm came nearer, he spent more and more time on the computer trying to perfect it. He worked all day on the day of November 2 trying to launch this worm. It was designed deliberately and intentionally."

The prosecutor stopped just short of voicing his private theory that Robert had actually panicked upon seeing that the *ftp* bug had been fixed and that he rushed to get the worm finished that day before other holes were plugged as well. Moreover, he believed that Robert's rush at the end accounted for some of the careless code in the program.

Rasch told the jury that he intended to call a collection of witnesses into the courtroom, victims from computer centers throughout the nation who would describe the horrendous events of November 2 and estimate the dollars lost in the effort to fix their systems. "You will hear testimony from the scientists, the engineers, who use these computers that when this thing was attacking them, they didn't know what it was doing. It was designed to hide its effects, to pretend to be an innocuous program. It disguised itself to frustrate the victims of its attack."

In explaining what a computer virus was, Rasch drew parallels to the biological flus the jurors could all relate to. "If you have just one virus you may not get very sick. But if you get many viruses, if you have hundreds, you will get very sick. Not only will you get very sick, you will get other people sick." Not only was Rasch comfortable with most of the computer arcana that was tossed around in the courtroom, but he did his share of tossing. And even though Guidoboni had two computer experts at his disposal throughout the lengthy preparation for the trial, he was still struggling a bit with some of the concepts. If the case was to be decided on pure technical machismo, Rasch would undoubtedly win.

The prosecutor reminded the jurors that the government did not intend to prove that Robert was an evil person, and that it did not need to prove that Robert intended to cause the damage he caused. For the jury to return a guilty verdict, Rasch said, the government needed only to prove that he intended to obtain unauthorized access and break in.

In contrast to Rasch's clipped and formal language, Guidoboni adopted a folksy tone. "I guess you can tell from looking at me that I am not Perry Mason," Guidoboni said. But where Rasch was self-assured, Guidoboni was almost apologetic. "I don't have a rabbit to reach down here and pull out of my hat," he said. "That's not the way criminal trials work in real life, so you're going to have to bear with me. We are going to do the best that we can." A few minutes into his speech, in fact, Guidoboni made Robert's confession for him. He acknowledged that his client had indeed written the worm. However, he added, evidence would show that "this worm caused no permanent damage and it was not designed to cause permanent damage. It didn't break any machines.

The virus didn't read anybody's private files and didn't steal any information and didn't put one dollar into Mr. Morris's pocket."

The defense attorney then listed Robert's credentials. Not only did he work hard at summer jobs while at Harvard, but he had made important contributions to the field of computer security. He had written papers that were published, and he had alerted the computer community to security holes in the past.

Guidoboni depicted the Internet as a loosely organized, almost ragtag collection of networks. "There was no drug czar, if you will, computer czar sitting up in Washington who has control, who established rules." Moreover, he said, the Internet is not a network that launches missiles. It's used for conducting research and sharing scientific data. It's also used for frivolous things like playing chess, sending love letters and exchanging recipes.

Guidoboni's speech had an odd air of desperation to it. He wasn't asking so much for a judgment of innocence as for forgiveness.

Guidoboni said he would present evidence showing that Robert's virus was purposely limited, its intention to spread slowly and quietly, "but you will hear that he made a mistake, a critical mistake." And in the end, "he alerted the computer community that the system wasn't secure and that they needed to take steps to fix it. This in turn caused a lot of embarrassment. But a simple mistake, together with embarrassment and some inconvenience, are not the equivalent of a federal felony offense."

Witnesses were what Mark Rasch promised and witnesses were what he delivered. The first witness for the prosecution was Dean Krafft, Cornell's director of computing facilities. Krafft had been pacing outside the courtroom, wondering aloud to another witness how he would be able to explain data decryption to the jury. Krafft described to a mystified jury the process of sorting through Morris's old files and finding early versions of the program. If nothing else, the jury got from Krafft's testimony that this was indeed to be a trial filled with technical terms.

Many of the witnesses for the prosecution knew each other, if not personally then by electronic mail, and if not by electronic mail then by reputation. Rasch and Meltzer had selected their witnesses carefully. They represented a good cross section of the nation's computer system managers. They traveled to Syracuse from universities, Army research labs and other government institutions. Local reporters joked that they were disappointed to see the witnesses weren't wearing pocket protectors, which they believed to be a computer nerd's signal appurtenance.

Mark Brown, the University of Southern California system manager who had battled Kevin Mitnick, was the second witness. He gave the first in what would become a blur of testimony about the devastation of November 2, 1988. Brown testified that when he logged in to his computer that night at about 11:00 P.M. California time, he saw dozens of strange programs running. Even after Brown killed them, they were back within minutes. The situation threw Brown and others in the computer center into a state of panic. "We'd never seen anything this widespread," he told the court. "We had no idea if it was modifying the data on the files." So at 6:00 A.M., to prevent infections from coming in and going out, Brown disconnected the university's computers from the Internet. To prevent reinfection, Brown had to turn off the computers, halting file transfers, cutting off electronic mail and bringing all research to a halt. In the end, 350 Suns and VAXes on the USC campus were infected.

And so it went, in painstakingly systematic fashion. Rasch and Meltzer called computer managers to the stand, and one after another they described the mayhem that ensued after the worm hit their institutions. In endless litany, witnesses told of realizing they were under attack, of taking computers off the network, of crawling under their desks and pulling the computers' plugs. The prosecutors asked each witness how many hours he and his colleagues spent ridding his computers of the worm. Estimates of the damages to each institution varied widely, from $200 to $54,000, and the total climbed steadily until it reached $150,000. And how much was their time worth? If anything was to sink in with this jury, it was that people who work with computers make good money: one witness valued his time at $21 an hour, another at $22.38. Was Robert Tap-in Morris authorized to use those computers? No, came the replies.

At times, the courtroom transformed into an introductory computer class, as witnesses attempted to explain complex and often baffling technical terms to the jury. The prosecutors submitted as evidence dozens of documents containing page after page of abstruse computer code, prompting a rebuke from Judge Munson. "For those of us educated in the forties, most of this is totally incomprehensible," the judge told the attorneys. "I don't see what good it would do for the jurors."

In his cross-examination of the prosecution's witnesses, Guidoboni had them state that the worm caused no damage. And he tried to establish that much of the time they devoted to working on the problem was spent in pure intellectual pursuit, analyzing the worm once it had already been eradicated. The most brilliant moment for the defense came in Guidoboni's cross-examination of Keith Bostic, the Berkeley system manager in charge of Berkeley UNIX. The prosecutor had Bostic tell of the catatonic state the Berkeley computers were in, and the exhausting decompilation effort. But Guidoboni asked Bostic whether the worm had in fact alerted the computer science community to security flaws in *sendmail* and *finger*.

"As a result of the worm," asked Guidoboni, "the *sendmail* program was a little bit better, wasn't it?"

"That is correct," Bostic responded.

"And it certainly was more secure."

"That is correct." By this time Bostic was smiling.

"And the same thing for *finger*, is that not correct?"

"That is correct," Bostic answered.

As long as Guidoboni was on a roll, there was another matter he wanted to clear up with Bostic. It was the question of authorized access. Didn't programs such as *finger* and *ftp* extend to anyone who used them a certain measure of access to all the computers on which the programs ran? Bostic tilted his head as if this were the first time he had considered the question. "Yes, it does," he answered. So might it be possible that, in a sense, Robert Morris was authorized to use any computer that could run those programs? Bostic agreed that he was.

After each witness testified, reporters followed him outside for more interrogation. Does Robert Morris deserve prison? "I certainly hope he is convicted and sent to prison," said one angry computer manager. "Otherwise, it will be open season on the whole network." The manager from the Army research lab had a slightly different view: "It would be bad to put someone like this in prison. He could learn things and be a real threat."

▲ ▼ ▲

Every day in court, Bob and Anne sat in the front row of the audience. Anne kept a composition book propped on her lap and took occasional notes. If anyone cut a conspicuous figure, it was the elder Morris. A mischievous gleam in his eye was dampened by a trace of sadness. He wore a gray three-piece suit and heavy black shoes bowed from wear. His complexion was gray and he smoked constantly outside the courtroom during recesses, enveloping whoever happened to be standing nearby in a curtain of smoke.

Over the three-day weekend, AT&T suffered a major failure in its long-distance service. For much of Monday about half the long-distance, international and toll-free 800 calls on the AT&T network didn't go through. Rental car reservation clerks and other toll-free operators sat idle. Television producers had trouble covering stories because they couldn't reach their bureaus. The problem was traced to an AT&T software glitch. A rumor circulated that a hacker in the New York area had broken into AT&T computers there and caused the disruption as a protest against Robert's prosecution. Mark Rasch even called colleagues in Washington to see if it had indeed been the work of a hacker. In the end, it turned out to be an error in the phone company's software. Yet the incident again brought home the unexpected importance of computers in everyday life. Any member of the jury who had tried and failed to make a long-distance call that day would have had a more vivid understanding of the Internet computer managers' frustration when faced with Robert's worm. In the Morris defense camp there was concern that this could make jurors less receptive to arguments in Robert's defense.

By the second week of the trial, more supporters of the defendant appeared in the courtroom. Peter Neumann, a security expert and former Bell Labs researcher, came from California to testify for the defense. Doug McIlroy was there as a character witness. Robert's friend Janet stayed through most of the trial. Meredith and Ben had shown up in time to see Robert testify. During the many long breaks, Janet, Robert, Meredith and Ben sat hunched over the day's crossword puzzle in the newspaper.

That week, three longtime friends of the Morris family arrived—Ken Thompson, Fred Grampp and Jim Reeds, all of whom also happened to be experts in the field of computer security. They had flown from New Jersey to Syracuse in a small private plane to do nothing more than sit in the courtroom and show their support for the family in general and the young defendant in particular. The defense already had more character witnesses than it could use, otherwise it would have called upon these three. It was like a twenty-five-year reunion of computer security's power hitters. Indeed, the scientists engaged in lively talks outside the courtroom might have been mistaken for a group of technical conference attendees on a coffee break. The Bell Labs scientists could recall the Arpanet crash of October 1980, when a simple hardware failure on a single computer brought the network to its knees. Of course, there were only several hundred computers on the Arpanet then.

When the Bell Labs trio arrived, Anne Morris began steering reporters in their direction, letting the journalists know just what sort of eminent scientists had taken the trouble to make the trip to Syracuse. In search of fresh story angles after days of tedious technical testimony, the reporters clustered around Thompson, perhaps the best known among computer scientists. His thinning brown hair brushing his shoulders, Thompson looked the most like someone whose eccentricities were indulged out of deference to his genius. Bob Morris stood by and watched in amusement, twirling a cigarette in his fingers. When Thompson remarked that he thought the release of the worm was an irresponsible act, Morris grinned and spoke up: "Have you ever done anything like that, Ken?"

It was peculiar that these of all people would be assembled in a court of law in the first place. The group provided more than just a technical reference point: they were the people behind the theoretical debates over technological morality and intellectual property. Their presence was a reminder of a long-standing tradition among computer security experts that one earns one's stripes in the field by defeating a computer's barriers. But that was in the old days, when the community was far smaller, when computers weren't ubiquitous, when the stakes were much lower, when breaching security was still considered to be all in the spirit of science.

Not the sort of people who allow idle time to slip by unproductively, the Morris family came to court each day with books to keep them occupied during recesses. The volumes they chose were light reading only by their standards: Bob kept Suetonius's *The Twelve Caesars* at his side, Robert had brought along a copy of Robert Graves's historical novel *Count Belisarius*, and Meredith was reading Umberto Eco's lastest novel, *Foucault's Pendulum*. Anne kept a careful watch on Bob, who was rather enjoying the journalists' attention. She adopted a more studied approach toward the assembled press: she spent many of the breaks quietly talking to reporters of Robert's various virtues. His brilliance she illustrated with a story about steamer trunks from Robert's childhood filled with back issues of *Scientific American*. His sensitivity she certified by telling of the fainting spell on the Washington Metro the day they first visited Guidoboni.

Paul Graham was put on the stand by the prosecution, and Andy Sudduth was called by the defense. When Robert saw his two friends enter the courtroom, he turned crimson, as if he were embarrassed for all three of them. Testimony from the two Harvard students provided some of the trial's few light moments. Paul was as animated as ever. He recounted the conversations he had with Robert as the worm was in the planning stages, and the panicked call he got from his friend just after it was clear that this would be no subtle experiment, but a monumental disaster.

Paul and Andy brought to the trial a further air of elite institutions. Andy told the court he had gone to Phillips Exeter Academy before entering Harvard, and that he had helped Harvard win the Head of the Charles rowing competition on the weekend just before the worm hit when Robert was in Cambridge allegedly poring over UNIX source code. Even the judge, a proud graduate of the University of Pennsylvania, another Ivy League school, was beginning to get caught up in the club-biness of it all. As if it mattered, he interrupted Paul's testimony to clarify some rowing parlance. The judge insisted the sport was called "crew," and it wasn't until Andy took the stand and called it "rowing" that the matter was resolved. At the end of Andy's testimony, the judge went out of his way to congratulate the witness on his 1984 Olympic silver medal.

The only element of mystery in the defense was whether Guidoboni would put Robert on the stand. But it wasn't really much of a question. As Guidoboni saw it, it could only help for the jury to hear the whole story from Robert himself. Not only would this be the first time since the worm hit that Robert made any public statement whatsoever, but his sincerity and utter lack of defensiveness would help the jurors see their way through the technical jargon to the core of the matter. Robert Morris had made a terrible mistake and shouldn't be punished for it.

But when he took the stand on the final day of testimony, Robert unintentionally came across as somewhat aloof, less endearing than he might have been. So intent was he on explaining the technical details that, rather than steal the jurors' hearts, he seemed slightly superior.

Guidoboni had a hard time getting Robert to express his feeling about the worm incident without provoking an instant objection from the prosecution. Guidoboni tried three times and each time Rasch stopped him. Finally, Robert managed to blurt out, "It was a mistake and I'm sorry for it," before the prosecutor could object to Guidoboni's fourth attempt.

Even Robert's honesty backfired on him a bit. When Rasch cross-examined him, Robert answered with a simple and direct "yes" to nearly every question. Yes, he planned and wrote the worm deliberately and consciously. Yes, he deliberately planned for the worm to break into computers on which he did not have an account. And yes, he deliber-ately made the worm's code difficult to understand.

When both sides were finished with their questions, the judge had a few of his own.

"Let me ask you something, Mr. Morris," he began. "You designed this program so that only once in seven times when the virus tried to intrude on some computer where it was already there that it would succeed again."

"Yes."

"Did it work?"

"I expect that it did work, yes."

"So the mistake you made was that the access was too rapid, it exceeded your expectations."

"Yes, that is right."

"So you misjudged the speed of the system."

"Yes, I certainly made an error in judgment."

▲ ▼ ▲

After the testimony ended, Guidoboni argued for acquittal, claiming that the government had not proven its case. At most, the defense lawyer maintained, prosecutors proved the elements of a lesser charge, a misdemeanor under the statute. The question of a lesser charge had arisen before, and this time the judge appeared to be ready to consider it seriously. But he could not give the jury the option of considering a misdemeanor charge along with the felony unless the defense asked him to. That meant the decision was up to Robert. It was a gamble. It meant weighing the chance of an acquittal against the chance of a compromise verdict. Having a misdemeanor on his record was certainly preferrable to a felony. Then again, if the jury didn't have a lesser charge to fall back on, it might choose the harsher conviction. When Munson adjourned the court until the following Monday, the matter was still unresolved.

The family had three tense days to kill in Syracuse. In search of distraction, having already toured the nearby Corning glass factory, everyone went bowling, then to the movies. Robert spent the weekend sleeping late.

Since the judge hadn't yet ruled on the prosecution's motion to disallow testimony related to Robert's intent, it was expected that his instructions to the jury would serve as his ruling. He took the weekend to think about it. Both sets of attorneys presented themselves in the judge's chambers first thing Monday morning and heard his decision on intent. To Guidoboni's disappointment but not to his surprise, the judge had not changed his mind on the irrelevance of Robert's intent to cause damage. As the judge read the law, the question of intent pertained only to whether the defendant intended to break in, not whether he intended to cause damage.

At that point, Guidoboni knew the case was over. His closing arguments to the jury were listless; his demeanor smacked of defeat. Prosecutor Ellen Meltzer, on the other hand, made her closing arguments with special vigor. She reminded the jury that they had heard Morris himself admit to the crime, to its deliberate nature and to causing authorized users to lose use of their computers. "Mr. Morris's computer worm was not a juvenile prank," she intoned. "In no way was it a legitimate Cornell research project gone awry."

Indeed, she said, there is nothing to thank the defendant for. "You do not thank a terrorist for increased airline security awareness and you do not thank a drunk driver—"

But before she could finish Guidoboni was on his feet, outraged by the comparison. The judge agreed. The word *terrorist* had special meaning in Syracuse. A year earlier, terrorists had blown up a Pan Am jumbo jet over Lockerbie, Scotland; among the passengers were thirty-five students returning home from a Syracuse University study-abroad program. Certainly in this city, of all places, Robert Morris could not be compared to terrorists who claimed innocent lives. The judge sustained Guidoboni's objection.

When Judge Munson delivered his instructions to the jury, in an implicit ruling on the prosecution's intent motion, he told the panel to disregard the testimony concerning the defendant's intention to cause damage. It was not relevant to the charge. He read the indictment to the jurors and explained the elements of the crime. If, in its deliberations, the panel should conclude that the government had failed to prove each of the elements, then the jury must return with an acquittal. But he said nothing about a lesser charge. Robert had decided to gamble on an acquittal.

During jury deliberations, Robert, Janet, Ben and Meredith started playing cards at the defense table until the guard told them to stop. Robert picked up his historical novel. Bob occupied himself with an old eighty-five-cent Penguin Classic edition of one of Xenophon's histories. Anne mingled with some of the reporters. Five hours later, the jury filed back into the courtroom and delivered its verdict. Robert watched intently as each juror was polled, and as each said, "Guilty." Reporters made a dash for the two pay telephones in the building. The judge set a date for posttrial motions and adjourned the court. Robert stood up and smiled meekly at Janet. Anne rushed up to the defense table but Bob hung back, staring straight ahead, the fingers of his left hand rubbing together as if touching an invisible piece of fabric. Anne motioned for him to join the rest of the family, which had formed a huddle at the defense table. Anne wanted to talk to Guidoboni about their options. The Morrises were people accustomed to having the privilege of options and a felony conviction seemed to be no exception. In the face of something essentially beyond her control, Anne was still considering where to go from here. But Guidoboni held up his hands and said, "Listen, folks, we don't have to go over this right now. There's plenty of time to talk about it."

In other circumstances, the computer scientist who worked for an American intelligence agency so hidden that even its name was once secret and his young son would quietly have lived out their lives as part of the country's best and brightest. Now the father was caught in the glare of publicity and the son was a felon. Someone who seemed destined to become one of the country's software stars was facing jail.

The lawyer gave Robert an apologetic handshake, snapped at the throng of reporters on his way out of the courthouse, and returned to his hotel. Flanked by family members, Robert left without saying a word. In his son's absence, Bob Morris stood in the plaza in front of the court, drawing reporters into a tight circle around him by responding to their questions in a voice just above a whisper. He said he still believed that his son didn't have a fraudulent or dishonest bone in his body.

Epilogue

by Katie Hafner

It was following his 1988 arrest in the Calabasas parking garage that the popular myth surrounding Kevin Mitnick first ignited. Local law enforcement officials rushed to see the U.S. magistrate and urged that she deny bail. They told her that Mitnick had tampered with one judge's credit rating already and had disrupted the phone service of one of his probation officers. The suspect, they said, had once fled to Israel and there was no telling where he might run this time. The magistrate denied bail.

Newspapers portrayed him as an electronic terrorist, capable of triggering a nuclear holocaust from a Touch-Tone telephone. " 'Dark Side' Hacker Seen as Electronic Terrorist" and "Computer Whiz is a Threat to Society," read the headlines. He was placed in maximum security prison, and his use of the phone was heavily supervised. Mitnick's manual dexterity struck fear into his wardens. One imagines devices on his hands similar to the cage over Hannibal Lecter's mouth.

Alan Rubin, Mitnick's court-appointed lawyer, attempted to discredit some of the myths. Not only had Mitnick never fled to Israel, he had never been out of the country. Rubin didn't dispute the tampering with a probation officer's phone service, but he did insist that Mitnick had never altered a judge's credit rating. After Digital assessed the damage

Mitnick had done at $160,000, the lawyer worked out a plea bargain with the Justice Department: Mitnick would plead guilty to two of the government's four felony counts and get a year in federal prison and psychiatric counseling. The prosecutor on the case supported the agreement because taking the case to trial would have required granting immunity to Lenny DiCicco, whom the prosecutor considered to be equally guilty. And for their own reasons, Digital officials were satisfied to see the matter avoid a public trial.

But the press already had painted such an unsavory picture of this defendant that the judge, influenced by what she had read, rejected the plea bargain. She pointed out that Kevin Mitnick had enjoyed leniency in the past, and it had been ineffective. A year in prison wouldn't be enough for this dangerous criminal, whose sophisticated stunts the public found terrifying. "We won't know the damage Mr. Mitnick has done until it's too late," the judge told the courtroom.

In a show of the same willingness to turn on a friend that Lenny had displayed, Kevin offered to cooperate with Federal prosecutors against Lenny. Then Kevin's attorney switched tacks. He convinced the judge that his client's computer behavior was something over which his client had little control, not unlike the compulsion to take drugs, drink alcohol or shoplift. It was a persuasive argument. This time, the judge accepted the one-year prison sentence. The prison sentence would be followed by six months in a rehabilitation program for everything from drug addicts to compulsive gamblers. Now Kevin wasn't just a computer hacker: he was a computer addict. Mitnick apologized for his actions in court and vowed never to repeat them.

Mitnick became a model resident at a halfway house, and was released from the program early, in the spring of 1990. At first, the conditions of his three-year probation prohibited him from so much as touching a computer, but once he demonstrated that he could control his behavior, he was allowed to search for computer work. He was still prohibited from using a modem. Fed up, Bonnie asked for a divorce. So Kevin moved to Las Vegas, where his mother and grandmother both lived.

▲ ▼ ▲

I met Mitnick—after a fashion—when he called Tom Snyder's radio show from Las Vegas in the summer of 1991. I was on a promotional tour for Cyberpunk. Throughout the writing of the book Mitnick had refused to speak with me unless I paid him. He became a ghostly presence during my time in L.A., as I visited his friends—and his erstwhile friends. They are a mistrustful bunch. The reporting entailed a lot of meetings at odd hours in the parking lots of fast-food outlets. To reach them, I paged them, or called from one pay phone to another. I got lucky when a couple of L.A. cops who had come to loathe Mitnick and his confreres handed me Mitnick's police file, which was several inches thick.

While conceding that "eighty percent" of what I wrote was accurate, Mitnick was apparently so irked by some of it that he called the show to air his grievances on national radio. He was polite and friendly, and even waited patiently while Snyder, who seemed oblivious to the high drama of this little contretemps, broke for a commercial.

The ensuing conversation was tense. Reciting verbatim from the book, Mitnick dwelt on the bias I had displayed favoring Lenny. I was polite back. I told him that if he had spoken with me before the book was published, he'd have a more compelling case.

Snyder interrupted with a question: "Kevin, you don't deny you broke into computers. You did that, right?"

"Yes, I did."

"Why'd you do it? What's the thrill?"

"It was basically the fascination of learning more about computer systems," Mitnick replied. "See, they didn't offer that type of in-depth exploring in an educational environment."

"But when you're breaking into a system," Snyder continued, "do you know you're doing wrong or do you feel you have a right to access these systems?"

"Oh, I know it's wrong," Mitnick responded. "But it's basically to learn more about the operating system and computer systems. It's basically the fascination of learning about the computer itself, not necessarily the information the computer contained."

When Snyder cut him off to take another call, and the elusive Mitnick disappeared back into the unknown, I didn't ask what he was doing with his days but gave him the benefit of the doubt. I imagined him reformed, living with his mother, walking the straight and narrow path of a crime-free life.

I was wrong.

▲ ▼ ▲

Returning to the subject after five years has been like walking into Kafka's closet. It is an image from *The Trial*, Kafka's waking nightmare of a novel about an accused man in search of his crime. In the midst of his travail Joseph K. opens the door to a small storage room and sees a man poised to flog two others with a rod. Some time later, K. opens the same door and the three men are still there, in precisely the same position. A lot can happen in five years. For my part, I had divorced, remarried, had a child, moved to a different state and written a book on an unrelated topic. But the people who inhabit Kevin Mitnick's universe are all exactly as I left them. They seem to have had none of life's experiences that make us wiser, more serene or resolved to do something different with our lives. They are people locked into the fringes of criminality and the law enforcement system.

When I returned to the story in early 1995, Susan Thunder was still nurturing her grudge against Roscoe. She posted messages to Internet news groups, quoting from Cyberpunk and inserting such epithets as "putrid" and "wicked" in front of Roscoe's name. For his part, Roscoe was still employed as a computer manager at the auto parts reseller in L.A. that he worked for when I met him six years earlier. As for Mitnick himself, it seems that even if he had wanted to stop breaking into other people's computers, he couldn't.

▲ ▼ ▲

At 11 o'clock on a Sunday morning in the spring of 1995, Susan Thunder and I were seated in the dining room of the Silver Saddle Ranch and Club, a low-rent dude ranch of sorts in the high Mojave Desert. She was scrutinizing a newspaper photograph of Tsutomu Shimomura, a young computer scientist whose digital sleuthing had brought Mitnick down a few weeks earlier. This arrest followed a two-year computer crime spree thought to be Mitnick's most egregious and prolonged to date. "Is Tsutomu gay or bi or what?" she asked. I told her I was relatively certain he's straight. "Oh," she replied. She was plotting her revenge against Shimomura, she said, and thought she might try something from a sexual standpoint, perhaps a "medium-term, possibly long-term" revenge program.

When I met with Susan, Mitnick was in jail in North Carolina, suspected of having spent more than two years breaking into computers everywhere from San Diego to the north of England. On the lam since late 1992, the thirty-one-year-old Mitnick finally was caught when he tangled with Shimomura, whose obsessive persistence matched his own.

After Mitnick's arrest, one of the first calls I got was from Susan. I was interested in her as a former cohort of Mitnick's. I had last seen her in Las Vegas in 1989, and knew to be wary of her. Our conversation six years earlier had been a disjointed mess and I had spent weeks trying to verify her stories, many of which hadn't checked out. But now I was intrigued by what she said. She was irate over her friend's ignominious downfall at the hands of Shimomura, a brilliant and quirky scientist who entered the chase earlier this year, after his own computers in San Diego were breached. He monitored the hacker intensively and within a week had led FBI agents to an apartment complex in Raleigh, where they found their fugitive, presumably in the midst of a late-night hacking session. "Tsutomu's way out of line," she said, although she has never met him. "You understand? Way out of line." I decided to go visit her.

I was more than a little suprised when Susan, now thirty-six, told me she held elected office. She was city clerk in a small town of abut 8,000 in the middle of the desert. When she first greeted me at City Hall, she was dressed to dramatic effect in a multicolored chiffon skirt and black velour hat with a large spider pin vamping down one rim. Susan takes little at face value and sees conspiracy and scandal around every corner. She told me proudly that she had been devoting much of her time to rooting out malfeasance among her colleagues in municipal government. As we spoke, she was busy collecting her evidence—picking torn papers out of a waste bin and placing them in carefully labeled envelopes.

But now she was preoccupied with the injustice visited upon Mitnick. She had turned his plight into her own life-affirming cause. One of her on-line monikers was now "friendkm." Over breakfast at the Silver Saddle, where Susan had procured me a room for the night, she plied me with chapter and verse on every topic from her days as a groupie (which included, she said, a systematic conquest of all four Beatles) to her triumphant computer hacking career.

It was in Las Vegas that Mitnick had reconnected with Susan Thunder. While Kevin was still in Lompoc prison, I had visited Susan in Las Vegas. She tossed dozens of invectives in Mitnick's direction. But following his release Mitnick apparently looked her up and they became friends, or rather, hacking compadres. Susan claims Mitnick was interested in her expertise in "military systems" but it's more likely that he was after her social engineering skills, which are admirable.

In 1991, Mitnick was employed at a local direct-mail company called Passkey Systems, where he programmed Digital computers. But the restriction against using a modem must have made his days pretty bland. So he and Susan looked for other diversions. One of their favorite games, Susan told me, was to station themselves outside McDonald's and use a ham radio to override the drive-thru order takers. They waited for black customers to drive up, then shouted, "We don't serve niggers here!" She tried to teach him how to play poker, but he wasn't interested. She took him to a Barry Manilow concert. And she "employed" him as what she vaguely refers to as a telecommunications specialist. When Mitnick was in the throes of a project and needed her help, he could be demanding. Susan occasionally received frantic calls from him, insisting that she drop everything to call some place to extract a password or other nugget of information.

In 1990, when Susan had first laid eyes on Mitnick after so many years, she was shocked. He had lost a lot of weight and was working out regularly. He had gone from slob to dude. "You know," she volunteered, "that hooker side of me always wondered if he was a good fuck." With that, she embarked on the story of her attempt to seduce Kevin. First she tried a little innocent flirting, but when subtlety didn't work, she asked him flat out: "Kevin, are you dating anyone?" He accused her of prying into his private life. "I'm not interested in dating anyone," he replied curtly. She gave up. "But I have no doubt that if I wanted to fuck him I could have," she said.

▲ ▼ ▲

A few weeks after our brunch at the Silver Saddle, Susan and I met in Las Vegas. She picked me up from the airport and careened through the city streets, past the Excalibur (King Arthur does Las Vegas), the MGM Grand and the more high-brow Luxor, a faux Egyptian pyramid outside of which sits a massive Sphinx.

Susan and I checked into the Gold Coast, a no-frills hotel and casino on the south side of town, and the next morning we toured the places she and Kevin had frequented during his time in Las Vegas. She took me to Kevin's former "office," a giant Kinko's on the east side of town that he used as his fax center. Kevin, Susan told me, had the run of the place. He did not like to stand in line, and usually just stepped behind the counter to retrieve his own faxes. But, she hastened to add, he always paid, and was always friendly to the Kinko's employees.

▲ ▼ ▲

I then went to visit Harriet Rossetto, the director of Gateways Beit T'Shuvah ("House of Repentence" in Hebrew) in Los Angeles. It was here that Mitnick spent several months after his release from Lompoc in 1989. Rossetto, a clinical social worker, started Beit T'Shuvah eight years ago when she saw that Jewish ex-convicts with addiction problems had no place to go after their release from prison. She decided to combine the teachings of Judaism with the classic twelve-step Alcoholics Anonymous model and opened a residential treatment program. Twenty-five men live there. She showed me around. Small groups of men, clearly more rough-hewn than your average citizens, clustered in the primitive kitchen and in an adjoining common room. I had trouble picturing Kevin Mitnick tough from behind a computer screen, perhaps, but a pretty congenial guy in person—at home in such a place. While there, Mitnick had shared a room with three other men. He was not allowed near a computer.

The hacker-as-addict theory has been ridiculed widely. I confess to having used it for a cheap laugh at dinner parties, and I arrived at Beit T'Shuvah prepared to be skeptical of the God-invoking Twelve Steps in general and of their application to a computer criminal in particular. But Rossetto makes a convincing case. She is careful to point out that her definition of addiction is broad. It embraces everyone from alcoholics to overeaters to sex addicts to obsessive shoppers. When she heard about Mitnick, it was her first encounter with this particular twist on the addiction theme. But she recognized an important symptom: the avoidance of painful feelings of vulnerability and loneliness through an activity. "Either you face the feelings and work them through or you avoid them and look outside yourself for relief," she said. It was not her job to see that Kevin be brought to account for his computer crimes, but to see that he reach inside himself.

Much has been made of the fact that when Mitnick was arrested in February 1995 he was in possession of 21,600 credit card numbers. But there is no evidence that he used any of them. As Rossetto sees it, this is well in keeping with addictive behavior. The closest cousin to Mitnick's affliction, she said, is gambling. It's not about money or winning: he's addicted to the action.

Rossetto doesn't stop there. She maintained that many people who work with computers are similarly addicted, but they carry out their addiction in a socially acceptable way. I am inclined to agree with her. I've often witnessed the spell a computer can cast. My own mother, who has been a programmer for twenty-five years, has described to me what it's like to sit in front of a computer screen and have four hours whiz by. I can imagine the tonic effect for someone like Mitnick. It satisfies and focuses his attention like nothing else. While in the throes of it, he is transformed. "He's not the zhlub," Rossetto said. "He's powerful and important. He's no longer the fat kid with glasses from a dysfunctional poor family."

When Mitnick first arrived at Beit T'Shuvah, Rossetto reported, he was frightened, mistrustful and private, and remained aloof from the rest of the group. But in sessions with Rossetto, he began to open up. "I saw how lonely he was," she recalled. "He was a throwaway child, who found contact and affection wherever he could."

One source of that contact was his father's brother, Mitchell Mitnick, a drug addict who, by bizarre coincidence, moved into the halfway house while Kevin was there. According to Rossetto, Kevin had been making some progress toward becoming part of the group, then up popped Uncle Mitchell. The two cordoned themselves off and formed something of a Mitnick clique. By all accounts, Mitchell is not a nice man. He once wrote Rossetto a bum check to attend Beit T'Shuvah's Yom Kippur services. And two years ago, he was charged with murder.

When Kevin left Beit T'Shuvah, Rossetto tried to convince him to pursue a job in the field of computer security. "I always believed he would have been willing to join the other side," she said, "that the wrong side of the law was his consolation prize choice." She remains convinced that had he gotten self-respect legitimately, the outcome of the story would be different. But Kevin believed that people were too afraid of him to give him a chance.

He was right. After moving to Las Vegas, when he tried to attend an annual meeting of users of Digital computers, conference officials barred him from the door. He refused to leave, and even made a lengthy appeal to the group's board of directors. They were unmoved.

Rossetto last saw Mitnick in early 1992. He had returned to L.A. after his half-brother, Adam, died of a drug overdose. Shaken, depressed and lonely, he visited her to discuss the possibility of returning to Beit T'Shuvah. She asked him if he was in trouble. "I'm slipping," he replied. She told him that she thought a second stay there would be a good idea. "I'll think about it," Mitnick said. She didn't seem him again.

"Somewhere, sometime, all of us have been on the dark side," a friend who is a respected computer scientist told me. That is, nearly everyone in the computer field has at one point succumbed to the temptation to cross the line and break in, if just for the adrenaline rush. It's an impulse that usually lasts about fifteen minutes. In Mitnick's case it has lasted fifteen years.

Throughout those years, as other liaisons came and went, Mitnick seemed to remain intensely loyal to one friend—Roscoe. Mitnick and Roscoe were in regular touch during his time underground. He called Roscoe an hour after his arrest in February to tell him he was in jail in North Carolina. Roscoe taped the call. Or, as Roscoe explained it to me, his answering machine picked up Mitnick's call and he left it on.

Roscoe and Mitnick had met in 1978 over a ham radio channel, when Mitnick, then fifteen, was being accused of harassing other ham operators over the air. Mitnick's telephone expertise and general love of mischief caught Roscoe's attention and the two became tight friends. He has lurked on the sidelines of Kevin's personal dramas ever since. He is big brother and confidant. Even Roscoe's taking up with Bonnie, who moved in with Roscoe shortly after she and Kevin divorced, has a curiously all-in-the-family air to it.

As Mitnick sat in jail in North Carolina, I sent Roscoe e-mail, asking if we might meet. I got this in reply: "Dear Katie, The past came back to life as if theater curtains were slowly drawing open. . . . Such was the effect of your simple 're: l.a.' after all these years! I'm flattered to hear from you. . . . At this juncture, I feel it only appropriate to call to your attention the fact that I'm not accustomed to dating married women. However, since I have known you from the past, I would be delighted to make an exception."

We met at a restaurant called Wild Thyme in South Pasadena. Roscoe had changed little from the slightly nerdy, mustachioed, stiff young man I first met in 1989. Although I have spent many hours in Roscoe's company and carried on interminable telephone conversations with him (like Mitnick, Roscoe is on the phone a lot), he remains a riddle to me. Whenever I speak to him, I catch a whiff of Eddie Haskell. He is the model of politeness to me, but I've heard that he curses to others about my prying.

It's unclear just when Roscoe's time on the dark side ended. Roscoe always seems to have more than a superficial grasp of what's going on in Mitnick's life. And yet, whenever Mitnick goes down, Roscoe simply goes on. One of the first things Roscoe reported to me was that a week before his arrest, Mitnick signed over the rights to his story to Roscoe, and now Roscoe was busy working up a proposal for a book about his buddy's time on the run.

For the last few years, Roscoe told me, he has been dabbling in something called "speed seduction." The concept, he tells me, is based on what he calls neuro linguistic programming. From what I can gather, speed seduction entails talking a very good line very quickly to entice very good-looking women into bed. When Roscoe tried the speed seduction method over the Internet, he was so successful, he told me, that he lost track of the number of women he was dating. He has published a booklet titled *Sensual Access: the High Tech Guide to Seducing Women Using Your Home Computer*, which promises to instruct readers on how to "reel in fish by the dozen" just by logging on to electronic bulletin board systems. The primer includes a section on "whale watching, or learning to spot fat and ugly pigs . . . before they talk you into a blind dinner date." Mitnick, who had been having some trouble meeting women in recent years, had expressed some interest in speed seduction techniques. When Roscoe gave him a quick tutorial, Mitnick was impressed. He had remarked on how closely it resembled social engineering.

Mitnick had moved back to L.A. in early 1992, after Adam's death. He lived in an apartment on the western edge of the San Fernando Valley, near Malibu, working for his father, Alan, a general contractor who lived in the same apartment complex. But things between father and son didn't work out, and Mitnick went to work for Tel Tec Investigations, a private investigation firm. Unfortunately for Mitnick, this wasn't just any private eye firm. Around the time that Mitnick joined, Tel Tec's owners were arrested on suspicion of tapping into TRW computers to get financial records. Roscoe claims that as part of their plea bargain, the Tel Tec guys offered up Kevin Mitnick.

Over tea at Wild Thyme, Roscoe told me that for more than two years he and Bonnie had been appealing a case against the Government. The lawsuit followed a September 1992 search of Roscoe's home and workplace, as well as Mitnick's apartment. The search warrant was based on evidence the government has gathered pointing to illegal access to a Pacific Bell computer. During the raid, FBI agents had seized computers, computer disks containing dozens of encrypted files and scores of documents. The agents even took Bonnie's purse. Claiming that the warrant was invalid and unfounded, Roscoe sued, demanding that the items be returned and that the warrant be quashed.

Two months after the raid, a federal judge issued an arrest warrant for Mitnick for having violated probation for his 1989 conviction. There were two reasons: unauthorized access to the Pac Bell computer and his association with Roscoe. At 5:30 A.M. in early December 1992, three federal agents showed up at the door of Mitnick's apartment. They found his mother and her boyfriend, but no Mitnick. His mother told the agents that her son had moved out.

▲ ▼ ▲

That Christmas Eve, someone claiming to be a probation officer called the California Department of Motor Vehicles central headquarters in Sacramento and requested that three photographs be faxed to L.A. The caller had the DMV lingo down, complete with a legitimate "requester code." Ed Loveless, a DMV investigator, knew something was up because two separate calls had come in requesting the same photographs. On a hunch, Loveless called the office from which the request ostensibly had been placed for confirmation. Sure enough, the call had been a fraud. When he checked the fax number the caller had given, he discovered that it was a Kinko's in Studio City.

First, Loveless dummied up a set of photos to fax. Next he dispatched two local investigators to conduct surveillance on the Kinko's and alerted the management to watch for someone picking up the fax. It wasn't an easy vigil. Sandwiched between strip malls on a busy, congested stretch of Ventura Boulevard, the Kinko's shares a parking lot with a supermarket and a video rental shop. The investigators waited inside the Kinko's, but got no word from management. Then one of them saw a man leave the store with a fax and followed him into the parking lot. The man saw the approaching agents, dropped the papers and took off on foot. The agents tried to chase him but lost him in the crowded parking lot, which was filled with last-minute Christmas shoppers. The documents were covered with Kevin Mitnick's fingerprints. Nine months later an L.A. judge issued a second warrant for Kevin's arrest, setting bail at $1 million.

One of the photographs Mitnick wanted was that of Justin Petersen, a small-time criminal and computer hacker who had been informing on Mitnick for months. The FBI had recruited Petersen to inform on suspected hackers, Mitnick among them. Petersen had approached Mitnick and begun encouraging both Mitnick and Roscoe to break into computers with him. Mitnick eventually caught on to Petersen. After consulting an attorney, he and Roscoe taped their phone conversations with him. But it was too late. Thanks in part to Petersen's work, the government was able to issue the probation violation warrant.

▲ ▼ ▲

Around the time of the Kinko's incident Mitnick became a true fugitive and his lifestyle descended accordingly: he moved from city to city and took on a false identity in each new place; he paid for goods in cash; he didn't own a car. He typically moved to university towns, rented a furnished apartment in the student district and took a job with a hospital or clinic, usually as a computer troubleshooter. He was in Colorado for a time, then moved on to Seattle. The fact that computers have gotten much smaller and more powerful helped make the operation mobile. Mitnick could get a lot of hacking done with an inexpensive laptop, a cellular phone and a modem. He apparently took precautions to hide his tracks by routinely cloning new numbers (reprogramming a phone with purloined serial numbers) and routing his calls through far-flung dialing areas. Getting onto the Internet was easy enough: he just dialed a local Internet service provider and logged on, presumably by gaining full privileges on the system and keeping a low profile by using a little-used account. From there, the world was his oyster. The Internet has a feature called "telnet," a remote log-in program that allows users to hop from computer to computer, all over the world.

Not long after going into hiding, Mitnick struck up a rather bizarre relationship with one of his hacking targets—Neill Michael Clift, an unassuming thirty-year-old software engineer in Stockport, England. Clift is the computer security hobbyist who spends much of his spare time studying security flaws in VMS. For years he has shared his findings in an informal arrangement with Digital, and this earned him Mitnick's close attention.

Mitnick's interest in Clift began in the 1980s, when Mitnick was intercepting internal electronic mail traveling among Digital employees. During his regular perusals of their messages, he discovered Clift as a rich source of security bug secrets and began to work on getting into Clift's machines directly. The night before Mitnick's first major arrest back in 1988, he had been busy beating a path into Clift's computers. It seems he just put a bookmark there, with the intention of some day picking up where he had left off.

Mitnick resumed his pursuit of Clift in early 1993. Posing this time as a well-known Digital engineer, he called Clift up and poured on the flattery. He explained that the company was going to be recruiting engineers and wanted to know if Clift would be interested. Clift told him to send him e-mail at a computer in Loughborough. Mitnick then began sending Clift e-mail, posing this time as a different Digital engineer. To draw Clift in, he offered him a lot of information that Clift recognized as proprietary to Digital. Clift was fooled. Before long, they had exchanged decryption keys, so that each could read the other's encoded transmissions. Mitnick asked for—and got—nearly every security flaw that Clift had discovered in recent months.

After some weeks of exchanging e-mail with the imposter, Clift asked him a few pointed technical questions. When his correspondent seemed slightly adrift, he grew suspicious. He probed the path the mail was taking, and saw it wasn't going to a computer at Digital at all, but was being forwarded to a machine at the University of Southern California.

Clift immediately suspected Mitnick. He knew Mitnick had been out of jail for some time. But it was too late. If this was Mitnick, he already had conned Clift out of scores of interesting security tidbits. Embarrassed and in fear of losing his job, Clift cut off the correspondence.

Then, in the summer of 1994, Mitnick rang Clift up—not just once but every day for weeks, both at Clift's work and home. Clift took the calls, partly because he wanted to hear just how much Mitnick knew. Surprisingly candid, Mitnick told Clift how he had gotten into his machines and bragged in detail about how he had tracked him down. He told Clift that he had full, albeit unauthorized, access to computers at Netcom and the WELL, two Internet service providers based in California, and was using them as his launchpad to other computers. He recited a litany of the many programs he had stolen from Clift and Digital over the years.

Mitnick's daring impressed Clift, but his technical skills didn't. "He spoke to me at the start hoping that I could teach him how to find security problems himself," Clift said. "He thought there was some kind of trick to finding them. He just didn't realize that it takes a great deal of hard work to find them and you have to be technically proficient." It appeared to Clift that Mitnick's skill lay in sniffing out work done by others.

From what he had heard about Mitnick through the years, Clift had expected to encounter a maniac. But Mitnick was friendly and relaxed, hardly an edgy criminal. Their conversations occasionally lasted three, four, even five hours. Mitnick interrupted the conversations to replace the batteries in the cellular phone he was using.

Mitnick seemed to enjoy the fact that he had pissed people off. "He obviously enjoyed the one-upmanship," Clift recalled. "He can think his way round what people know or do but probably can't see how much he upsets them or how much trouble he causes them." Clift told Kevin how much he had feared for his job. Kevin apologized and said it was nothing personal.

Mitnick's spirits never seemed down, except when he was in financial straits. Money got very tight, he told Clift. But when he got paid a couple of hundred dollars for an interview with a small British magazine, he spent most of it on a 1,500-page manual on the inner workings of the VMS operating system. Mitnick inquired of Clift which parts he should read. When Clift suggested he read the entire book, Clift recalls, "that didn't sit well. The internals book came to signify his investment in technical skill." When Clift accused Kevin of being interested in nothing more than how to break into systems, Mitnick nearly shouted into the phone: "Well, I bought that fucking book didn't I?"

At the same time, Kevin was still using the Loughborough computer to exchange e-mail with Clift and Michael Lawrie, a friend of Clift's. In an attempt to learn more about him, Clift and Lawrie had decided to give Mitnick his own account on the machine. It was a risky decision, because Mitnick crashed the computer at will. According to Lawrie, he routinely altered files, stole other users' identities and killed users' jobs and data. Lawrie gradually got fed up and sent Mitnick a message telling him he would be happy to see him rot in jail.

The phone conversations with Clift continued. Clift had come to look forward to them. He had done his share of opening up to Mitnick as well. But what Clift didn't tell Mitnick was that he had enlisted the help of the FBI. Somehow Mitnick found out. And when he did, he was furious and hurt. He had erected a friendship, at least by his definition, and had taken a great risk in doing so. "You are a paranoid bastard," he wrote to Clift. "Thanks a lot for contacting Digital and informing on me about Israel. You really underestimate my contacts. I know more than I let on. Too bad we can't be friends, that would have been nice, but all you want to do is help them bust me."

▲ ▼ ▲

In late 1994, just as his telephonic marathons with Clift were ending, Mitnick once again came within a whisker of getting caught. He was in Seattle, living under the name Brian Merrill, and working at a local hospital, earning $1,900 per month as a computer troubleshooter. Two investigators looking into a spate of fraudulent cell-phone calls trailed him one night. The local cellular service company had seen a sudden surge in the use of cloned cell phones. The investigators traced the calls to Merrill's apartment and, using a scanning device, listened to a lengthy, spirited conversation he was having with someone in Colorado. The two men were discussing a particular computer system they were interested in cracking. The two chortled a great deal while they talked about sorting through files and modifying records.

A couple of months later, still unaware that Brian Merrill was Kevin Mitnick, the Seattle police, accompanied by Secret Service agents, returned with a search warrant. After staking out the apartment for three hours, they broke down the door. No one was home. They seized a roomful of electronic gear—a laptop computer, modems, cell phones, battery packs, manuals, chips and chip programming equipment—and left the warrant on the kitchen table. By the time word reached Seattle from the FBI in L.A. that this was the hideout of America's most wanted cybercriminal, Mitnick had disappeared.

Just where Mitnick went after leaving Seattle and before arriving in Raleigh a couple of months later isn't clear. I grilled Roscoe on this point, but he claims to know little about his friend's precise whereabouts during his time on the lam. When I first asked Susan about it, she said she thought he was in Germany. "The papers were in order," she intoned in her best conspiratorial voice. A few weeks later, she confessed to having fabricated that story—at Kevin's request of course. "That was the line we had agreed I would use if anyone asked me," she said. I told her that when Shimomura's machine was cracked on Christmas Day, the attack had apparently originated from a computer at Loyola University in Chicago. But she didn't hear me, for I had uttered the S word.

Susan was worrying the topic of Shimomura again. "This guy is an arrogant motherfucker," she said. "How can he say that Kevin isn't very good? If he was better than Kevin, his system should have been impenetrable. I'm sure he's pissed that Kevin got into his system. It's humiliating. But that's no reason not to acknowledge the skills of your competition. If this guy was such a hotshot security expert, how come Kevin got into his computers?"

I had been wondering the same thing.

▲ ▼ ▲

Tsutomu Shimomura, an elfin man with jet-black hair that flows well past his shoulders, was born in Japan thirty years ago to biochemists who shuttled their two children between Japan and the United States. He dropped out of high school at age fourteen to work at Princeton's astrophysics department, where he did image processing. He enrolled at Cal Tech as an undergraduate, where he found as his mentor the physicist Richard Feynman. The two spent hours discussing the deep connection between computation and physics, one of Feynman's favorite topics.

Before he could graduate, Los Alamos National Laboratory offered him a postdoctoral position. He was nineteen. At Los Alamos, he worked on processor architectures and computational models.

Five years ago Shimomura went to the San Diego Supercomputer Center on a year's sabbatical from Los Alamos and stayed, mostly because San Diego suited him better than the relative isolation of New Mexico.

Shimomura is driven, and has little tolerance for second-rate minds. "He can draw conclusions from things so quickly that if you can't keep up with him, he writes you off," said Gerard Newman, a colleague at the center.

When Shimomura first arrived at the center, he was twenty-five years old and extremely demanding. "Staff people were complaining about having to cater to him." recalled Sid Karin, the center's director. But others came to Shimomura's defense and told Karin to keep him on. "They said, 'This guy may be a pain in the butt but he's worth it,' " Karin recalled. By all accounts, Shimomura can be a terrible pest. "He just drops by in the middle of the afternoon, interrupting whatever you're doing and sits down and talks to you for two hours," Karin said. "It can be a real imposition. I've heard people say, " 'Tsutomu, what part of no don't you understand?' "

Among his friends, Shimomura is famous for his obsessions. He has been known to expound for hours on the virtues of the best in-line skating wheels and the most effective ski wax. Then there are the computers in his life. He owns about thirty, of which half a dozen or so are usually up and running at any one time. He also has a devilish knowledge of cell phones, and has been known to eviscerate them just to see how they work. He uses programming commands not listed in the user manual. And, dancing on the edges of the law, Shimomura has used his expertise to convert cell phones into powerful scanners capable of listening in on others' conversations.

Over the years, Shimomura has developed an interest in computer security and the solving of complex problems the field requires. When the Supercomputer Center had a problem with break-ins a few years ago and called in the FBI, Shimomura caught their attention and they began to call on him as a resource. He also has done security consulting for the Air Force and the National Security Agency.

Perhaps no one was more surprised at what happened last Christmas Day than Shimomura himself.

One of the first things an electronic prowler typically does is delete system logs so that his victim cannot go back and retrace what happened. But the person who broke into Shimomura's computer was unaware that Shimomura had taken an extra precaution: a copy of his logs file was mailed off regularly to a safe spot on another computer on the network. Andrew Gross, a UCSD graduate student employed at the Center who regularly monitored Shimomura's log files, was at home for the holidays in Tennessee when he decided to take a look at his e-mail on the day after Christmas. When he saw that Shimomura's log files, ordinarily a cumulative record that should only grow larger, were shrinking, he knew they had been tampered with.

Gross reached Shimomura's cell phone. Shimomura was in a car on his way from San Francisco to Lake Tahoe for several weeks of skiing. Gross told him what he had seen. "I think we have a problem," he said. Both flew back to San Diego at once.

The intruder had used a method called Internet Protocol (IP) address spoofing. IP spoofing depends heavily on the fact that computers on a given network are often programmed to trust one another. That is, they are instructed not to grant privileged access to an outside computer unless that computer is designated as a trusted machine—not unlike an admonishment not to speak to strangers, only to trusted friends. A spoofing attack exploits a flaw in the networking software that allows a computer to be fooled into thinking it is communicating with a familiar computer.

The Internet works by breaking data into groups of digital "packets" containing addressing information. The IP spoofing method essentially falsifies the return address information. The intruder apparently logged on to a computer at Loyola University in Chicago (a stranger to the target machine). He then had that computer pose as another of Shimomura's personal machines in San Diego, using an electronic address unique to that computer. Once the rogue computer at Loyola had taken on the identity of a trusted machine, it told the target machine to trust every other host on the Internet. This allowed the imposter to enter Shimomura's other computers and have free run of the system. Dozens of Shimomura's files were copied.

"We were more than mildly surprised," said Gross. "Between Tsutomu and me we've covered his machines for pretty much everything imaginable and then some."

Gross and Shimomura both knew about the technique, but had never seen it done. There was even some doubt in the community as to whether IP spoofing, which involves delicate timing, would work at all.

"But it turns out," said Gross, "that on days when the Internet is less traveled, it's possible." Christmas Day, perhaps the one day of the year when families manage to convince the resident wirehead to log off, was apparently no random choice.

Shimomura, like everyone in computer security, was well aware of Kevin Mitnick's reputation. He knew a warrant was out for Mitnick's arrest. A couple of months earlier, he had gotten word from a friend that someone presumed to be Mitnick had broken into her computer; the attacker was trying to steal some cell phone code that Shimomura had worked on that turned a cellular phone into a scanner. At around the same time, a computer of Shimomura's named ariel had gotten a few pokes, but they weren't terribly sophisticated. Shimomura had even gotten a letter from Justin Petersen, the person whom the FBI had set up to bust Mitnick, offering to help him hunt Kevin down. But the relatively complex nature of the Christmas Day attack made Shimomura doubt that Mitnick could have written a program to carry out the spoof. "Either he found it somewhere or he got it from somebody," Gross said.

Once they figured out the attacker's method, they went out gathering forensic evidence, looking at what the person had stolen, in search of clues to his intent. The first thing the trespasser had gone after was the cellular phone code. He also took Shimomura's home directory, which contained his e-mail and several security tools. After Shimomura and Gross had set up some additional protection for the machines, each returned to his respective vacation. At that point, Shimomura viewed the incident as a giant annoyance. "It was something I didn't want to deal with," he said. "I wanted to go do something else." He was eager to put the ordeal behind him.

On December 27, a chilling message was left on Shimomura's voice mail: "My technique is the best," came a male voice in an odd, slightly Cockney accent. "Damn you. I know sendmail technique. Don't you know who I am. Me and my friends, we'll kill you." Then another voice came on: "Hey boss, my Kung Fu is really good." Three days later came a second voice-mail message: "Your technique will be defeated. Your technique is no good." Shimomura took the messages and put them on the Internet for all to hear.

Shimomura might have dropped the matter if it hadn't been for what Bruce Koball found in his account on the WELL, a Bay Area conferencing system that offers access to the Internet. On January 27, Koball, a technical consultant and organizer of an annual conference called Computers Freedom and Privacy, received a notice from the WELL that, according to an automated program called Disk Use, CFP's account was taking up too much disk space and he should clear it out. This struck him as odd, because he seldom used that account. When he took a look at the account, he saw that it was indeed bloated—with 158 megabytes of files—and the owner of the files was Tsutomu Shimomura. Koball called Shimomura, and Shimomura confirmed that those files had been stolen from his computer. The intruder somehow had acquired what are called "root" privileges on the WELL, which means he had the run of the place. He apparently had stashed his booty in the CFP account because it was used so seldom. The frightening part about his having full access to the WELL was that, if he felt like it, he could bring the system to its knees. So far, he hadn't.

The intruder was logging on to the WELL routinely, apparently comfortable that his wanderings were going undetected. Here was an opportunity to monitor the interloper closely. Shimomura seized it. On January 31, Gross flew up to San Francisco and set up shop in a back room at WELL headquarters in Sausalito. Shimomura joined him a few days later and the two monitored the intruder's every electronic footprint. They captured each character as the intruder typed it. "It was obvious that this was not your average break-in," Gross said. With most break-ins, someone has a few files they might be interested in, spends most of the time haphazardly joyriding around the system, then leaves. But this intrusion was far more widespread. The WELL was by no means the only target. The thief had also taken control of computers at InterNex and Netcom, two other Bay Area Internet service providers.

When it became clear that the attacker's launchpad for the sessions was Netcom, based in San Jose, the sleuthing team packed up and moved the operation down the peninsula. From Netcom's headquarters, the team saw the intruder name files ("japboy" was one favorite file name, while "fucknmc"—as in "Neill Michael Clift"—was usually his password of choice) cruise through places he already had been and poke at places he still wanted to get into. One of the Netcom files he had was a customer account record containing 21,600 credit card numbers. He routinely searched through the e-mail of about a dozen people, including the mail of my coauthor, John Markoff, who was also now following the chase closely. When the monitors saw that the intruder was looking for the text string "itni," their doubts disappeared. They were dealing with Kevin Mitnick.

Occasionally the trespasser went looking for one of his friends and engaged them in real-time conversations using a private "talk" feature of the Internet. "There were two or three people he was communicating with," Gross said.

▲ ▼ ▲

A crucial piece of information came a few days into the monitoring at Netcom. Kent Walker, an assistant U.S. attorney in San Francisco, had issued subpoenas of telephone company calling records. The records showed that the calls were coming from a local Netcom dial-in site in Raleigh. They were originating from a cellular telephone hooked to a modem. As soon as possible, Shimomura was on a plane to Raleigh. By 1 A.M. on February 13, he was in the passenger seat of a Chevy Blazer driven by a Sprint cellular technician, his lap piled with scanning and homing equipment: a surveillance device he had rigged out of an Oki cell phone, a palmtop computer to control the Oki and the Sprint technician's cellular scanner, which had a directional antenna for detecting signal strength, like a sophisticated geiger counter. Shimomura describes that part of the chase as trivial. "It's like finding a lightbulb in the dark, or an avalanche beacon in the snow," he said. "You walk toward where it's brightest."

Within thirty minutes, Shimomura had homed in on the Players Club apartments, a three-story complex near the airport. When he turned things over to the FBI to make the arrest, Shimomura advised the agents to move swiftly, to reduce the time Mitnick would have to destroy evidence. At 2 A.M. on February 15 the agents knocked on the door of apartment 202. It took Mitnick five minutes to open the door. When he did he demanded to see a search warrant. They had one, but for the wrong apartment. Prosecutors had called a federal magistrate to get a valid warrant, but the agents already were inside. Mitnick was under arrest.

▲ ▼ ▲

Susan Thunder told me that she once saw the following graffito on a bathroom stall: "Your ass is the only thing you can sell and still have after you've sold it." The same might be said of software. It's the only thing you can steal and still leave behind, because what you're taking is a copy. This is a common line of defense among attorneys who represent computer criminals. When I visited Kevin's lawyer, a genial, middle-aged criminal defense attorney named John Yzurdiaga, I was sure he'd trot out that line. But he didn't. In fact, Yzurdiaga, who has the beaten-down look of a man who has spent a lot of time around a lot of rough characters, seemed more than a little mystified and slightly overwhelmed by the particulars of the case, and confessed that he didn't know the first thing about computers.

I had gone to L.A. to talk to Yzurdiaga about getting an interview with Kevin, who was awaiting trial in North Carolina. After politely dashing any hopes I had of getting to his client ("It's far too premature," he said), Yzurdiaga bought me lunch. Over gazpacho and salad in a trendy restaurant in downtown L.A., I raised some questions that had been nagging at me. Who was Mitnick talking to in Colorado from Seattle that night? Everyone I had spoken with seemed to agree that the taunting threats on Shimomura's voice mail last December were not left by Mitnick. Eric Corley, a friend of Mitnick's who publishes 2600, a magazine for hackers, told me he knows who left the messages and it wasn't Mitnick. And who wrote the IP spoofing program?

After dispatching his lawyerly duty and proclaiming his client's innocence, Yzurdiaga, a man who has defended people accused of far more heinous acts, threw up his hands. He looked sincerely baffled. He said he couldn't understand just how bad these crimes could be, no matter who committed them. After all, no one was harmed physically. Nothing was sold. Nobody profited. And what is a suitable punishment? Even Shimomura thinks that a protracted jail sentence is, as Shimomura puts it, not an elegant solution. Roscoe's lawyer, a brash, opinionated man named Richard Sherman, dispersed with elegance altogether. "You know what they should do to Kevin Mitnick?" he said when I visited him. "Kevin Mitnick should have his pants taken down for six months so everyone can see what a little tweeny he has."

▲ ▼ ▲

When the Internet was built, it wasn't intended to be secure. It was intended to be useful. What we now see as a systemic shortcoming was the Internet's best feature. There were few locked doors. Visions of a civil, free, open electronic community blossomed. Then human nature intervened and those on its dark side revealed less romantic possibilities. The best and worst of us turn up in cyberspace, no differently than, say, if we were to colonize the moon. The day we accept technology's encroachment on our lives is perhaps the day we will stop overreacting to Kevin Mitnick.

After his arrest, it was inevitable that newspapers would tell the story of Shimomura versus Mitnick in the black and white terms of a fifties Western. Mitnick is the bad guy because we want him to be the bad guy.

In point of fact, it's likely that he is being blamed for a lot of things he didn't do. More to the point, revisiting the story of Kevin Mitnick has made me wonder if our preoccupation with Mitnick is an easy distraction, diverting our attention from those lurking in cyberspace whose intentions are far more malicious than his. It's already known that there's a fair amount of industrial espionage, and the presence of paramilitary fanatics on-line is difficult to overlook. Mitnick, it may turn out, is just the one who keeps getting caught.

▲ ▼ ▲

Disregarding Yzurdiaga's categorical thumbs down, I flew to North Carolina to try to see Mitnick. I had heard security on Mitnick was tight and that no one but the Raleigh public defender had been to see him. I wasn't on his visitation roster, I knew he didn't like me much and my chances probably weren't as good as those of a total stranger. But it was worth a try. I wrote Mitnick offering him the chance to correct any errors of fact. There was no response. So I decided to deliver to him a copy of the book myself. At least, that was the ostensible mission. For years, he had been like a phantom in my life. Now at least I knew where to find him.

When I arrived in Raleigh, I dawdled a bit before setting out to the jail, which is in the town of Smithfield, about thirty miles east of the city. I drove to the Players Club apartment complex, where Mitnick had been living when he was arrested. Set in the lush Carolina pines, the neocolonial apartments are clean and new, the grounds immaculate. From the balcony of his $550-per-month, one-bedroom unit, Mitnick had had a view of a nearby state forest. I could imagine his relief at arriving here, with the possibility and promise of nearby Research Triangle Park, the area's own Silicon Valley. My mind wandered to something I had read in a newspaper article: the search of Mitnick's place after the arrest had produced a book on the best companies to work for in America and forty-four job application letters.

I drove past farms and trailer parks. The closer I got to Smithfield, the more convinced I became of the futility of my journey. And, having skirted the lawyer, I couldn't help but feel a bit naughty. When a police car pulled out of a parking lot and followed close at my tail for several miles, I could imagine what Mitnick must have felt every time he saw a cop. As I entered Smithfield, a billboard advertising Zima Gold greeted me: "Kiss Your Expectations Goodbye."

Smithfield is a sleepy, amiable place. I wouldn't have been surprised to see Andy and Opie ambling down the main street. The window of the town's computer store was rather inappropriately adorned with a toy swimming pool and plastic dolls in bathing suits, propped next to a few PC programs. The local library, with hardly anyone in it, was filled with racks of well-thumbed paperbacks. Around the corner from the jail the local movie theater was showing *Bad Boys*.

The jail itself, part of a complex that includes the county courthouse and sheriff's office, is smack in the middle of town. I pushed the intercom button and explained to a disembodied woman's voice that I was here to deliver a book. "Are you immediate family?" she asked. "No," I said, and tried to explain, but she cut me off. "We don't accept books for federal prisoners," she said. "They have to go through the mail, or the marshal's office has to approve it."

I walked next door to the sheriff's office, found a pay telephone and called the marshal's office. "Well," said the deputy marshal I reached, "if the U.S. attorney's office has nothing against you giving him a book, neither do I." I called the U.S. attorney's office. "I don't have a say in who can visit," the assistant U.S. attorney in charge of the case told me. "Call the defense attorneys." I called the public defender's office in Raleigh but was told that Mitnick's attorney was in a meeting. I called the marshal back. "I'll call the captain," he said. "Give me five minutes." The wall had cracked.

I walked into the sheriff's office and within a few minutes an accommodating man in uniform was whisking me upstairs. The bars closed behind us. "Shouldn't I sit in there?" I asked, pointing to the visitors room. "Oh no, don't bother with that," he replied, and directed me straight into the warden's office. "Come right in here." Southern hospitality at its finest. Bars, floor to ceiling, painted yellow and blue, were everywhere. Jailers sat around, passing the time of day. The captain told them to retrieve Mitnick. One of the jailers stepped into the hallway, cupped his hands together and yelled up at a glass-enclosed warden's booth: "NIT-WICK." The others corrected him: "Mitnick!"

"Mitnick!" he repeated.

Mitnick emerged a few moments later. He was wearing a bright orange jumpsuit that was several sizes too large over a turquoise T-shirt, an ensemble that struck me as incongruously cheerful. His brown, wavy hair, pulled back into a small ponytail, was in a frenzy. He had gone three or four days without shaving. Otherwise, his face had a healthy ruddiness to it.

As he approached me, he looked suspicious and baffled. "Kevin, I'm

Katie Hafner," I stammered, and held up the book like an I.D. card. The skepticism melted from his face and he smiled. "I recognize you from your picture," he said, evidently referring to the book's dust jacket. "You are the last person I expected to see. I didn't know who they were calling me out for. And you're the LAST person I expected to see."

Then he stopped himself and backed away a bit. "I can't talk to you. Under no circumstances. Orders from my attorneys." He stopped himself again. "How did you get IN? Even my friends can't get in." I mumbled something about the book. "I can't say anything to you," he repeated. "I have charges pending." Then he laughed. "You are persistent, aren't you?" I suppressed the urge to reply in kind.

I gave him the book and told him to look it over and let me know what he'd like to change. "Just what you want to be given. The book you hate," I joked. "How did you get in?" he repeated. He obviously was impressed by the feat. I hadn't social-engineered my way in, exactly, but he no doubt suspected it. He was smiling and began to laugh again, a genuine, friendly laugh. I laughed too. I wasn't surprised by how much I liked him.

None of the wardens, seated in an adjacent office, seemed to be paying any attention to our hallway meeting. "If my attorneys say I can talk to you, I will," he said. He disappeared back into the cell block, then reemerged a couple of minutes later, wearing the same gloomy look that he had when he came out the first time. He pressed the book back into my hands. "They told me to give you the book back and they want to talk to you." A young warden showed me into a room with a telephone and the two of us spent several minutes wrestling with the problem of a calling-card call. I nearly suggested we call Mitnick in for assistance. When I got through to Yzurdiaga, who was on a speaker phone with another attorney helping with the case, he was livid. The two attorneys ordered me to leave the book with the jailer and leave the building at once, as if my very presence implicated their client. But I wanted to say good-bye in person. I asked the wardens to call him out again. "He says you can have it," I said when he returned. I handed him the book and he held out his hand. "It was nice to meet you," he said as we shook hands. With that, he turned on his heels, lifting his hand in a friendly, slightly sorrowful wave. The large steel door opened to reclaim him, and he disappeared.

When I returned my rental car at the airport, the Budget clerk, a young man in his twenties eager for small talk, asked where I had been. "At the county jail in Smithfield," I replied, and I explained the reason for the visit. He seemed fascinated but mystified. "Have you heard about this guy, Jim?" he asked his colleague, an older man who had been nodding solemnly as I spoke. "Yep. He was arrested right around here," and pointed in the direction of the Players Club apartments. "He stole a whole buncha credit cards." I began to explain the Mitnick hadn't used any of the credit cards, that the whole case wasn't so simple. But then the shuttle came. I said good-bye and stepped onto the bus.

▲ ▼ ▲

Shortly before the trial in Celle, Pengo moved into a spacious apartment on a quiet street in Kreuzberg and continued freelance programming. After his court testimony in January 1990, he awaited the official decision on his own fate. Three months later, after a series of difficult discussions between Pengo's lawyer in Bayreuth and the German authorities, word came from Karlsruhe that all charges against Pengo had been dropped. There was no probation and no fine.

At the recommendation of the Digital Equipment Corporation lawyer who had attended the Celle trial, Digital had considered filing a civil suit against Pengo for his possession of the company's proprietary software. Pengo's lawyer finally negotiated a settlement. Pengo signed a document promising never again to break into Digital's computers. He was not required to pay restitution to the company.

In late 1990, Emery Air Freight Corporation, whose network user identification Pengo had used so generously, sent Pengo a letter demanding $13,000. After receiving no response from Pengo, Emery's lawyers filed a civil suit against him in Berlin. Once again, Pengo's lawyer intervened, took up negotiations with Emery and settled out of court for $2,300, just enough to cover Emery's attorneys' fees. Pengo has yet to receive a bill from his lawyer.

Markus Hess is still a programmer in Hannover.

Cliff Stoll, divorced from Martha Matthews, was paid a large advance to write a book on astronomy, but eventually dropped the project and wrote a detractor's view of the Internet instead, called *Silicon Snake Oil*.

▲ ▼ ▲

While the final decision was being made in Pengo's case, Robert Morris was sentenced in Syracuse, New York. Reporters filled the courtroom. The entire Morris clan made the trip for the event. Bob was wearing new black shoes and a baseball cap bearing the name "Basil's Deli," a lunchtime hangout near NSA headquarters.

Tom Guidoboni's partner, David O'Brien, made an impassioned plea. Taking care to refer to his client as "Robert," O'Brien called the twenty-four-year-old defendant "a decent kid" who had already taken sufficient responsibility for his actions. Guidoboni then took issue with the government's estimate of $160,000 in losses from the worm, and argued that Robert should be shown some leniency because he had taken responsibility for his actions. The crime, Guidoboni maintained, equated with a "six" under the Federal sentencing guidelines. That was the magic number that translated into no jail time.

When his turn came, the federal prosecutor, Mark Rasch, had little to say except that he believed some time in jail was appropriate. There had been a last-minute dispute within the Justice Department over the appropriate punishment. Unable to make up their minds for more than eight months after the incident about whether to prosecute, government officials now appeared to be having difficulty deciding how to punish him. In an unusual move, the department decided not to file a sentencing recommendation in the case.

Like a game-show host breaking for a commercial, Judge Munson called a five-minute recess and said he would come back with a sentence. When he returned, he asked if the defendant had anything to say. Robert replied that he didn't. The judge then delivered a short speech. He had been inundated with letters, he said. Some called for leniency, while others branded Robert Morris a grave threat to a free society. Munson complained that he could hardly step outside his front door without encountering someone who wanted to give an opinion. Even middle-aged women at his country club were buttonholing him to express their sympathy for the young computer scientist. Prison, he said, would not be a suitable punishment for this crime. He sentenced Robert to three years' probation, a $10,000 fine, and 400 hours of community service. As he adjourned the court, he added, "I'd still like to hear about that trap door." Had Robert been within reach, the judge might have tousled his hair. A few days later, Guidoboni announced that Robert planned to appeal the conviction.

Like the act he had committed, Robert's sentence elicited deeply divided opinions from both computer scientists and the larger community. Many believed the decision not to include a jail term in the sentence was fair. Robert had made a terrible mistake and had paid for it already in accruing a debt of more than $150,000 in fines and legal fees. Others reacted to the light sentence with particular rancor. Eugene Spafford, the Purdue computer scientist turned ethicist, was quoted in the *Journal of the* ACM as calling on the computer industry to boycott the products of any computer company that would hire Robert Morris.

For his own part, Robert kept a low profile. Beyond his statement on the witness stand that he had written his program as a poorly conceived experiment, he chose not to say what else motivated him to write it. After the sentencing he got a long hug from his mother and a firm handshake from his father. But he still refused to talk with reporters. While his family went downstairs to face the throng of reporters, Robert stayed in the courtroom and discussed the terms of his probation with his attorney and court officers. Later he left via a side door, unobserved.

To fulfill his court-ordered community service, Robert worked at the Boston Bar Association. Today he is working as a programmer at a software engineering research firm in Cambridge, Massachusetts, and learning ancient Greek as his father once did. On March 7, 1991, the U.S. Court of Appeals for the second circuit upheld Robert's conviction.

Notes on Sources

PART ONE. KEVIN: THE DARK-SIDE HACKER

The history of phone phreaks that appears on page 18 ("When Susan and Roscoe met . . .") was drawn largely from personal interviews, the chapter titled "Blue Boxes and Phone Phreaks" in *The Biggest Company on Earth* by Sonny Kleinfeld (Holt, Rinehart and Winston, 1981) and from *Fighting Computer Crime* by Donn B. Parker (Charles Scribner's Sons, 1983).

The description of the U.S. Leasing incident that begins on page 33 ("The first opportunity for revenge . . .") was based on personal interviews and on court transcripts.

The history of Digital Equipment Corporation that begins on page 37 ("This wasn't the reaction . . .") was based in part on the book *The Ultimate Entrepreneur: The Story of Ken Olsen and Digital Equipment Corporation* by Glenn Rifkin and George Harrar (Contemporary Books, 1988) and on a speech titled "Digital Equipment Corporation: The First Twenty-five Years," delivered by Kenneth H. Olsen on September 21, 1982, to the Newcomen Society in Boston.

The description of the COSMOS incident that begins on page 48

("It was at the Shakey's Pizza Parlor . . .") was based largely on transcripts of a preliminary court hearing in the case.

The description of the USC incident that begins on page 68 ("For all Mark Brown knew . . .") was based on personal interviews and on official police records.

The description of the National GSC incident that begins on page 74 ("Richard Cooper wanted to know . . .") was based on official police records.

The description of the Santa Cruz Operation incident that begins on page 86 ("Steph Marr had been around computers . . .") was based on interviews with employees of the company and on reports filed by officers of the Santa Cruz Police Department.

The description of the events involving Pierce College and Kevin Mitnick that begins on page 93 ("The first call from Pierce . . .") were based on personal interviews and on reports of the Los Angeles Police Department.

The description of Kevin Mitnick's brush with Security Pacific Bank that begins on page 100 ("For anyone with Kevin's record . . .") was based on personal interviews and on reports of the Los Angeles Police Department.

The description of Digital Equipment Corporation's entanglement with Kevin Mitnick and Lenny DiCicco that begins on page 105 ("Mark Brown noticed immediately . . .") was based on personal interviews with Digital employees and others, on electronic mail messages that passed among Digital employees at the time and on FBI reports and court documents.

The description of Kevin Mitnick's final arrest in 1988 that begins on page 128 ("It was the stunt . . .") was based on personal interviews and FBI reports.

Digital officials turned down our interview requests and declined to comment on this and other sections of the book.

General Sources

Douglas Colligan, "The Intruder," Technology Illustrated, October/November 1982.

"The Electronic Delinquents," transcript of the April 22, 1982, broadcast of the ABC News program "20/20."

John Johnson, " 'Dark Side' Hacker Seen as Electronic Terrorist," *Los Angeles Times*, January 8, 1989.

Karen Kingsbury, "An Obsession with Computers," *Los Angeles Daily News*, December 19, 1988.

Karen E. Klein, "Magistrate Refuses Bail for Valley Computer Whiz," *Los Angeles Daily News*, December 24, 1988.

Bill Lawren, "Computer Crime," *Penthouse*, July 1982.

Jeffrey Perlman and Debi A. Hastings, "Computer Raiders Hunt Secrets," December 15, 1981.

Eddie Rivera, "The Phine Art of Phone Phreaking," *L.A. Weekly*, July 18–24, 1980.

Ted Rohrlich, " 'Phone Phreak' Sentenced to 150-Day Term," *Los Angeles Times*, June 11, 1982.

Ron Rosenbaum, "Secrets of the Little Blue Box," *Esquire*, October 1971.

John R. Wilke, "At Digital Equipment Slowdown Reflects Industry's Big Changes," *The Wall Street Journal*, September 15, 1989.

PART TWO. *PENGO AND PROJECT EQUALIZER*

In general, facts surrounding the espionage activities of the German group were gathered from personal interviews, court testimony and the final seventy-one-page verdict issued by the West German court.

The message from CERN system manager Alan Silverman that appears on page 154 was reprinted in *Die Hacker Bibel*, Part 2.

The log of the hacker's attempts to gain access to the White Sands Missile Range that appears on page 180 was taken from *The Cuckoo's Egg* by Cliff Stoll (Doubleday, 1989).

The section on Laszlo Balogh of Pittsburgh was based on interviews with Howard Hartmann, Linda Doebler, Cliff Stoll and Yale Gutnick.

General Sources

Thomas Ammann, Matthias Lehnhardt, Gerd Meissner and Stephan Stahl, *Hacker für Moskau* (Wunderlich, 1989).

Stefan Aust, *Der Baader-Meinhof-Komplex* (Hoffman und Campe Verlag, 1985)

Dieter Brehde, "Der Tod des Hackers," *Stern,* June 8, 1989.

Malcolm W. Browne, "World's Biggest Accelerator Surges to Life," *The New York Times,* August 8, 1989.

"Hacken für den Weltfrieden," *Süddeutsche Zeitung,* January 12, 1990.

Jane Kramer, *Europeans* (Farrar, Straus & Giroux, 1988).

Steven Levy, *Hackers: Heroes of the Computer Revolution* (Anchor Press/ Doubleday, 1984).

James M. Markham, "France's Minitel Seeks a Niche," *The New York Times,* November 8, 1988.

Robert Shea and Robert Anton Wilson, *The Illuminatus! Trilogy* (Dell, 1975).

Ulrich Sieber, *The International Handbook on Computer Crime* (John Wiley & Sons, 1986).

Mary Stolberg, "Informant Key in Computer Parts Theft Trial," *The Pittsburgh Press,* January 21, 1983.

Cliff Stoll, *The Cuckoo's Egg* (Doubleday, 1989).

Sherry Turkle, *The Second Self: Computers and the Human Spirit* (Simon and Schuster, 1984).

Jürgen Voges, "Hacker wollten Weltfrieden sichern," *Tageszeitung,* January 12, 1990.

PART THREE. RTM

In general, the events in this section draw on interviews with friends, associates and relatives of Robert Morris, and with law-enforcement officials.

The description of events surrounding the paper on breaking the M-209 encryption scheme that begins on page 267 ("In the mid-1970s . . .") comes from an interview with two of the paper's authors.

General Sources

John Perry Barlow, "Crime and Puzzlement," *Whole Earth Review,* Fall 1990.

Harold L. Burstyn, "RTM and the Worm That Ate Internet," *Harvard Magazine*, May–June 1990.

Dorothy Denning, "Concerning Hackers Who Break Into Computer Systems," Digital Equipment Corporation, Systems Research Center. Paper presented at the Thirteenth National Computer Security Conference, Washington, D.C., October 1990.

Peter J. Denning, "The Science of Computing: The Internet Worm," *American Scientist*, March–April 1989.

Ted Eisenberg, David Gries, Juris Hartmanis, Don Holcomb, M. Stuart Lynn and Thomas Santoro, "The Cornell Commission: On Morris and the Worm," a report issued by Cornell University, February 6, 1989.

David Kahn, *The Codebreakers* (Macmillan, 1967).

Gina Kolata, "When Criminals Turn to Computers, Is Anything Safe?" *Smithsonian*, August 1982.

Jonathan Littman, "The Shockwave Rider," *PC/Computing*, June 1990.

John Markoff, "Author of Computer 'Virus' Is Son of N.S.A. Expert on Data Security," *The New York Times*, November 5, 1988.

John Markoff, "Cornell Suspends Computer Student," *The New York Times*, May 26, 1989.

John Markoff, "How a Need for Challenge Seduced Computer Expert," *The New York Times*, November 6, 1988.

John S. Quarterman, *The Matrix: Computer Networks and Conferencing Systems Worldwide* (Digital Press, 1990).

Jon A. Rochlis and Mark W. Eichin, "With Microscope and Tweezers: The Worm from MIT's Perspective," *Communications of the ACM*, June 1989.

Donn Seeley, "A Tour of the Worm," University of Utah Department of Computer Science, Technical Report, November 1988.

John F. Shoch and Jon A. Hupp, "The 'Worm' Programs—Early Experience with a Distributed Computation," *Communications of the ACM*, March 1982.

Eugene H. Spafford, "The Internet Worm: Crisis and Aftermath," *Communications of the ACM*, June 1989.

Ken Thompson, "Reflections on Trusting Trust," 1983 ACM Turing Award Lecture, *Communications of the ACM*, August 1984.

Michael Wines, "A Youth's Passion for Computers, Gone Sour," *The New York Times*, November 10, 1988.

Acknowledgments

Hundreds of people agreed to be interviewed for this book. Many sat patiently, explaining again and again some of the more technical aspects of the subject. Many others helped us to reconstruct events, hauling out their old calendars and notebooks and computer printouts. We are especially grateful to Renate and Gottfried Hübner, Bob and Anne Morris and Gil DiCicco for spending time with us to talk about their sons. Thanks, too, to Bonnie Mitnick for speaking with us about Kevin Mitnick who, despite repeated requests, refused to be interviewed.

We are deeply indebted to Paulina Borsook, Mark Seiden and Deborah Wise for their editorial body and fender work. The manuscript was also read in its various stages of completion by Everett Hafner, Steven Levy, Katherine Magraw, Annabelle Markoff, Andy Pollack, Peter Preuss, Debbie Yager and Susie Zacharias, all of whom made helpful comments. The manuscript benefited greatly from the keen eye of copy editor Stephen Messina.

We received invaluable research help from Thomas Ammann, Keith Bostic, Dieter Brehde, Dave Buchwald, Eric Corley, Udo Flohr, Tom Guidoboni, John Johnson, Dan Kane, Ekkehard Kohlhass, Phil Lapsley, Jon Littman, Mike McAndrew, Doug McIlroy, Gerd Meissner, Peter Neumann, Dennis Ritchie, Eddie Rivera, Alan Rubin, Ulrich Sieber,

Steve Steinberg, Cliff Stoll, Ken Thompson and Michael Wines. For helping to piece together various incidents involving Kevin Mitnick, we are grateful to Bob Ewen of the Los Angeles district attorney's office and Detective Jim Black of the Los Angeles Police Department.

Ellen Kosuda, ace researcher, provided invaluable help with the epilogue. Teresa Carpenter offered guidance every step of the way. Thanks, too, to Kathleen Cunningham, Ivan Orton, Brian Salt, Noel Chliappa, Steve Bellovin, Michael Lawrie and Neill Clift. Amy Goodwin graciously took on extra baby-sitting. For their friendship and hospitality we'd like to thank John Kelley and Lisa Van Dusen, Sara Charno, Silke Grossmann-Brehde, Keith Hammonds, Andrea Klotz, Mara Liasson, David Olmos, Ralph and Sonya Raimi, Heiko and Mechthild Rogge, Seth Rosenfeld, Marc Rotenberg and Gail Schares. Thanks, too, to Paul Saffo and the folks at the Institute for the Future in Menlo Park for office space and a Macintosh.

Fortune smiled on us when we were introduced to Bob Bender, the senior editor at Simon & Schuster who took this book on. He thought this was just a swell idea from the start and he didn't change his mind. His every suggestion was invaluable. And thanks to our agents John Brockman and Katinka Matson for making the Bender connection.

A final note: Roscoe and Susan, two of the characters in the Kevin Mitnick section, cooperated with us in the understanding that their true names would not be revealed. We respect their right to privacy.

Index

COSMOS incident and, 48–59, 64, 71–72
DiCicco's first meeting with, 64
DiCicco's relationship with, 81, 105, 126–28
DiCicco turned informant against, 129–37
divorce of, 344
education of, 63, 81–82, 84–85, 100
as fugitive, 353–54, 356–57
goal-orientation of, 111–12, 127
Hafner's attempts to visit, 364–67
Hafner's on-air confrontation with, 345
in hiding, 75–78, 100, 342
in jail, 73, 343
jobs held by, 64, 74–75, 100–102, 127
lack of technical skills of, 355
malicious streak of, 26, 65–66, 96, 102, 105, 111, 128–30, 342
marriage of, 82–85, 91, 127, 133
1988 arrest of, 343–44
Pacific Bell "minis" project and, 79–80
Pacific Bell's memo on, 91–93, 131–32
paranoia of, 81, 112, 126, 127
patience and perseverance of, 32–33
physical appearance of, 26
Pierce College incident and, 93–100, 103–4
plea bargain of, 342–43
psychiatric counseling of, 343–44
Radio Shack computers used by, 64–65
Roscoe's first meeting with, 23–24
Roscoe's friendship with, 351–52
Santa Cruz Operation break-in and, 86–91, 92, 96, 103
searches of homes of, 52–53, 88–89
in Seattle, 356–57

social engineering skills of, 26, 66, 113
software theft as viewed by, 124
Susan's friendship with, 347–48
Susan's lack of rapport with, 15, 26
Susan's quest for revenge and, 33, 40, 41–42, 51–52
USC computers used by, 65, 67–73, 105–10, 123–24, 126, 129, 131, 135
U.S. Leasing break-in and, 33–40, 47, 55
VMS operating system stolen by, 105–37
XSAFE stolen by, 124–25, 130, 131
Mitnick, Mitchell, 350
Mitre Corporation, 181–82, 186, 246
modems, 21
Morris, Anne, 269, 270–71, 272–273, 276, 283, 286, 291–92, 322, 324–25
computer virus incident and, 311–12, 313, 318, 319, 326, 335, 336–37, 340, 341, 346
Morris, Ben, 269, 270, 271, 272, 273, 311, 336, 340
Morris, Meredith, 269, 270, 271, 272, 273, 326, 336, 337, 340
Morris, Robert (Bob) (father), 238, 261–76, 277, 282, 283, 285, 293, 299, 322, 324–25
at Bell Labs, 262–68, 269, 270, 275–276, 291
computer virus incident and, 261–62, 296, 311–13, 315, 317, 318, 319, 326, 329, 335, 337, 340, 341, 345, 346
cryptology and, 267–68, 271
education of, 262, 284, 292
home life of, 269–73
iconoclasm of, 264
marriage of, 269

tiger teams, 266, 276
Time, 295
time-sharing, 265–66
TIPs (terminal interface processors),
278
Torek, Chris, 307, 308
trapdoors, 37
trashing, 27, 49
Très riches heures de duc de Berry,
270
Trial, The (Kafka), 345 46
Triam International, 191, 202,
204
Trojan horse programs, 107, 115,
119, 126, 182, 198, 256, 258,
306
TRW, 27–28, 68, 74, 75, 76, 96,
97
Turing, Alan, 267
"20/20," 59
Tymnet, 110, 149–50, 151, 152,
181, 186, 188

UCBVAX, 256–57
UCLA (University of California at
Los Angeles), 53, 278
Ultrix, 117, 119
Union Carbide, 185
University College, London, 186
University of Bremen, 187, 188
University of California at Berkeley,
118, 282
computer virus incident and, 253–
254, 256–59, 306, 307–8, 309,
310, 335
see also Lawrence Berkeley
Laboratory
University of California at Los
Angeles (UCLA), 53, 278
University of California at San
Diego, 254, 259
University of California at Santa
Barbara, 278
University of Illinois at Champaign-
Urbana, 255

University of Maryland, 255
University of Southern California
(USC), 65, 67–73
Brian Reid hoax at, 109–10
computer virus incident and,
334
VMS theft and, 105–10, 123–24,
126, 129, 131, 135
University of Utah, 278
UNIX, 86, 88, 119, 122, 181, 195,
247, 268, 328
Berkeley, 181, 189, 197, 254,
256, 258, 286, 288, 296, 301,
303, 305, 335
development of, 266–67
Digital's use of, 117–18
Hess's familiarity with, 164, 165,
167
Morris's expertise in, 282, 284,
285, 289, 292
security flaws in, 275–76, 288,
291, 297, 298, 299, 301, 306–
307, 327
source code for, sought by Soviets,
178, 185, 189, 193, 197
UUCP program in, 282
UNIX conference (1988), 307–8
USC, *see* University of Southern
California
USENET, 287, 295
U.S. Leasing, 59
ease of breaking into, 34
U.S. Leasing incident (1980), 33–40,
47
data destroyed in, 36–37
Digital's response to, 37–38
Susan's quest for revenge and, 33,
34–35, 40
trial in, 55, 56–59
UUCP, 282

VAX, 65, 81, 99, 107, 115, 122,
151, 154–55, 176, 184, 289,
302
meaning of acronym, 65